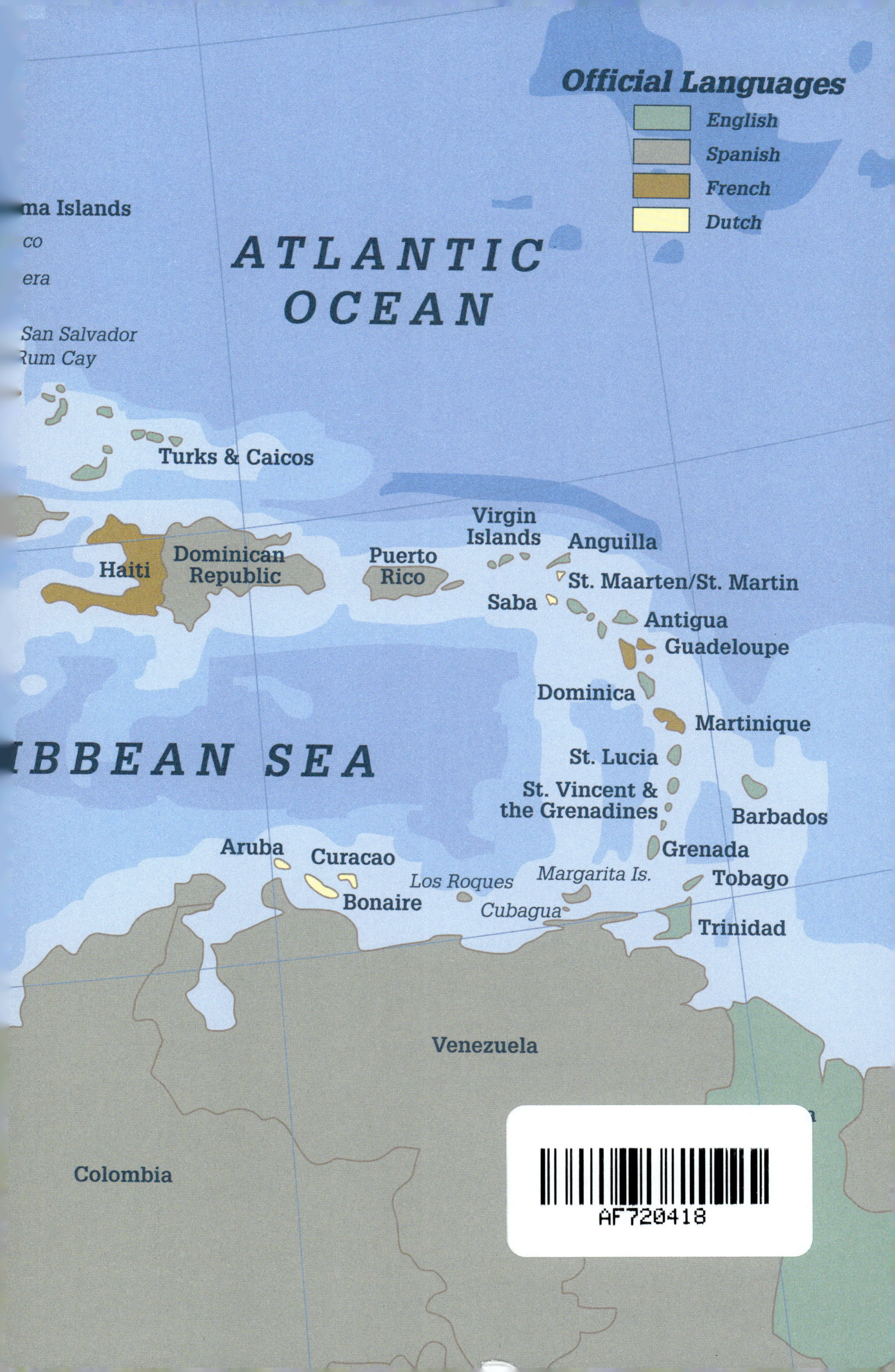

Official Languages
English
Spanish
French
Dutch
ATLANTIC
OCEAN
Islands
San Salvador
Turks & Caicos
Haiti
Dominican
Republic
Puerto
Rico
Virgin
Islands
Anguilla
St. Maarten/St. Martin
Saba
Antigua
Guadeloupe
Dominica
Martinique
SEA
St. Lucia
St. Vincent &
the Grenadines
Barbados
Grenada
Aruba
Curacao
Los Roques
Margarita Is.
Tobago
Bonaire
Cubagua
Trinidad
Venezuela
Colombia

REEF FISH Behavior

FLORIDA CARIBBEAN BAHAMAS

NED & ANNA DELOACH

with PAUL HUMANN

NEW WORLD PUBLICATIONS, INC.

Jacksonville, FL USA

For my mother
Mary Belle DeLoach
who taught me to love natural history.

Cover - *Butter Hamlets in a spawning clasp.*
Title page - *Seaweed Blennies locking mouths in a territorial dispute.*

CREDITS:

Editor: Ken Marks.
Photography Editor: Eric Riesch.
Layout & Design: Eric Riesch, Ned DeLoach.
Art Director and Drawings: Michael O'Connell.
Print Consultant: D'Print Pte Ltd, Singapore.
1st edition ISBN: 978-1878348-28-9; 2nd edition ISBN: 978-1878348-68-5
Library of Congress Control Number: 2019905871
First Edition, 1999; Second Edition, 2019.

Publisher: New World Publications, Inc., 1861 Cornell Road, Jacksonville, FL 32207, (904) 737-6558
www.fishid.com, orders@fishid.com.

Acknowledgments

Having the private use of a dive boat was indispensable during our work on the first edition in Bimini. Close friends John and Laura Smith of Jacksonville, Florida provided the *Pocahontas* in 1995, Scuba Bimini allowed us to use the *Blenny* in 1996 and 1997, and our South Bimini neighbors, Dr. W. J. "Pete" Morse and Helen Swallow, made the *Big Hellie* available to us in 1998. Their generosity provided the freedom to study the fishes at our own pace. And a hearty thank you to our dear friend Richie Collins who kept our engines running. We are especially grateful to Bill and Phoebe Foster of York Harbor, Maine for the extended use of Wen-Mar, their island home on South Bimini. The Rosenstiel Marine Library staff will always be remembered for their unfailing courtesies and expertise. The staff of the Reef Environmental Education Foundation (REEF) provided a wealth of helpful fish survey data as well as sharing a sea full of personal fish behavior observations. We'll always be indebted to the hands-on team who put the first edition together: Nancy DeLoach, Fred McConnaughey, Eric Riesch, Ken Marks, Dr. Emily Schmitt-Lavin, Michael O'Connell and Sean Keeler.

Much of our work on Bonaire was made possible by the support of Martien & Ingrid Van der Valk, Michelle Coolen-Van der Valk & Paul Coolen and dive manager Augusto Montbrun and the staff at Buddy Dive Resort. Judie Clee and Chris Flook always made things happen during visits to Bermuda. Amy Slate has long been a dear friend and our go-to facilitator in Key Largo.

We wish to express our deepest respect and gratitude to the scientists whose life's works are the heart of both editions. Their contributions are listed under "Reference" beginning on page 432. A special thanks to James Bohnsack, Pat Colin, Michael Domeier, Anne-Marie Eklund, Chris Koenig, Christy Pattengill-Semmens, Brice Semmens and Ben Victor who personally shared details of their past and present projects. A special thanks to editors Ken Marks and Eric Riesch whose knowledge of fish behavior has been invaluable to the project. The insight of Paul Humann our friend, mentor and partner touches every page.

PHOTOGRAPHY:

The photographs in this book show species in their natural habitats. Many outstanding underwater photographers contributed their work. Their skill, dedication and time underwater have help make the novel collection of photographs as complete as possible.

Charles Arneson–154, 213; **Andrew Bruckner**–372, 373; **Mike Bryant**–380; **Nick Caloyianis**–385, 386; **Mickey Charteris**–10, 21t, 90t, 147br, 302, 303, 304t&b, 356, 370, 409; **David Cubbin**–389b; **Jim Church**–111; **Deb Devers**–36; **Alex Fogg**–355; **Rick Frehsee**–288; **Debbie Fugitt**–90b; **Mike Haber**–45tr; **Pam Hardie**–4; **Jim Hellemn**–297; **Paul Humann**–12t&b, 13t&b, 18, 30bl&br, 43mr, 60b, 72t, 76b, 94, 96, 100, 101t&b, 103tl&tr&br, 104bl&br, 106b, 114t&b, 115t, 117t, 122tl&ml, 123tl&tr, 136t&br, 137br, 139b, 140, 145t, 148, 172, 173bl&br, 180, 188t&b, 189t, 199b, 217tl&bl&br, 224tl, 226, 227tl&tr, 230br, 255t&b, 260, 273tl&tr&mr, 275, 279, 286, 290, 292, 301tr, 317tl&tr&bl, 320bl&br, 330r, 363, 374, 376, 384b, 390, 391tl, 393t&m, 395l, 403t, 417, 423; **Steve Kovacs**–1, 48, 49, 54t, 199t, 344t, 347t; **Ken Marks**–97, 272; **Woody Mayhew**–261, 263; **Susan Mears**–50; **Ellen Muller**–61tl, 62, 147tl&tr&ml&mr, 214, 266t, 267tl&tr&bl&br, 358; **Doug Perrine**–20, 34, 60t, 216, 357, 388; **Scott Michael**–17; **Eric Riesch**–43tr, 59br, 61bl, 113, 173mr, 186, 212, 235l&r, 273bl, 274, 318mr; **Douglas Seifert**–280; **Wayne Shoemake**–35; **Walt Stearns**–85, 91, 125, 282, 283, 284, 285; **Wesley Strong**–382, 384t; **Carlos Villoch**–389t; **Cary Yanny**–6; **Dick Zingula**–39t&b. The remaining photographs by **Ned DeLoach.**

Ned & Anna DeLoach

Authors' Note
1999

Reef Fish Behavior is intended as an overview of what is presently known about the nature of reef fishes inhabiting the waters of Florida, the Caribbean and Bahamas. It is also meant to serve as a companion reference to *Reef Fish Identification Florida, Caribbean, Bahamas*—a comprehensive guide to the fishes in the same region. Both volumes are the result of an expanding and productive partnership between the scientific and the recreational diving communities.

The information included in the text came from two primary sources—four summers of underwater observation and photography in Bimini, Bahamas and many hours of research in the library at the Rosenstiel School of Marine and Atmospheric Science, University of Miami. Extended stays in Bonaire, Belize, the Florida Keys and shorter visits to Roatan, the Cayman Islands, Puerto Rico, Walker's Cay (Bahamas), New Providence (Bahamas), St. Lucia, St. Vincent and the Grenadines, and Barbados rounded out five years of research and regional travel. Video images, recorded by my wife Anna, were used extensively to analyze behavior described in the text. Anecdotal observations gleaned from conversations with fish aficionados in the scientific, marine management, conservation, aquarium and diving fields add insight and spice to an otherwise data-dominated narrative. Entries beginning with "we" were personal observations made by Anna, my partner Paul Humann or me. In the text, I chose to do the following: capitalize common fish and invertebrate names to set them apart from descriptive adjectives; enter a species' scientific name only the first time it appears in a section; and, we followed the recommendations of *Common and Scientific Names of Fishes* published by the American Fisheries Society.

Anna and I thoroughly enjoyed every facet of gathering material for this book but our four extended stays in South Bimini were magical times, when, as John Steinbeck wrote, "the world was spinning in greased grooves." While in Bimini we had the luck of living at Wen-Mar—a rustic, thick-walled white concrete cottage built in the 1940s at the very edge of the sea. Our times there revolved around a small, brightly-lit sitting room that faced west toward the Gulf Stream. On clear nights we could see the glow of Miami from the room's glass doors and, while enjoying coffee in the cool of morning, we watched Brown Pelicans dive and Spotted Eagle Rays jump in our front yard. Afterwards, I would settle into an ocean-facing futon surrounded by stacks of copied research papers and dog-eared references and read about fishes or hammer away at my laptop until noon.

We dived seven days a week, except during prolonged westerly blows. Our borrowed boat (the *Pocahontas*, the *Blenny* or the *Big Hellie* – depending on the year) invariably pulled away from the dock at one with Capt. Anna at the helm. Thirty-minutes later, we were happily underwater. From among the dozens of excellent dive sites that edge the islands, we consistently returned to the same six. The locations, ranging in depth from 14 to 55 feet, allowed us to observe several different fish communities on a semi-regular basis over extended periods of time. When we were studying new behavior or trying to capture a difficult image, we often returned to the same location daily for weeks. During these periods, when we were "on to something," we worked the hardest, stayed underwater the longest and were the happiest.

Although we observed interesting fish behavior throughout the day, we found fish watching best at sunset when the large predators prowl and many fishes court and spawn. Our most productive dusk dives had a marked tendency to occur on the evenings of the seventh through the ninth day after the full moons of early spring to early summer.

Anna carried a Sony 3-chip video camera, and I lugged along a Nikon F3 with a 60 mm macro lens and a Nikonos III with a 20 mm lens both connected to a single Ikelite Substrobe 200. We typically made three dives each afternoon, the last beginning 45 minutes before sunset and ending with just enough light left on the horizon to help Anna navigate home. Even before showering, we had the day's videotape turning in the VCR and reviewed our afternoon's exploits while sipping rum and taking notes.

When we began our research we had little idea how much, or what, was known about reef fish behavior. At the Rosenstiel Marine Library, we dived into the journal-packed stacks with the same enthusiasm that kept us underwater for hours on end. We quickly discovered that, unlike terrestrial animals that have been observed and documented for centuries, the study of coral reef animals is in its infancy. With little choice before the birth of scuba, just a half-century ago, marine biologists based much of their early research on preserved specimens fished from the reef. Their inability to observe the fishes and invertebrates in their natural habitat left large gaps in our understanding of the behavior and ecology of reef animals. Today's diving naturalists, whether scientifically trained or inquisitive amateurs, are pioneers chronicling the last great undocumented natural history on earth. It is an exciting time to be a diver.

Following Page: *Author's Note 2019*

Authors' Note
2019

Ned & Anna DeLoach

Since the publication of the first edition two decades ago, fishwatching has taken us halfway around the world to document marine species in the Indo-Pacific region; however, during that period we invariably made it back to the Caribbean for a month each year to gather additional material for our favorite pet project—the next edition of *Reef Fish Behavior*. So here it is, off the press 20 years to the month following the publication of the first edition.

During the interceding two decades notable advancements have been make to our understanding of reef fish behavior. Most important for our study has been the utilization of DNA markers for tracking fish larvae during their early oceanic life. The results of this consequential work has helped transformed our understanding of larval transport and its crucial role establishing and maintaining reef fish communities. During the same period, an impressive body of lab and field experiments, primarily from the Pacific, has revealed that cleaning symbiosis—a hallmark of underwater lore—once believed to be clear-cut example of cooperation between species is in reality rife with conflict and high-stakes drama. We also take a long look at the present plight of grouper and snapper spawning aggregations, and find that the ongoing endeavor in the Cayman Islands to reestablish the Caribbean's iconic Nassau grouper population showcases a collaboration between science and local conservation initiatives functioning at its best. On a less encouraging note, no-take marine reserves—prohibiting the extraction of all resources—have not fared as well as hoped, primarily from a poverty of political will. In 1999, less than one percent of the world's oceans were protected in a proper manner. Even after a goal of protecting 20 percent of our oceans by 2020 was proposed in 2007, by 2017 only two percent had been appropriately safeguarded.

Research for the second edition, emphasizing fish reproduction, cleaning symbiosis, feeding strategies and early life history, has taken us to 20 destinations in Florida, the Bahamas, Bermuda and the Caribbean. However, the majority of our underwater time has been spent in Bonaire. Although the small, arid island, off the coast of Venezuela, has a substantial fish population comparable with other regions, the island's magical attraction for us is its 25-mile leeward coastline offering easy access from shore to a variety of underwater habitats. The freedom to observe animals on their, and our own, schedule has been a boon to this edition.

Contents

Reproduction

The slow spawning rise of Honeycomb Cowfish, which carries them 25 feet above the bottom, ends with an explosive release of gametes.

When searching for sex on the reef, there is no better time to be underwater than dusk. Unfortunately, divers traditionally are hanging up their wetsuits or preparing for dinner when the captivating mating rituals of reef fishes go into high gear. Following extended courtship, during the last hour of daylight, numerous male/female pairs of reef fishes make side-by-side rocket blasts toward the surface culminating in explosive, simultaneous releases of thousands of free-floating gametes (eggs and sperm). The visible portion of the milky clouds left behind is sperm; the thousands of transparent eggs are virtually invisible. As the gametes disperse the parents dive for the bottom abandoning their progeny to the currents. This reproductive strategy, known as pelagic or broadcast spawning (external fertilization), ordinarily involves pairs; but a few species court and spawn in groups that can number into the thousands. Approximately 75 percent of reef fishes spawn in open water.

A second spawning strategy, demersal egg laying, generally begins at sunrise with females laying eggs in nests prepared, tended and guarded by males. Most nests are hidden away in difficult to get to places. A male's spawning color pattern is generally the best indicator that a nest is nearby. Divers rarely have a chance to see internal fertilization and live bearing, a reproductive strategy common to sharks and rays.

Group-spawning Blueheads

Toadfishes, family Batrachoididae, are the only Caribbean bony fishes known to care for free-swimming offspring. As many as 60 one-inch juveniles have been counted around the den of a guardian male. One male was observed tending three different broods during a five-month period. Because toadfishes lack a dispersal mechanism, they are restricted to the continental shelf of their birth and are absent from oceanic islands.

Sperm is little more than DNA with a tail requiring little effort to produce; in contrast, reproducing a daily cache of eggs packed with fat and protein to nourish thousands of yolk-sac larvae demands a substantial amount of foraging. This energy imbalance between the sexes coupled with an exemption from parental care makes polygamy (multiple mates) the reproductive strategy of choice for the majority of pelagic spawning males. The males' penchant for promiscuity and the danger of predation involved with traveling in search of mates combine to make harems the preferred social system of small, site-attached pelagic spawners. Of course, harems have their costs—it takes plenty of energy for territorial males to defend feeding grounds and isolate mates from rivals. Also food and shelter, nearly always in short supply, must be shared.

Although the spawning rises of pelagic spawning species are brief, the extended courtship leading to the dramatic finales are often packed with melodramas revolving around a basic theme: A revved male relentlessly chases, cajoles, corners, nudges and shows-off to a seemingly indifferent female. To an untrained eye courtship escapades often appear as little more than fish milling near the bottom. But after observing one or two fish trysts, the signs become unmistakable. Clues, both subtle and conspicuous, include color and pattern

Sergeant Major laying and fertilizing eggs on a rock bottom.

A pugnacious male Bearded Toadfish guards his stay-at-home offspring inside his den.

changes, fin displays, body twitches, and most noticeably, females, heavy with eggs and ready to spawn hovering, in mid-water as males swim rapidly about chasing away competitors.

Males arrive at spawning sites ready to shed sperm. For females it is a different matter requiring additional time for their premium cargos of eggs to hydrate (the absorption of ovarian fluid allowing the capsules to float). When ready, females customarily initiate spawning by rising a few feet above the bottom. Not all pelagic spawners wait until twilight to mate. Members of a few common fish families, notably parrotfishes and wrasses, spawn during the day. The timing of a species' daily spawning period, which typically lasts from one to two hours, varies between regions in the Caribbean, but occurs locally at the same time each day. If there is such a thing as a typical courtship and pelagic spawning scenario, the Redband Parrotfish, *Sparisoma aurofrenatum,* provides a good example. From two to seven females gather at a preferred time between late morning and early afternoon at a traditional spawning site usually along the outer edge of the reef. While the swollen females graze, the male blazes about with his fins folded back and his tail arched upward. With his fins spread and body twitching he briefly approaches each mate to assess and encourage the female's readiness to spawn before speeding out of the area. Displays intensify with each subsequent visit until willing females rise 8 to 10 feet above the bottom and pause. The male circles, and from time to time ceremoniously voids a plume of waste as he analyzes the situation. When a hovering female begins a slow steep rise from her hovering position, the male dashes to her side and together the pair streaks toward the surface and simultaneously release gametes, separate and dive for the bottom. A male often spawns with two or three partners in succession before swimming away, only to return minutes later to repeat the process with other females in his harem. Spawning often continues in this manner for nearly an hour.

During reproductive activity species vary in their reactions to divers. After courtship begins, nothing seems to bother wrasses, flounders and small sea basses, but parrotfishes are slow to spawn with observers nearby. At dusk, the beam from a diver's handlight often distracts courting fishes. When we approached a pair of Scrawled Filefish, *Aluterus scriptus,* during their slow, spiraling spawning ascent, the fish broke apart. One partner fled, leaving the other alone in mid-water. Confused, it settled to the seafloor and released a stream of gametes next to a sea fan.

A number of species, including most members of the wrasse, parrotfish and sea bass families, are sequential hermaphrodites, born with the ability to change sex during their lifecycle. This strategy allows a species to maintain a desired sex ratio under varying social and environmental conditions. Sex changes, once completed, are irreversible. In the Caribbean, nearly all sex-changing species are protogynous hermaphrodites, switching from females to males after reaching a certain size or age or when social situations, such as the loss of the harem's dominant male, warrant. Common Snook, *Centropomus undecimalis,* are one of the few Caribbean species documented to be protandrous hermaphrodites, changing from males to females. The exact factors triggering sex reversal have yet to be determined for most species.

Spawning Graysbys

Graysby, *Cephalopholis cruentata,* common, moderate-sized, protogynous sea basses, have a straightforward pattern of sex reversal. All individuals are born female. After reaching a length of eight inches during their fourth or fifth year, females change into males and establish harems. However, social systems are much more complicated for wrasses and parrotfishes. Nearly all the species in the two families have two visually distinct types of reproductive adults. The most numerous are the young, sexually mature individuals in the initial phase that, depending on their species, can be either males or females but are generally females. When in the initial phase both sexes resemble each other in size, color, markings and basic behavior, which often allows initial phase males to more easily approach spawning activities without being chased by dominant males. Initial phase males can be either primary males (born males) or secondary males (sex-changed females). Terminal males, the largest, most colorful and least numerous individuals in a population, are typically secondary males; but in a few species, primary males also transform into the terminal phase.

Although the social systems of parrotfishes and wrasses appear chaotic, their strict, size-oriented pecking orders promote stability. The majority of terminal males join harems as young females and, with growth, rise through the hierarchy until positioned next in line to change sex and take control. The ability to transform quickly into terminal males is essential for establishing dominance in a harem and for discouraging neighboring terminal males from taking advantage of a sudden power vacuum.

Bluehead, *Thalassoma bifasciatum,* one of the Caribbean's most abundant and thoroughly studied reef fishes, do not form harems. Instead, the terminal males take temporary control of traditional spawning sites, usually along the periphery of forereefs, where they spawn for a one- to two-hour period each afternoon. When ready to spawn, females migrate from their feeding grounds

Young initial phase Bluehead

A female shows an early transitional color pattern as she begins to change into a male.

to favored spawning sites, each controlled by one of the larger terminal males in the population. During a research project that removed all the terminal Blueheads from an isolated reef, the larger females immediately displayed aggression among themselves. Within minutes, a few initiated male courtship behavior. Later the same day, the ranking females attempted to spawn as males and exhibited a slight darkening of their heads—a unique terminal phase characteristic. After three days the same fish displayed obvious transitional color patterns, intermediate to the two phases. Recently sex-changed Bluehead males grow 50 percent faster than females, but it takes a minimum of four months before they are large enough to experience significant reproductive success.

It would be death defying for non-ranking sequential hermaphrodites to buck the system and change sex before social order dictates. Offenders would be driven from their feeding grounds and unable to establish territories of their own or to join other harems. However, there always seems to be a twist on the reef. Within the small harems of the Rock Beauty, *Holacanthus tricolor,* one of two sex-changing angelfishes in the Caribbean, the ranking females regularly switch sex without the loss of the harems' dominant males or the opportunity to control nearby territories. These bachelor males, as they are called, possibly escape the wrath of the much larger dominant males by keeping low profiles. Even though they are unable to reproduce for a short period, their premature

A late-stage transitional pattern. By this time the individual is probably functioning as a male.

Terminal phase Bluehead

Male Sailfin Blennies display oversized fin to attract females and intimidate rivals.

transformations better position bachelor males to change quickly into a dominant role if the opportunity arises, and, similar to transitional Blueheads, their rate of growth is accelerated—no small matter in a world where size in relation to cohorts means everything.

On reef systems with small communities of Blueheads, Yellowtail Parrotfish, *Sparisoma rubripinne,* and Striped Parrotfish, *Scarus iseri,* terminal males of each species control spawning. As populations grow, the number of initial phase males increases substantially putting competitive pressure on the terminals' dominance. On these reefs, both harems and roving bands of initial phase males and females exist together. To boost their reproductive success, the young males not only interfere with pair spawning terminal males, but also spawn in large autonomous groups. Several aggregations, some with as many as 50 young males, gather upcurrent of traditional spawning sites and waylay egg-swollen females on their way to spawn with terminal males. Resembling females so closely makes initial phase males difficult to distinguish allowing small males, known as streakers, to hang around harems waiting for opportunities to streak in and join the spawning rises of terminal males. Sneakers, on the other hand, maintain low profiles in order to surreptitiously court and spawn with females inside the territories of terminal males. When recognized, the impostors are immediately chased, but often

A male Threespot Damselfish exhibits reproductive pattern to attract mates.

A typically drab Hairy Blenny changes into a carnival of color when breeding.

cover their escapes by disappearing into a passing initial phase herd. Controlling spawning territories is no easy task; on average, site-holders chase interfering males three times each minute during the two-hour reproductive periods. Occasionally terminal males can no longer fend off the competitors that overrun spawning sites and mate with the waiting females. As populations grow, competition can become so heated that terminal males are forced to abandon their prime sites and reestablish territories on the periphery of initial phase group-spawning activities.

Females do have some say in these male-dominated domains. In an attempt to acquire a worthy set of genes for their eggs, many choose to mate only with the largest and most conspicuous males (sexual selection). To help the females make their decisions, terminal males often display their brightest colors during courtship. Most polygamous species are either permanently or temporarily dichromatic (the sexes exhibit different color patterns). Fishes with terminal and initial phases, such as many parrotfishes and wrasses, are generally considered to be permanently dichromatic. Most temporarily dichromatic species change colors and patterns only during reproductive periods. These temporary spawning patterns are common among damselfishes and many large sea basses. Ordinarily monogamous species, without the need to attract multiple mates, are monochromatic (the sexes exhibit the same permanent color pattern). Some species are dimorphic, with males obviously larger or developing physical features absent in females. Three-inch male Sailfin Blenny, *Emblemaria pandionis,* rise above the sea floor flaunting dorsal fins larger than their bodies in spirited displays to attract prospective mates. Male Lancer Dragonet, *Paradiplogrammus bairdi,* and the males of several flounder species display enlarged fins as markers of hereditary fitness.

Although it seems obvious that bright colors play an important role in sexual selection, there is little evidence supporting the idea. To date, only the vigor of a male's courtship displays has been directly correlated to reproductive success. Female Blueheads have been found to select male partners because of their control of preferred spawning locations rather than for their flamboyance. When dominant males were removed from a reef or switched to nearby locations, their former female partners remained faithful to their previous spawning sites even when they were appropriated by less colorful males. Bright male colors and distinctive features also play a significant role in intrasexual competition. Several members of the wrasse family display brilliant colors when engaged in territorial disputes. Sailfin Blennies, Lancer Dragonets and flounders regularly erect their oversized fins when confronting rivals.

Some polygamists willingly mate with any obliging female that happens along, while others show a preference for the largest females, apparently because of their greater egg production. Studies have shown that male Blueheads and Bucktooth Parrotfish, *Sparisoma radians,* allocate more sperm during spawning rises with larger females. And in proportion to body weight, initial phase males have larger testes than terminal phase males, a significant advantage when sneaking and streaking or competing in group spawns where competition is intense. However, terminal phase Bluehead males are estimated to be from 20 to 50 times more reproductively successful than the smaller males.

Reproduction

A male Spanish Hogfish escorts one of his smaller haremic females on a spawning rise.

Because of constant competitive pressure, terminal males of most species can defend successfully no more than seven females at any one time. Late one afternoon we observed the break-up of an unusually large harem of 13 Spanish Hogfish, *Bodianus rufus,* by a small, probably recently sex-changed male that took advantage of the unmanageable situation. Unlike other Caribbean wrasses, Spanish Hogfish are monochromatic, wearing the same purple-and-gold coats throughout life. The harem's dominant male was identified easily by his large size, but the sex of the smaller harem members was visually impossible to determine. Two similar-sized harem members squaring off with their jaws flared, provided our first clue that something unusual was taking place.

Courtship began when the females gathered loosely at the side of a reef. Moments later the terminal male made a close pass, prompting three females to rise above the reef and curve their bodies into an S-shape, signaling their readiness to mate. The unenthusiastic male casually pair spawned with each before wandering away. When he returned several minutes later, five additional females were waiting to mate, but the aloof male obliged only two before once again leaving. As soon as the terminal male was out of sight, the smaller male made his play. Flaring his fins and jerking his body, he repeatedly nudged the

A large dominant male Spanish Hogfish (foreground) approaches members of his large harem hovering above the reef indicating their readiness to spawn.

The passionate spawning clasp of egg-trading hamlets.

waiting females. When the terminal male returned and attempted to spawn, the young male streaked in and joined the rising pair. The large male instantly turned on the upstart and chased him down to the reef, where he disappeared among a cluster of similar-sized females. By the time this problem seemed in hand, a second large terminal male appeared on the scene. The harem-master instantly gave chase. In his absence the young male quickly spawned with two females, and, with the dominant male still occupied, he and three females fled as a group toward deeper water.

A few small sea basses have both functioning ovaries and testes, which allows them to spawn as males, females or both. These simultaneous hermaphrodites pull off this odd lifestyle by egg trading. Unlike most pelagic-spawning females that typically discharge their entire cache of eggs in a single spawn, egg traders parcel out their energy-expensive cargo a portion at a time. This behavior helps to ensure that their partners will reciprocate by releasing eggs, which they, in turn, fertilize. Palm-sized egg-trading hamlets, *Hypoplectrus* spp., offer one of the most entertaining peep shows on the reef. An hour before sunset, a solitary hamlet leaves its small feeding territory for a nightly rendezvous with a customary mate. The courting couples display fins, alter colors, twitch and chase about the sea floor. The partner taking the female role always acts as the aggressor. Just as the sun sets, the pair rise above traditional prominent features such as a tall coral mound or gorgonian where they intertwine their bodies like a pair of clasped hands for a count of three before releasing a small puff of gametes as they pop apart. Each pair spawns from 2 to 20 times before separating for the night. On occasions a partner will refuse to make the final trade, shortchanging the waiting male. Such inappropriate behavior ends with an aggressive chase.

Resident spawning aggregations of Blue Tang assemble late in the afternoon.

Of course, not all reef fishes are hermaphroditic or polygamists. Most species are gonochoristic, remaining the same sex for life. And although monogamy (mating with a single partner) was once believed to be rare on the reef, new evidence indicates otherwise. It is difficult to substantiate the fidelity of mated pairs over time, but stable relationships lasting several breeding cycles have been documented for some species of butterflyfishes; Gray Angelfish, *Pomacanthus arcuatus,* French Angelfish, *P. paru,* egg-trading hamlets and Harlequin Bass, *Serranus tigrinus,* pipefishes and the egg-laying Cocoa Damselfish, *Stegastes variabilis.* Mouth-brooding male jawfishes and cardinalfishes, with limited capacities to incubate young, have marked tendencies toward monogamy. Because mated pairs of butterflyfishes and angelfishes feed together throughout much of the day, they are the most conspicuous twosomes regularly sighted by divers. Gray and French Angelfish even seem affectionate, rising to circle returning mates. Having only a single reproductive partner limits a male's ability to distribute his genes, but ready access to a mate does offer advantages. However, complete monogamy appears to be rare. Scattered reports have noted male butterflyfishes straying into adjacent territories to spawn and permanently paired angelfishes joining in group sex. Even egg-trading Harlequin Bass and hamlets occasionally spawn with other mates.

Taking promiscuity to extremes, several species of pelagic spawners migrate to traditional spawning sites along outer reef shelves. There they participate in mass spawning events that, for some species, can, depending on local population size, involve thousands of individuals. There are two types of mass spawning groups: resident spawning aggregations involving small grazers, plankton-pickers or mobile invertebrate feeders that travel a few hours or less to local spawning sites inside or near their home ranges, and transient spawning aggregations composed of the large predatory groupers and snappers that, in many cases, migrate for days and, possibly, more than 100 miles to traditional spawning grounds. Species known to form resident aggregations include the Bluehead, Striped Parrotfish, Yellowtail Parrotfish, Ocean Surgeonfish, *Acanthurus tractus,* Blue Tang, *A. coeruleus,* Spotted Goatfish, *Pseudupeneus maculatus,* and egg-laying Brown Chromis, *Chromis multilineata.* With additional fieldwork, other species will undoubtedly be added to this list.

A resident aggregation of Blue Tangs estimated between 6,000 and 7,000 individuals was observed for a four-year period by researchers in Puerto Rico. The gatherings always occurred in late afternoon around the full and new moon at a site in 60 feet of water. The cloud of fish would repeatedly rise, 10 or more feet, then slowly settle back to the bottom. Eventually subgroups of 4 to 15 males each following a single female broke from the main group and made a high spawning rise that carried them near the surface.

Nearby 20,000 Ocean Surgeonfish spawned in a similar fashion each afternoon from December through March. Cued by the moon, large, generally solitary groupers and snappers migrate long distances once or twice each year to reproduce at traditional spawning grounds for multi-day periods. During the week before the full moons of two consecutive winter months (usually December and January), Nassau Grouper, *Epinephelus striatus,* become restless and move out of their home ranges. Decades ago, before inordinate fishing pressure on their aggregations significantly reduced population numbers, groups of up to 500 fish, known as grouper runs, were sighted swimming along outer reef shelves toward spawning grounds. As the fish arrive they congregate

Transient mass spawning aggregations of Nassau Grouper.

Reproduction

A transient spawning aggregation of Dog Snapper, numbering between 500 and 1,000 individuals, gathers in 90 feet of water along an outer reef bank.

across a wide expanse of moderately deep bottom. Spawning takes place in late afternoon typically for three to four consecutive days. Shortly before sunset the fish, wearing a dramatic black and white spawning pattern become restless. Small subgroups consisting of a single female and up to 20 males, dash upward for 20 to 30 feet at steep angles, leaving visible gamete clouds at the apex of their rises. Because Nassau Grouper have never been observed pair spawning outside aggregations the annual events probably represent the species only opportunity to reproduce.

Mutton Snapper, *Lutjanus analis,* Dog Snapper, *L. jocu,* and Cubera Snapper, *L. cyanopterus,* normally form large, transient spawning aggregations during consecutive months in late spring and early summer. However, one team of researchers studying a gathering of Nassau Grouper in Belize in January unexpectedly discovered an aggregation of Dog Snapper estimated from 500 to 1,000 individuals milling about the bottom in 90 feet of water 600 feet from the grouper. The snapper did not exhibit spawning patterns, but a few were noticeably darker than normal. When approached by divers, the scattered group dropped over a ledge to 160 feet, then ascended, forming a comet-shaped cloud. At 30 feet the rising snapper compressed into a compact ball before disappearing back into the depths. A few groupers displaying their distinct black-and-white spawning colors were mixed in with the throng. Although the rising fish appeared to make a spawning rush, a gamete release was not seen. In the Dry Tortugas, Mutton Snapper have been observed releasing gametes in the late afternoon.

The question of how fishes know when and where to gather has prompted considerable research and debate. One hypothesis suggests that sites and times are selected to maximize survival of eggs and larvae. Possible benefits to offspring could include rapid dispersal of eggs, protection from predators,

Spawning rise of Tiger Grouper from transient spawning aggregation.

placement of larvae in optimal conditions to find food in the open ocean and increased chances for offspring returning to their reefs of origin. Research in Little Cayman, using current drifters, has substantiated that Nassau Grouper, gathered at the local spawning aggregation, select evenings to spawn when conditions are most favorable for retaining offspring in nearby waters. Many pelagic-spawning species migrate to the outer, downcurrent edges of reefs or island shelves, where they spawn at dusk presumably to protect their progeny

Yellowtail Snapper and Sergeant Major occasionally feed on the gamete clouds of pelagic spawners.

from the many egg predators on the reef. Water column feeders, such as Brown Chromis and Creole Wrasse, *Clepticus parrae,* which pick plankton above the reef throughout much of the day, move closer to the bottom as light fades. Their departure creates a window of opportunity for clouds of fresh spawn to dissipate unmolested. A few pelagic spawners, notably parrotfishes and smaller wrasses, spawn during daylight hours when feeding in the water column is intense. However, even though their conspicuous courtship behavior signals impending spawns, few planktivores feed directly on their clouds of gametes. Only Yellowtail Snapper, *Ocyurus chrysurus*, and Sergeant Major, *Abudefduf saxatilis,* regularly lurk around active spawning grounds.

Spawning fishes leaving the secure cover of the bottom to place their gametes above the reef brazenly expose themselves to nearby piscivores. Under these circumstances, speed would seem essential. Studies have demonstrated a correlation between the body size of spawners and their rate of ascent. Larger species less vulnerable to predation, such as Stoplight Parrotfish, *Sparisoma viride,* often climb over 15 feet. Whitespotted Filefish, *Cantherhines macrocerus,* immune to most predators because of size and deep bodies, ascend high off the sea floor in slow, graceful, head-to-tail spiraling spawning rises. At the opposite end of the size scale, three-inch Peppermint Basslet, *Liopropoma rubre,* barely rise above their coral cover to release gametes. These tactics evidently work: few successful attacks on spawning fish of any size have been documented.

If spawning on outgoing tides offers survival benefits for offspring, the strategy is not adopted by many species. Small and medium-sized site-attached pelagic spawners seldom coordinate reproduction with the lunar cycle, which in turn controls the tides. Of 22 pelagic spawners from eight families documented, only five species, all large predators that migrate to distant spawning grounds, were found to have lunar spawning cycles. However, in regions with pronounced tidal flow, there are indications that several species alter their reproductive schedules to spawn with outgoing tides. In areas with minimal tides, the same species reproduce at established times.

Most reef fishes have long reproductive seasons with one or more annual peaks. These peaks tend to occur during the cool-water months of late winter and early spring. Whether spawning in pairs or groups, daily, seasonally or annually, pelagic spawners rendezvous faithfully at traditional sites during specific times. Generally, pair-spawning individuals and resident-spawning groups migrate to customary sites typically near three- to five-foot structures such as tall coral heads, sponge clusters or gorgonians. Some of these sites have been documented to be in continuous use for at least three decades. It was originally proposed that raised features act as launching platforms, facilitating higher spawning ascents. However, when elevated structures are absent from an area, no decrease in spawning activities occurs. The spawning grounds of transient aggregations are generally associated with distinct promontories jutting seaward from the insular shelf. Several of these sites are known to have been active for more than 75 years and are often used by different species at different times. Dyes and surface floats used to track the dispersal of eggs from the resident spawning sites of Blueheads produced little evidence of eggs leaving inshore waters from either location. It was the researchers' conclusion that, at least in some instances, reef fishes do not spawn at superior dispersal sites or at the best times for their offspring to be swept to sea. Today, many researchers lean toward the idea that prominent topographical features act simply as landmarks to help widely dispersed individuals locate each other during spawning periods.

A long-range study of isolated Bluehead populations demonstrated that tradition, rather than resource assessment, is the strongest influence in the spawning site selection process. For 12 years, encompassing four generations, the same sites were used continually for daily spawning even though the area

The slow spiraling spawning rise of Whitespotted Filefish.

had an abundance of alternate sites. When entire populations were removed from three separate reefs and replaced with like numbers and sex ratios of Blueheads, it was discovered that the previously used sites were no more likely to be selected than any other suitable sites in the area. However, when the manipulated populations were themselves replaced a month later, the new populations demonstrated a strong tendency to reuse the sites selected after the first manipulation. This led the researchers to suspect that, in the absence

A hybrid cross (left) between an Atlantic Creolefish and Coney (below) both members of the sea bass family.

of tradition, resource assessment, possibly for optimal egg transport, plays a dominant role in mating site selection.

Like nearly all reef fishes, Blueheads settle out of the plankton to reefs where they remain for their maximum three-year life span. The tiny juveniles, quite susceptible to predation, remain close to shelter greatly limiting their ability to migrate or to assess their surroundings. Therefore it would seem almost certain that individuals learn the routes to mating sites by following members of the existing population rather than by trial-and-error. After researchers replaced a population, it took three to four days before the new residents began to spawn. It can be hypothesized that if interruptions on a small scale cause such confusion, then they should create even greater disruption for large spawning aggregations. Many scientists fear that when transient aggregations are repeatedly overfished, too few experienced adults remain to teach succeeding generations how to locate spawning grounds. This could have a devastating effect on the large species of groupers and snappers that are believed to concentrate their entire reproductive efforts during limited mass-spawning events each year.

With so many species of pelagic spawners placing gametes in the water column, it seems inevitable that occasional hybrids would occur. And, in fact, cross-fertilization among fishes in the wild is fairly common. The mixed offspring are almost precisely intermediate in body structure and color patterns to parental species. The more closely related, the better the chance two species have of crossbreeding. It is extremely rare for members of different genera to mix, but four cases of such natural crosses between western Atlantic fishes have been documented. The most peculiar occurred in Bermuda and Brazil between two members of the sea bass family. The two species, Atlantic Creolefish, *Cephalopholis furcifer,* and Coney, *Cephalopholis fulva,* were formerly classified in separate genera (as *Paranthias furcifer*) and do not even share the same habitat. Atlantic Creolefish, reported to form loose spawning aggregations, are plankton-pickers that feed high in the water column. Coneys inhabit bottom territories where they pair spawn within harems. The three other recorded hybrids are crosses between Yellowtail Snapper and one of three species of snapper in the genus *Lutjanus* (Schoolmaster, *L. apodus,* Gray Snapper, *L. griseus,* and Lane Snapper, *L. synagris*). The Yellowtail Snapper/Lane Snapper hybrids were once classified as a valid species, the Cuban Snapper, *L. ambiguus.* In an experiment in the late 1980s a Lane Snapper and a Yellowtail Snapper were crossed in the laboratory. When examined 16 months later, the offspring had the same morphology and coloration as individuals previously described as Cuban Snapper. The research points to the apparent ease with which Yellowtail Snapper cross in the wild with *Lutjanus*, bringing into question the validity of *Ocyurus* as a legitimate genus.

It seems unlikely that the sperm and eggs of different species, released separately, would find each other in the currents. The hybrids were more likely engendered when Yellowtail Snapper joined the spawning activities of the three *Lutjanus.* Like the Nassau Grouper involved in the spawning ball of Dog Snapper in Belize, Yellowtail Snapper could accompany the spawning rises of *Lutjanus.* We have twice observed Yellowtail spawning in groups of 10 or 12 individuals during the late afternoon one summer. The fish circled a few feet off the bottom before suddenly charging straight up 15 to 20 feet, where they left visible gamete clouds. This behavior could be easily assimilated in the spawning activities of other snappers.

Because of the close physical and behavioral similarities between Striped Parrotfish, *Scarus iseri,* and Princess Parrotfish, *S. taeniopterus,* researchers have long expected to find a natural hybrid of the species. Interestingly, Princess Parrotfish are one of the few pelagic spawners that reproduce in the early morning; the look-alike Striped Parrotfish spawn in the more traditional late afternoon. This is probably the factor limiting cross-fertilization between the two.

Although most hybrids are born sterile or die before reaching sexual maturity, crosses between Queen Angelfish, *Holacanthus ciliaris,* and Blue Angelfish, *H. bermudensis,* occur regularly where the ranges of the two species overlap in Florida and the western Bahamas. Their hybrid offspring, once incorrectly classified as the Townsend Angelfish, *H. townsendi,* are fertile and frequently backcross (mating with either parent species), creating a variety of intermediate color forms, known as a hybrid swarm. In rare cases, crosses between species can lead to the formation of new species. The importance of this evolutionary process to reef fishes is still being investigated.

When visiting reef communities divers are often astounded by the tranquility.

Bluestriped Grunts compare the size of their mouths in a behavior, known as gaping.

But lurking just beneath the apparent harmony is a pervasive tension engendered by a primal competition for reproductive success. Open hostilities originate from territorial disputes for resources and mates, or more commonly, from attempts to maintain a position in the local pecking order, where size dictates dominance. When matters come to a head they are settled with the only weapon fish have at their disposal—mouths. But first there is the matter of injuries. Having an open wound in the sea can quickly lead to life-threatening infections. To limit the risk most fish attempt to resolve disputes with a behavior, known as gaping, where rivals compare the size of their open mouths with little physical contact. As the thinking goes: An unobtrusive comparison of mouth size functions as a proxy for body size. In instances where the mouth of one of the rivals is larger than the other's, even by a hair, the smaller opponent relents, and the two go their separate ways. However, when adversaries are equal in size the strategy quickly breaks down and the fish lock jaws. Usually, one of the fish concedes within seconds and the pairs move apart. On rare occasions rivals sling each other about in violent, protracted struggles. In cases of territorial disputes between neighboring bottom-dwellers, pairs often remain stationary on the bottom locked jaw-to-jaw for extended periods. On a morning dive in Dominica,

Young Spanish Hogfish attempting to establish rank within their harem.

Territorial Spotted Scorpionfish locked mouth-to-mouth in a lengthy standoff.

French Grunts engage in a long struggle.

Brown Chromis locked in combat.

A Smooth Trunkfish challenges a rival.

Striped Parrotfish attacks its image.

we noticed a pair of beefy Spotted Scorpionfish, *Scorpaena plumieri,* each the size of my forearm, lying across the bottom attached to one another in what appeared to be mindless rage. At the end of our dive, an hour later, the pair was still locked together, exactly where we originally encountered, them in what seem to be blind, eye-to-eye fury.

Benthic Egg Laying (Demersal Spawning)

Although pelagic spawning predominates on tropical reefs, benthic egg laying is widespread. As a general rule, egg-laying species are small and numerous. Damselfishes, blennies and gobies are the most common egg-laying families on Caribbean reefs. Surprisingly a few large species in the triggerfish family also prepare and tend demersal nests. As benefits, egg-layers do not have to make dangerous openwater ascents to release their eggs, search for mates or make risky migrations to distant spawning sites. However, substantial energy costs are required of the parental males to prepare nests, fin-fan clutches, pluck out bacteria and dead eggs, and keep aggressive egg-bandits at bay throughout daylight hours. Even after dark when the threat of predation from fishes subsides fireworms, brittlestars and crabs commonly menace unguarded egg nests. The males' care pays dividends. Unlike the eggs of their pelagic-spawning counterparts that hatch in little more than 24 hours as tiny, defenseless yolk sac larvae barely able to swim, the larvae of egg-layers emerge after six or seven days larger, stronger and better prepared to fend for themselves during their multi-week metamorphosis in the open sea. Also, demersal eggs are larger, negatively buoyant, opaque, and carry more yolk than their more numerous counterparts released in open water. Larvae usually hatch during the first hour after sunset when most egg predators have bedded down. Many egg-layers spawn with the lunar cycle. Of these, the majority follow biweekly timetables believed to correspond with the new and full moon. However, only limited correlation has been found between hatching times and optimal tidal flows that would quickly carry the hatchlings away from the reef.

Egg-laying species are most active just after daybreak when males dart about coaxing females to their nest sites normally located inside holes, crevices, shells, under rocks and encircling gorgonian stems. The majority of these nests are small and difficult to locate. The nests of Sergeant Majors are notable exceptions. Males of the common Caribbean damselfish prepare nest sites in

Male Sergeant Major (above) wearing reproductive colors, spawns with a compliant female. **Right:** *Purple Reeffish spawning in the algal turf of a reef crevice.*

Reproduction

Male Longfin Damselfish guards nest.

Genital papilla of Sergeant Major.

Well developed Sergeant Major eggs.

A spawned-upon Spinyhead Blenny.

Sapo Cano toadfish with nest of offspring.

Recently hatched brood of a Sapo Cano.

the open on hard smooth surfaces of coral rock, mooring blocks, pilings and metal structures. Be cautious, approaching closer than five feet frightens guardian males away from their nests leaving the contents vulnerable to nearby predators waiting for just such an opportunity.

Prior to multi-day bouts of courtship and spawning, which occur sporadically throughout the year with peaks in the summer, male Sergeant Majors pick nesting areas (measuring two to three feet across) clean of debris and algae with their mouths. When finished, suitors turn dark bluish gray and begin performing repeated bouts of eye-catching signal jumps to attract mates. Over the next several days from one to four females during numerous visits contribute up to 20,000 elliptical eggs each to a single nest. The total number of eggs deposited by a female during a single daylight period is referred to as a clutch. Eggs emerge one at a time from an extended triangular organ, known as a genital papilla, located behind the anus. Each capsule, the size of flax seed, is secured to the bottom by a fibrous holdfast allowing the tiny orbs to sway when aerated by the sweeping motion of an attending male's fins. The nests slowly expand in size, as new rows of eggs are added adjacent to those most recently laid. Following each pass the male fertilizes the latest batch with an invisible mixture of sperm and seminal fluid (milt) that streams from his extended genital papilla.

After a few days, with their contributions complete, females abandon their offspring to the care of guardian males. Males groom and protect their charges, throughout daylight hours, until all the clutches hatch six to eight days later. During incubation it is possible to estimate the approximate age of a clutch by noting their changing colors. Fresh eggs are reddish purple. By day four the eyes appear as dark dots inside increasingly transparent capsules. On day five a curl of vertebrae can be seen inside each swollen sphere and dark eye pigment dominates giving the nests a silvery gray sheen.

The egg nests of most damselfish species are difficult to find. Males in genus *Stegastes*, including the Longfin Damselfish, *Stegastes diencaeus,* and the Dusky Damselfish, *S. adustus* prepare small four- to seven-inch nests inside crevices, but can occasionally be found on exposed blades of fire coral. The best way to locate a nest site is to watch for males displaying courtship patterns during early morning hours. Although the spawning patterns of a few damselfishes aren't apparent, the whitish upper heads, backs and tails of courting Longfin and Dusky Damselfish stand out in marked contrast to their dark bodies.

Mass spawning episodes of Brown Chromis are fairly common and something to see. They begin during the day with a large number of males flocking to a section of reef tract or coral slope where the three-inch fish energetically set about fashioning crude nests in shallow algae-covered depressions. With the arrival of females two days later hundreds of fish turn into thousands. The territorial males go wild signal jumping, chasing and fertilizing eggs. Usually only one female at a time enters a nest site. Skimming over the surface the size of a platter for a minute or two they scatter eggs, too

Left: *Nests and developing eggs and embryos of reef fishes.*

Resident aggregations of Brown Chromis gather in profusion to spawn.

small to see. As they retreat the males are right behind spreading milt. This same sequence with other females continues for hours. Telling the sexes apart is difficult. Other than the males' slightly darker backs and the females' somewhat lighter complexion and slightly swollen bellies the sexes are indistinguishable. As one would suspect the lengthy spawning bouts attract plenty of egg predators. In Bonaire, besides the usual contingents of hungry wrasses and Sergeant Majors hanging around, we observed a pack of two-dozen Scrawled Filefish, *Aluterus scriptus,* mob the nests. Even Atlantic Trumpetfish, *Aulostomus maculatus,* join the frays, picking off distracted Brown Chromis rather than eggs.

A few species of triggerfish, most notably Queen Triggerfish, *Balistes vetula,* and Sargassum Triggerfish, *Xanthichthys ringens,* spawn in open water. Other family members lay fist-sized gelatinous masses containing thousands of eggs in shallow depressions blown in the sand. The egg-laying triggerfishes are not only the largest reef fishes that participate in demersal spawning, but the only species known to provide maternal protection. Lacking the agility and energy of smaller egg-laying fishes, large triggerfishes have a difficult time keeping darting egg-raiders at bay. Having more than one guardian on patrol is a great help. Fortunately, the eggs of triggerfishes hatch quickly, possibly within one or two days. One afternoon we watched three Black Durgon, *Melichthys niger,* probably members of the same harem, guarding an egg nest in a sand channel between coral ridges. The triggerfishes were kept constantly on guard by scrappy packs of wrasses and several damselfishes circling just out of reach of their beaks. When we approached, the triggerfishes moved several feet away and watched nervously. In a shallow sand basin no more than a foot across, we found a flattened gelatinous egg casing encrusted with sand. Before the wrasses attacked we backed away allowing the Black Durgons to remain in control. On another occasion, for two consecutive days, we observed 15 to 20

Black Durgon use teamwork to defend their sand nest from numerous nimble egg-predators.

large, normally solitary Ocean Triggerfish, *Canthidermis sufflamen*, gathered in the water column above a bed of Manatee Grass, *Syringodium filiforme*, in 55 feet of water. While most of the large graceful oceanic triggerfish cruised high above, a few individuals blew shallow, two- to three-foot-wide nests in bare patches in the seagrass bed below. The fish displaying dark faces and backs were wary and curtailed nest building when we approached. We never observed them spawn or found eggs in the nests.

Male jawfishes and cardinalfishes take parental responsibilities to extremes by incubating eggs inside their mouths. During the eggs' five- to seven-day incubation, male Yellowhead Jawfish, *Opistognathus aurifrons*, only momentarily leave their clutches unattended within their burrows while they pick bits of zooplankton above their burrows. Because of their small size male gobies and blennies are quite secretive about where they prepare nests. Most are located deep inside recesses impenetrable to egg thieves. Occasionally the nests of larger blennies, such as the common Redlip Blenny, *Ophioblennius macclurei*, spill out of a recess into the open. In coastal waters several blenny species use discarded bivalve shells for nests.

A male cardinalfish uses his mouth to incubate his clutch.

Male jawfishes regularly aerate and mix their clutches.

Live-Bearing

The majority of bony fishes are oviparous—releasing gametes for external fertilization. Cartilaginous fishes are typically viviparous—bearing live offspring. Male sharks, rays and skates fertilize their mates internally (copulation). They transfer sperm into the cloaca (a ventral opening used for reproduction and excretion) of their mates with claspers, organs modified from the inner edge of each pelvic fin. The presence of claspers can be used to distinguish males from females. The ends of claspers, often equipped with hooks and spines, flare when inside a female to help secure a hold. Sperm passes through grooves in the claspers, fertilizing a few large eggs.

The developing embryos of viviparous species receive nourishment in various ways, but most absorb food directly from their mother's uterine wall, much like mammals. Members of the hammerhead and requiem shark families reproduce in this manner. In a version, known as cannibal viviparity or intrauterine cannibalism, Sand Tiger, *Carcharias taurus,* produce two pups—one developing in each of two separate uteruses. The first young devours its sibling before consuming its mother's continued production of eggs. A hardy, three-foot pup is born after an eight- or nine-month gestation period. Ovoviviparous species, stingrays for example, carry their fertilized, yolk-rich

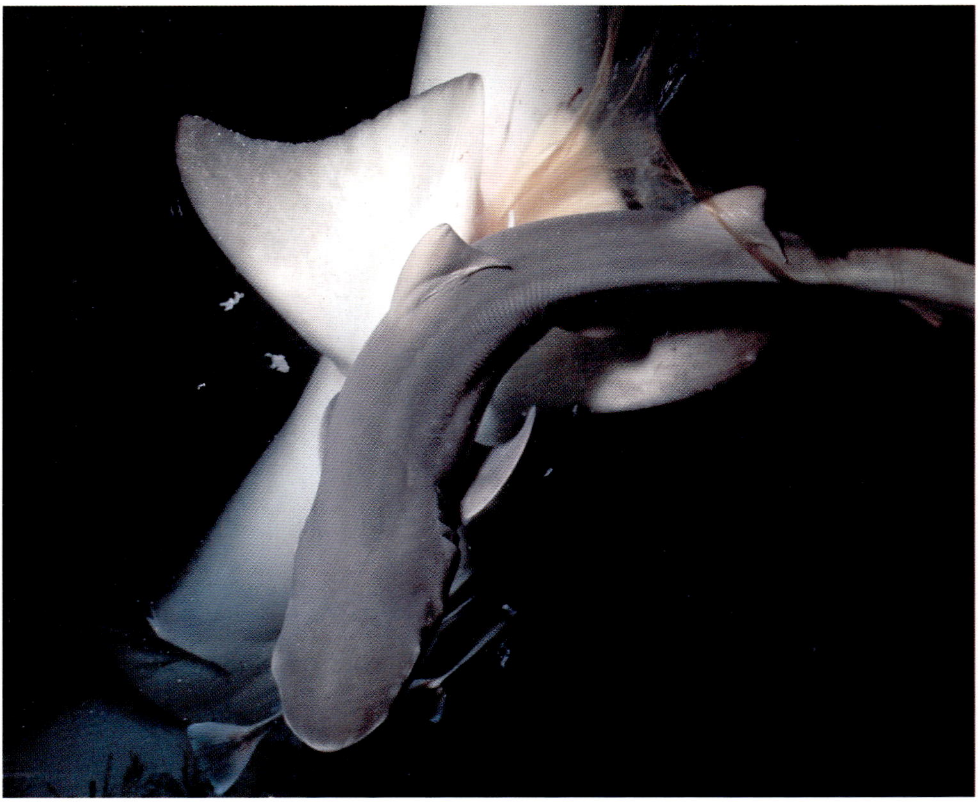

The live birth of a Lemon Shark.

A female Black Brotula packed with embryos.

eggs throughout gestation. After the embryos consume the yolk, they eat food secreted from the uterus. Skates are the only rays that do not bear live offspring. Females lay durable, leathery cases, commonly known as mermaid's purses, that they attach to bottom structures by tendrils. Each packet contains a fertilized egg and several months' supply of yolk.

Little is known about the reproductive behavior of sharks. Bite marks on the pectoral and dorsal fins of females attest to the aggressive nature of the encounters. During breeding season, males swim behind potential mates with their snouts close to the females' cloacas to assess their reproductive status. When ready to copulate the males bite the females' sides and fins, a courtship behavior believed to elicit a compliant response from their mates. With the females' pectoral fins firmly clenched in their mouths, the males turn on their sides and insert sperm through a clasper. Six to 22 months later, fully formed pups, able to fend for themselves and large enough to avoid most predation are born in the shallows.

A few species of brotulas in family Bythitidae are the only Caribbean bony fishes known to be viviparous. The 70 or so species of these curious little fishes inhabit a variety of habitats worldwide, including freshwater caves, estuaries, reefs and the deep sea. Black Brotula, *Stygnobrotula latebricola,* secretive three-inch reef dwellers, are occasionally sighted in pairs at night, inside crevices and under overhangs of shallow reefs. Brotulas swim by undulating fins encircling the length of their bodies. The males' copulatory organs, located behind the anus and surrounded by two pairs of pseudoclaspers, are apparently derived from the anal fin. Embryos, closely packed like cordwood, develop inside the females' ovaries. It is not known whether the offspring disperse in the currents or remain near their birthplace.

Life Cycles & Reef Fish Communities

Larval Pancake Batfish, an uncommon member of the batfish family.

The majority of reef fishes begin life in gamete clouds launched high above the reef. This hedge against the concentration of hungry predators and parasites below is the only parental assistance offspring receive before being carried away by currents on one of the most perilous odysseys in the animal kingdom. If one considers that to sustain population numbers, each adult only needs to reproduce itself once or twice during a lifetime, and that a single female typically produces tens of thousands of eggs annually, it's easy to comprehend the extraordinary odds against surviving their multi-week, oceanic passage though the larval stage. Weeks later and in some cases hundreds of miles from their origins, the few survivors navigate toward shore at night with the help of an inshore tide, and settle to the shallow sea floor. Over the next hours to days, the seafaring settlers metamorphose into site-attached juvenile reef fishes that, with growth and continued luck, will one day shed their own gametes above the reef. This endless cycle of prolific spawning and unpredictable recruitment is responsible for structuring fish communities on coral reefs around the world.

Amazingly, before 1980 reef ecologists were generally unaware of the great "pelagic orphanage" and its profound importance. Instead of gathering detailed natural history data, early researchers, often former bird ecologists, simply adopted the classic niche theory, the accepted model of terrestrial ecology. They believed that under water, as on land, most animals live in "closed" communities saturated with an endless supply of their own offspring. Constant competition for limited food and space, and the forces of predation ultimately shape the final assemblies. Pioneering reef ecologists were so enamored of their pet hypothesis that nearly everyone failed to note that marine fisheries researchers in Europe and the United States had, for eight decades, been working with a different set of population dynamics. Their "open" non-equilibrium model focused on a correlation between adult stock size and random waves of larval recruits arriving from the open ocean. As early as 1914, Norwegian fisheries biologist Johan Hjort indicated the potential importance of recruitment to adult populations of marine organisms. In 1950 and 1966, Gunnar Thorson stated that marine invertebrates have a planktonic phase and a "settled" juvenile phase. He also pointed out that such species are subject to large-scale fluctuations in abundance. However, these and other early insights played no role in molding the favored ecological theory of the day.

Thinking back, it is easy to understand how everyone was seduced by the niche theory. It seems to make good sense when observing a reef fish community in action. To an unsophisticated eye, a never-ending turf war appears to be taking place. Every niche in a reef's craggy face houses an animal ready to protect itself, its territory and its food supply. It simply was assumed that the local offspring of spawning reef fishes quickly found protected corners in their home reefs or were eaten or driven away. During the late 1970s and early 1980s, when evidence supporting an open system no longer could be ignored, marine ecologists scrambled to get on the bandwagon even though the new hypothesis was, and remains, difficult to demonstrate. Although field studies in the pelagic realm are logistical maelstroms that more often than not produce perplexing variations in nearly every set of data, ecologists have made great strides in improving larval collection methods and gaining a better understanding of physical oceanography since the beginning of the century. More and more findings indicate that a substantial amount of larvae are retained near their home reefs by tidal fronts, current gyres, enclosed bays and larval behavior, leading to the conclusions that reef systems vary in their degree of openness. For the most part, the new research supports the concept that while some larvae are carried great distances by currents, a substantial number settle near where they were spawned.

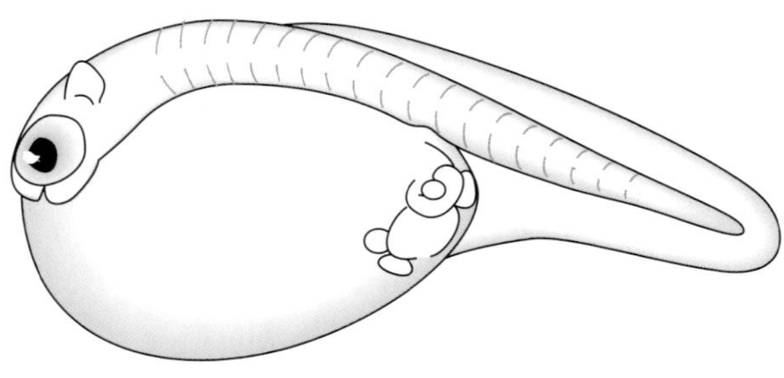

A typical yolk-sac fish larva.

The daily otolith aging technique introduced by Giorgio Panella in 1971 promoted everyone's research efforts. This revolutionary tool makes it possible to learn the age of a fish from embryo to adult by counting concentric layers of protein-rich and protein-poor materials embedded in tiny calcium carbonate structures, known as otoliths. The three tiny stones, used for balance and sound detection, are found within the semicircular canal at the base of the braincase of all bony fishes (cartilaginous fishes, such as sharks and rays lack otoliths). After preparation for observation, the structures' matrix of protein appears similar to the growth rings of a tree. Although the legibility of time-markers varies among species, daily increments are sufficiently distinct to be deciphered on most larval and juvenile fishes. The technique provides a wealth of indispensable information, including the duration of a fish's pelagic larval stage and even the ability to determine the day it was spawned, as well as estimating a species' maximum age, age at reproductive maturity, and the analysis of generational trends.

Despite these advances in knowledge, much of the early life history of reef fishes remains a mystery. It is known that within 24 hours the translucent, slightly buoyant, nearly spherical, fertilized eggs ranging in diameter from 0.6 to 1.2 mm released by broadcast spawning species hatch into tiny, weak swimming yolk-sac larvae with long tails and distended bellies packed with a two- or three-day store of fat. By the time their on-board supply of food is gone the tiny larvae's gut, eyes and ability to orient have developed sufficiently for them to feed on their own. Growth accelerates as medial fins form and the swim bladder develops enabling them to position themselves vertically within a preferred current flow strata that better influences their ability to remain close to home. As the vertebrae develop the tail fin becomes functional allowing greater swimming stamina and speed. In contrast, the somewhat larger eggs laid and fertilized on the bottom by typically small species, such as damselfishes, gobies and blennies, require several days of incubation allowing the embryo to go through the yolk sac stage inside the egg. At hatching the more physically mature larvae than those spawned into the open sea are ready to swim and feed—a substantial advantage for survival at the cost of far fewer offspring produced.

The clear, nutrient-poor, blue water bathing tropical reefs so prized by divers are food deserts for larval fishes. Fortunately, a fraction of tropical oceans transports random clouds of concentrated food, known as patches, near the surface. If these scattered patches didn't exist, food particles would be so widely dispersed that larval fishes wouldn't be able to survive. These patchy sources of food concentration have multiple origins including deep vertical upwellings, large cyclonic eddies, and meandering fronts bordering shorelines created when water masses of different densities collide. As one flow slips beneath the other everything that floats or swims collects along the upper surface creating a smorgasbord for hungry fish.

Poorly understood, thin surface layers of concentrated life, varying vertically from inches to several feet, often occurring in bays and estuaries, can extend like a blanket over the sea for miles around and persist from hours to weeks. Such layers that can incorporate 50 to 75 percent of the biomass from surrounding waters are considered another important component of larval survival in the pelagic.

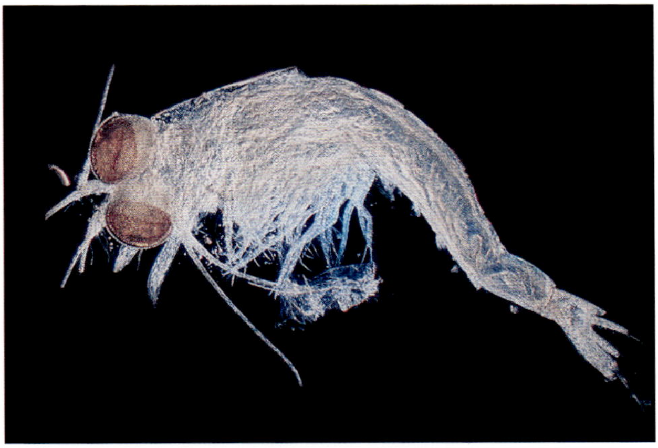

Microscopic pelagic crustaceans, members of the zooplankton community.

In the tropics deep vertical upwellings begin on warm, calm days when cool water barrels up from deep below. The sudden mixture of nutrient-rich water and sunlight creates a wild bloom of single-celled plants or simple chains of plant cells, known as phytoplankton. Single-celled diatoms, encased in intricate silica shells, are the most prevalent pelagic algae. These simple plants, forming the base of the food web, are a favorite food of oceanic copepods, the most abundant group of animals on earth, and the primary food of larval fishes. When the tiny pelagic crustaceans begin to gorge, they multiply by the millions. Soon, mollusks, salps, and other pelagic herbivores add their mass to the ever-thickening soup. The banquet is joined by macro-carnivores whose numbers are dominated by flat, elongate chaetognaths, commonly known as arrow worms, with formidable mouths ringed with teeth and grasping hooks capable of ripping chunks from egg-yolk larvae. These microscopic oceanic animals, collectively known as zooplankton or zooplankters, quickly attract jellyfishes and siphonophores trailing deadly mops of stinging tentacles. Schools of fast-swimming fishes including tuna regularly cut through the moveable feast. Even though this savage pelagic world should seem a nightmare to our cherubic fish larvae, their instincts indicate otherwise. For the next few weeks, this drifting patch of snapping-jaw hell will be home sweet home, stocked with a full cupboard of food only a tail-thrust away.

In the beginning, young fish larvae consume a variety of foods, including phytoplankton, larval copepods and mollusks and, in some cases, siblings. When potential prey is sighted the larval fishes curve into an S-shape and scull forward with their pectoral fins until close enough to engulf the victim with quick open-mouthed lunges. As they mature, body mass increases by a third each day, and each species begins to select a more specialized diet.

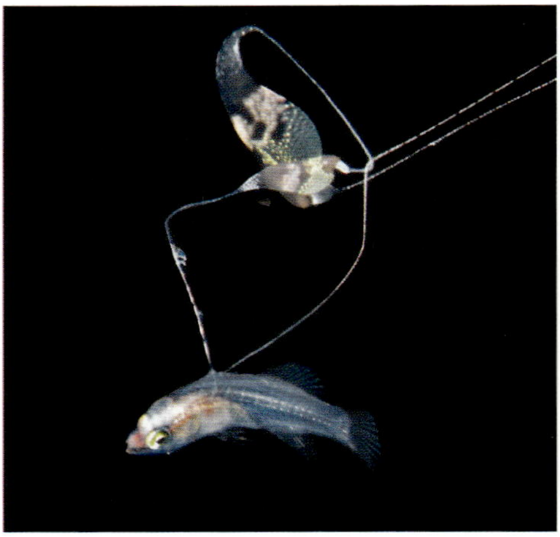

Larval basslet

Adult Candy Basslet

Larval burrfish

Adult Web Burrfish

Larval Squirrelfish

Adult Squirrelfish

Fortunately, evolution hasn't forsaken the little fishes during these dangerous times. As the yolk sac disappears, larvae develop elaborate morphological adaptations that attain a state of complexity that delights the imagination. Spikes, elongate spines and spinelets, long fins with streamers, enlarged heads sporting enormous jaws are a few adaptations designed to deter and deceive predators. Silver or transparent bodies aid in this death-defying game of openwater hide-and-seek.

The size of drifting patches, without defined borders or an even distribution of food, varies horizontally from meters to miles depending on the stability of the surface waters. When storms brew, wind and waves break patches apart; during periods of calm, they retain their size or even enlarge as they occasionally coalesce. Nearly all larval fishes inhabit the upper 600 feet, with members of different families inhabiting specific depth ranges during the day. At night, these strata often lose definition as larvae move toward the surface to feed. The purpose of these daily migrations is uncertain. Possibly the cooler water below preserves energy by lowering metabolism, the larvae descend into the dark to escape visual predators during the day or to avoid the sun's ultraviolet radiation. Whatever the reason, it must be important. The 600-foot journey for a two-millimeter larva is equivalent to a 120-mile trip for humans.

A typical larval voyage in the Caribbean lasts between 14 and 30 days but many continue for 120 days or more. In an exceptional case some porcupinefishes have a pelagic phase lasting more than a year. Before settling to a suitable habitat on the sea floor, some larvae undergo a second dramatic morphological shift; others change little more than color. The Queen Triggerfish, *Balistes vetula,* one of the largest species retaining pelagic features at settlement, measures two inches after spending 75 days in the plankton. One of the smallest, a reef drum called the Cubbyu, *Pareques umbrosus,* is only four millimeters when it ends its pelagic existence after 14 days. A few species, including puffers, chubs, bigeyes, Gray Triggerfish, *Balistes capriscus,* and Sergeant Major, *Abudefduf saxatilis,* develop into juveniles prior to settlement. Most semi-pelagic species such as jacks never settle. Instead, juveniles mature into adults in the plankton, often in association with rafts of *Sargassum*, jellyfishes or openwater predators like barracudas and sharks. During inshore night lighting dives we have observed marked developmental and size differences among different species of settling fishes, and to a lesser extent between members of the same species. For instance, most quarter-sized settling surgeonfish retain much of their pelagic translucence; however, we noted a single settling Blue Tang, *Acanthurus coeruleus,* with its complete juvenile coloration. A few species, such as members of the bigeye family and the Atlantic Trumpetfish, *Aulostomus maculatus,* are surprisingly large, regularly measuring more than two inches in length, other species are so small they can hardly be seen. Whereas some species, such as Brown Chromis, *Chromis multilineata,* arrive as full-fledged juveniles easily recognized to species, others settle with little or no opacity and many of their pelagic characteristics still in place.

Left: *A comparison between larval and adult reef fishes.*

A member of the bigeye family, among the largest species at settlement.

A recently settled Jackknife Fish—one of the smallest reef fish at settlement.

The elongate, glassine, leaf-shaped larvae of eels, bonefish and tarpon, known as leptocephali, belong to the superorder Elopomorpha. Leptocephalus, prevalent components of the larval diaspora, arrive inshore following extended sojourns in the sea, which can last for three months to a year, still showcasing pelagic features so otherworldly that it is difficult to imagine what adult form they will eventually acquire. Leptocephalus' long flexible bodies move with a typical eel-like swimming style, known as anguilliform, which sends undulating waves along their flexible bodies.

Scientific understanding of larval fish behavior in the great oceanic orphanage has undergone a revolutionary change since the end of the last century. Larvae are no longer thought to be simply passive drifters in an exclusively "open" system carried off to distant shores as captives of capricious tides and currents. A growing body of evidence indicates that although a significant amount are carried away from their natal reef, a large percentage, possibly the majority, remain in local waters until competent to settle near where they were spawned. This new self-recruitment model seems practical if one considers there is no place like home, especially if the alternative means being cast off to an uncertain future in the open ocean with suitable settlement sites few and far between. Science is just beginning to offer explanations of how tiny millimeter-sized larvae pull off this once unimaginable feat. The same path of research will also help explain how some of the waifs whisked off into the open sea eventually find a new far-flung home.

The capacity to alter their destiny may begin shortly after hatching with the inflation of the swim bladder allowing larvae to inhabit depths with horizontal flow patterns conducive for remaining in home waters. Days later, after their tail fins develop, larvae become strong, active swimming machines capable of influencing their dispersal. Swimming speed and endurance continue to increase throughout their pelagic existence. By the time they are ready to settle, larvae have been recorded swimming equal to or faster than

Right: A comparison between post-larval and adult stage reef fishes.

Settling post-larval Redspotted Hawkfish

Adult Redspotted Hawkfish

Settling post-larval scorpionfish

Adult Spotted Scorpionfish

Settling post-larval lizardfish

Adult Sand Diver

the mean current flow in most areas. In-water observations indicate that fish larvae spend much of their time navigating at a brisk pace along established trajectories guided by an unknown combination of sensory cues. Early research has focused on the possible use of sight, hearing, and smell. All three senses are known to develop early and improve with growth, but how this helps larvae remain near home or locate distant sites to settle is far from certain.

The conspicuously large eyes of fish larvae turn out to be more important in capturing prey and avoiding predators in low light conditions than for navigating in the open ocean. Researchers are exploring other possible uses of vision for orientation such as functioning as a sun compass, tracking celestial cues or monitoring wave directions. The larvae's eyes appear to be at their best when used at limited range and probably only come into play during settlement when searching for a preferred habitat or for locating members of their own species. Coral and rock reefs produce individual sound signatures created by a combination of such things as lapping waves, snapping shrimp, and grazing parrotfishes and sea urchins. These diagnostic sound patterns, effective for less than a mile, may be imprinted on embryos for use as a homing device. Reefs also generate fields of distinct smells in the form of dissolved chemical cues carried for a few miles by tides and currents before dissipating. Adjacent settlement habitats, such as seagrass beds and mangroves, also create unique smells. Specific odor halos, possibly imprinted at birth, are presumed to help larval fish remain in the vicinity of their home waters. Of course, the effectiveness of all navigational traits are at the mercy of ever-changing patterns of weather and oceanic processes.

Some species settle on high-profile reefs; others search for sand flats, rubble fields, shallow patch reefs, seagrass beds, estuaries, mangroves or surf zones. A few can be quite choosy: larval Nassau Grouper, *Epinephelus striatus,* find sanctuary in algae-covered clumps of dead finger corals inside shallow mangrove lagoons. Certain species are now known to seek out or to avoid members of their own species during the settlement process by sensing dissolved chemical cues. The same ability also may be used to steer clear of predators. The new arrivals make easy targets. To protect themselves while completing their transformations into juveniles, wrasses, eels and other species burrow down into the sand. Others evade predators by schooling, hiding inside the reef or within rubble fields. It has also been noted that ready-to-settle larvae tend to be choosy and do not always remain at the first location they encounter, instead returning offshore and later seeking a more suitable settlement site.

Although many species settle randomly throughout the year, recruitment reaches its peak in the summer when patches spread along narrow fronts paralleling the shore. During the dark nights around the new moon, when the threat of predation is least, concentrations of larval fishes, known as pulses, ride the flood tide toward the shallows. We have been underwater at night as pulses passed over the reef. One moment the water was clear, and in the next instant the beams from our handlights attracted a blizzard of darting small creatures as strange little fishes rained down. Surgeonfishes the size of quarters with polished heads and bodies so clear you could count their ribs,

A juvenile Atlantic Tripletail—one of several fish species that never settle.

A recently settled post-larval goatfish seconds before burying into the sand.

bounced along the bottom; transparent eel-like fishes squirmed into the sand; and tiny, silver-burnished butterflyfishes paused as if collecting their bearings before disappearing into the reef. Ten to 15 minutes later the patch moved on, leaving behind a new generation of fishes.

If the proper bottom is not encountered, some species can prolong their presettlement mode for days or even months. Surgeonfishes have been reported to settle anywhere from 18 to 65 days after hatching, and a few reef fishes put themselves in a holding pattern for up to 100 days. But even with this capacity, it is possible that a large percentage of larvae are never carried near shore and are lost at sea.

Researchers are quite familiar with the world of larval fishes from their long history of collecting offshore marine life in plankton tows. None other than Charles Darwin himself—co-discoverer with Alfred Russel Wallace of evolution by natural selection—trailed nets behind his ship during his famous 1831 to 1836, round-the-world voyage aboard the H.M.S. *Beagle*. Over the centuries, their efforts to document larval marine life have produced an admirable accounting of larval morphology recorded with painstaking precision, often accompanied by exquisitely detailed drawings. However, it wasn't until the invention of scuba that scientists had the means to observe the offspring of reef fishes in their natural surroundings. But, even after recreational diving became popular, few divers knew about, or had the inclination to search for the illusive life forms in the open ocean.

During the past decades a few innovative charter boat operators have made it possible for divers to safely visit offshore waters at night when the larvae rise to the surface to feed. Most enthusiastic are underwater photographers equipped with cameras capable of capturing the intricate details of the tiny exotics, which are small, fast and erratic swimmers. To give a perspective of what it is like to drift in the sea with a camera after dark we include a 2017 article from *Alert Diver* magazine portraying one of our early offshore night dives.

Night Drifters

from Alert Diver Magazine *2017*

I'm not sure how far off Palm Beach, Florida we are when the engines slip into neutral and a big inflatable ball dangling 40 feet of line trimmed with lights disappears into the night. I guess it's about four or five miles, and if Captain Dean's calculations are right, we're positioned near the ever-shifting edge of the Gulf Stream where marine life congregates. While the idling boat gently bobs, a dozen fidgety divers sit in full gear, waiting for the signal to bail out of the back like paratroopers.

First off the deck, Anna and I make a beeline for the glowing down line—our touchstone for the next hour and a half as we drift in the mysterious universe of larval sea life. We're amid a great oceanic diaspora of the externally fertilized offspring of reef fishes and invertebrates. The few survivors of the multiweek metamorphosis eventually settle to the seafloor, where they develop into their final adult forms.

Within minutes everyone is underwater and caught in the imperceptible grip of a three- to four-knot current that will carry us eight miles north by dive's end. In the interim we're adjusting lights and acclimating our eyes to a hazy ocean liberally flecked with organic detritus, known as marine snow. This is no game for slackers; larval fish are small, fast, erratic and tricky to track down in a featureless ocean. But when you find a jaw-dropping and possibly never-before-photographed animal, the thrill is as exhilarating as hooking a giant sailfish—although the "sailfish" we find in our watery fairyland can fit in the palm of your hand.

In less than 10 minutes Anna has found a larval flounder that was drawn to her light like a moth to a flame. Not knowing its behavior, I hold back and watch one of the fanciest fish I've ever seen flutter in her beam before flopping onto its side like a dead leaf. As I inch closer, the fish rights itself and sails away. I'm close behind, attempting to work my way alongside, focusing, framing and hoping it will slow or turn back on itself. Unsure how far I've traveled, I glance around, locate the down line and turn back. The fish is gone.

I spiral, sweeping my light in a circle. Somehow the beam catches a flicker of fins heading toward the surface. I'm after it and only catch up when the fish comes to a halt indignantly facing in the opposite direction. I slip to its left, but the fish slips right. I pivot right, and it tacks left, flashing its tail in my face as it speeds away. In fits and starts the chase continues until the larva slows, makes an about face and swims toward me at a leisurely pace. "Steady, steady," I coach myself as the flounder passes within a hand's width, fins flying.

All it took was that one outrageous larval fish, and Anna and I are hooked just like our South Florida colleagues, who seldom, if ever, miss a scheduled night drift. The majority are veteran Blue Heron Bridge shooters—naturalists to the core and well-versed in the trials and rewards of wildlife photography. As a group they have taken to night drifts like ducks to water, and they are

Right: *Larval Spotfin Flounder,* Cyclopsetta fimbriata.

Larval tripodfish, Bathypterois *sp.—a deepwater species.*

not the first to fall under the spell of offshore larvae. By the time the Florida folks made their first drift in 2014, divers in Hawaii had been photographing larval fish at night for 20 years. Inspired by legendary underwater photographer Chris Newbert, their night drifts, known in the islands as blackwater diving, introduced larval art to the world. Our Florida friends are happily carrying on the tradition 5,000 miles away.

Back on board, the boat is buzzing. Images of unfamiliar animals attract the biggest crowds around glowing camera screens. As you might expect, "What is it?" is the most frequently asked question about this world where crabs don't look like crabs, and fish appear as illusions. Photos of unidentified animals are forwarded to two gentlemen half a world away who acquired their taxonomic expertise cataloguing morphological sketches of specimens swept up in plankton tows.

Scientific understanding of the oceanic orphanage has been transformed since the end of last century. It is no longer thought to be an exclusively open system, randomly transporting passive larvae to distant shores. Recent studies have shown larval fishes to be strong swimmers with sophisticated instincts for remaining in local waters. But exactly where they go between spawning and settlement remains a mystery. Meanwhile, underwater photographers from Florida, Hawaii and beyond are adding bits of understanding to the sea of knowledge whose currents sweep us all along like night drifters having the time of their lives.

Observing larval fishes in the wild doesn't necessarily require a long boat ride. A few years back we found that our video lights would on some evenings attract post-larval fishes during night dives. So we began swimming out to the reef dropoff in Bonaire—famous for shore diving—where we would hang in the dark with our video beams blazing. On most evenings an assortment of larval fishes and invertebrates were drawn to our halo of lights. Many of the new arrivals exhibited hints of juvenile colors while others still retained the translucent bodies of openwater larvae. The following *Alert Diver* article from 2018 relates some of the trials and exhilaration of inshore night drifts.

Larval Ribbonfish, family Trachipteridae—a deepwater species.

Journey's End

from Alert Diver Magazine *2018*

Diving in the open sea at night isn't new to us. Typically we travel well offshore where enticingly exotic larval fishes and invertebrates spend the formative weeks of their lives. These tiny orphans of the reef equipped with translucent bodies, over-sized fins, spines and streamers are unlike anything we've ever seen. Unfortunately, because of logistics, dive operators are understandably reluctant to drop us far out in the ocean after dark. As an alternative Anna suggests swimming from the beach to see if our lights will attract larvae ready to settle inshore at the end of their pelagic larval voyage. Why not, we're in Bonaire with the most accessible shore diving in the world?

It's long past sunset when Anna and I swing tanks onto our shoulders, wade into the sandy shallows and slowly submerge. With our lights switched off we swim directly seaward navigating by the light of the moon. Crossing the interface where the sand gives way to a plunging coral slope we snap on three video lights illuminating an empty sea liberally pocked with stringy bits of detritus. We pause hovering 20 feet below the surface and 40 feet above the bottom watching for fleeting flecks of larval fishes and invertebrates until our eyes cross.

Larval Lookdown, Selene vomer

Pseudogramma sp., *a cryptic member of the sea bass family.*

We're still there 30 minutes later when the leading edge of a current begins pushing north. As the flow increases a cloud of worms and tiny crustaceans numbering in the thousands materialize as if out of nowhere like a Mongol horde swarming our lights and darting into our ears. A scattering of larval animals follow close behind. First to arrive are the mantis shrimp, as delicate as lace disoriented and spinning crazily followed by ribbon-thin bonefish larvae streaking through our lights. As quickly as the swarm arrives it disappears leaving us staring at the detritus once again. Disappointingly, the evening's results were not as productive as hoped, but from our experience drifting offshore we know that some nights are better than others for finding larvae.

These late-stage larvae we're looking for go through radical transformations toward the end of their stint at sea losing much of their transparency and specialized adaptations. At the same time they develop into proficient swimmers with the ability to use smell, hearing and other cues to navigate toward inshore habitats of choice where they spend the remainder of their lives. Before heading for shore settlers congregate in holding patterns along frontal slicks where water masses of different densities meet causing one to slip beneath the other forcing everything that swims or floats toward the surface. There they wait for an incoming night tide to carry them toward shore under the cloak of darkness when chances of predation are less. Dazed and defenseless, the arrivals dive for cover or bury in the sand emerging a

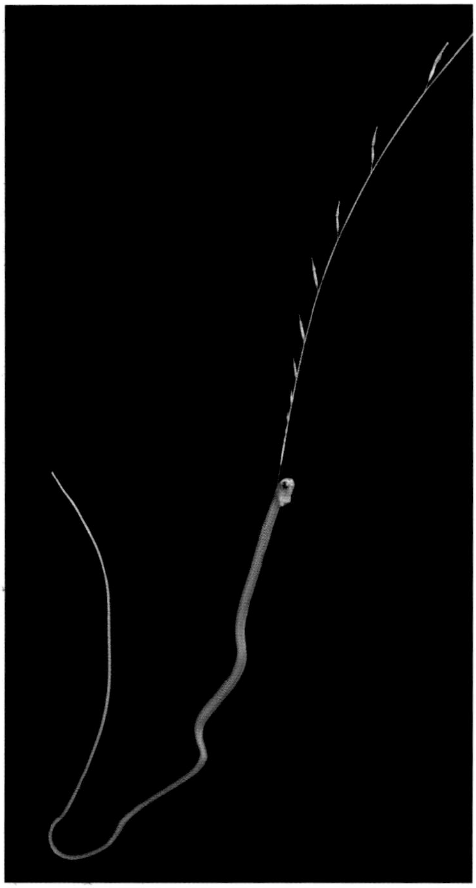

Larval pearlfish, member of family Carapidae. Their elongate predorsal filament is believed to mimic the stinging tentacles of siphonophores.

day or so later remolded into competent juveniles.

According to a local chart the next night tide isn't scheduled until after eleven, well past our usual bedtime. With a pioneering spirit we wade out from the beach just before eleven and hang for an hour without seeing anything of interest. The next evening, we abandon the chart and slip into the water at the reasonable hour of eight only to find ourselves battling a southbound current. Just as we're thinking about heading back to shore the first wave of crustaceans hits, bringing with them a splendid red trumpetfish only two inches long, followed by a pea-sized puffer that puffs up, then a blenny, a translucent lizardfish and a dozen surgeonfish along with crabs and shrimp galore.

A few months later we're back in the Caribbean this time visiting Dominica and excited about the chance of going out to night light from the resort's dock. But a week of pesky swells keeps us out of the water in the evenings. In frustration we hitch a ride on the night diving boat heading for a shallow inshore patch reef. Instead of going with the group we position ourselves above a mooring block and switch on the lights. Within minutes we are inundated by minnows—hundreds of misguided missiles the size of your finger, going nuts in the lights colliding with us hard and often until we feel like paintball players in a losing cause. Adding to our torment the crustaceans burst onto the scene joining forces with the maddened minnows. Once we discover that we can live through the onslaught, we bear up and begin looking for larvae. In the midst of it all Anna spots a breathtakingly spectacular long-armed octopus with a see-through mantle the size of a jellybean. Somehow she gets my attention. Together we watch the young cephalopod calmly parachute down through the chaos, settle on the sand, and promptly disappear beneath the surface.

Species of coral reef fishes vary considerably in the extent of their geographic ranges. Even closely related family members can differ widely in the geographic area they inhabit. For example the range of the Clown Wrasse

Halichoeres maculipinna, covers the entire Caribbean Basin, the Bahamas, both coasts of Florida, the northwestern Gulf of Mexico and also extends north to North Carolina and Bermuda and south to Brazil. At the same time the similar-sized Mardi Gras Wrasse, *H. burekae,* only inhabits deep reefs in the Gulf of Mexico. Interestingly species, known as endemics, with restricted distribution around small isolated island and reef systems in the Pacific had wide variations their pelagic larval durations ranging from less than 30 day to more than three months.

It would seem that the amount of time a species is capable of spending in the open sea would have a direct relation to its distribution, but such relationships are not as clear-cut as they might appear and have been substantiated only for a few species of wrasses. Another way to think of a species' range of distribution is by metapopulations, a mosaic of widely spaced settlement sites separated by open sea and bottom unsuitable for recruitment. These systems of distinctly separate populations, linked in various degrees by larval dispersal, allow species to colonize reefs across vast regions.

Juvenile Atlantic Moonfish, Selene setapinnis*—an openwater fish that never settles.*

After decades of study, the only thing reef ecologists agree on is that "larval fishes arrive at different reefs at different rates and at different times." Recruitment varies not only among fishes, but also among family members and even within the same species. Although similarities in the timing and magnitude of annual settlement often occur across a region, recruitment prediction at any scale remains haphazard at best. Seasonal pulses bearing the greatest influx of larval recruits are loosely synchronized with the moon's phases and with spawning dates. However, these patterns appear to be disrupted by fluctuating meteorological and hydrological events and unknown biological processes in the plankton. And making interpretation of what is going on even more difficult, every region has its own system of hydrology and topography influencing dispersal patterns of local offspring. Settlement peaks of the common wrasse species, known as Bluehead, *Thalassoma bifasciatum,* were found to consist of individuals spawned over an extended period of weeks, which, for unknown reasons, aggregated before being swept toward shore.

Larval Swordfish, Xiphias gladius, *feed near the surface at night.*

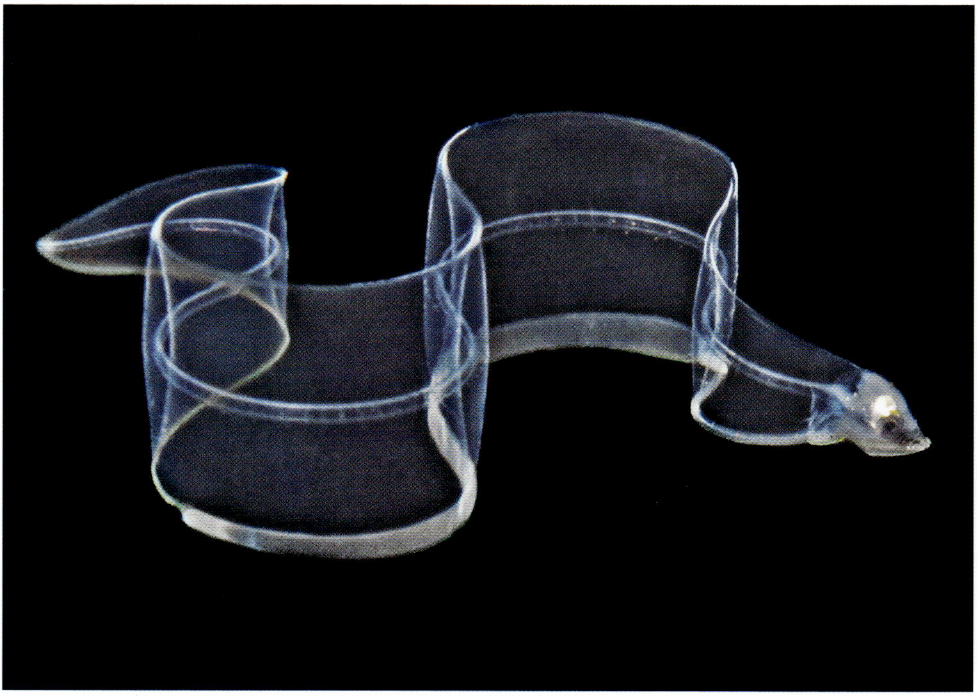

A Leptocephalus—a translucent larval eel—can remain in the pelagic up to a year.

Occasionally, a species settles in mass-recruiting episodes. In Panama, Queen Triggerfish, *Balistes vetula,* larvae traditionally arrive in low numbers. During a single week in 1985, however, the species settled at a rate 50 times higher than the total for any previous year. For less common long-lived species a dramatic influx such as this may increase the size of a local population for a decade, while poor recruitment often causes localized extinction for extended periods. Further complicating matters, high rates for one species have no bearing on the success of other species. Even when consistent across a region, settlement is often erratic among neighboring sites. During a 24-hour period the recruitment of a common damselfish varied 12-fold between two patch reefs separated by only 50 feet. And two species of recently settled damselfishes were found to have a 10-fold variation among patch reefs just 400 yards apart. These local fluctuations may be the result from aggregating behavior, forced concentration by local hydrodynamics or the random size and movement of patches.

Reef Fish Community Structures

Reef ecologists agree that the distribution and abundance of reef fish populations are controlled to some degree by larval supply, but to what degree remains in question. Those researchers favoring the recruitment limitation hypothesis believe that the numbers of arriving larvae will have a direct relationship to adult abundance. Their case is built on the concepts that settling larvae rarely exceed the carrying capacity of a reef and that a species seldom colonizes every area of suitable habitat. Others argue that competition for space, shelter, food and the rate of predation after settlement hold the keys to understanding population dynamics. Studies investigating the effect of predation after settlement tend to bolster this view, but such research is difficult to substantiate. General estimates place mortality between 25 and 90 percent for the first month after settlement. Research on Caribbean reefs, where piscivores often make up a large percentage of biomass, estimated that juvenile grunts experience 90 percent mortality during their first month. Bluehead juveniles suffer 22 percent mortality for each of the first three days after settling on the reef, and adults lose 13 percent of their population each month. Reefs where large predators had been eliminated by spearfishing were found to have a higher abundance of small species such as Blueheads.

Few studies have analyzed the long-term effects of larval recruitment on the structure of reef fish communities, and to date, these studies have produced varying results. For example, in Barbados recruitment of Redlip Blenny, *Ophioblennius macclurei,* does not appear to be limited by larval supply but rather by competition between settling larvae and residents for available space. However, recruitment rates and the post-recruitment survivorship of damselfishes in Australia and of both the Bluehead wrasse and damselfishes in Panama show no correlation to the density of resident populations. It is possible that the behavior of different species predisposes them to space

Juvenile Balloonfish the size of ping-pong balls, settled in such an uncommonly large number one year that the fish gathered into groups of dozens.

limitation or recruitment limitation. Although research forges ahead in the Pacific and Caribbean, such an intricate set of interacting processes is at work that it might well be some time before this complicated riddle is solved.

Over the years, we have occasionally monitored exceptional settlement episodes. While visiting Dominica in 2014, we found recently settled Balloonfish, *Diodon holocanthus*, the size of ping-pong balls, so abundant they gathered in tight-knit groups of dozens. When we returned to the island two years later the 10-inch adults were everywhere. Instead of checking off one or two sightings on survey dives as had been the norm, we tallied dozens every time we went underwater—the assumed effects of the bountiful 2014 settlement. In the early 2000s we observed a similar influx of Atlantic Trumpetfish in Bonaire. Once again, follow-up surveys recorded a substantial increase in species sightings for several years following the mass settlements.

Not all larval fishes settle in what will become their adult habitat. In the Caribbean, mangrove forests, seagrass meadows and inshore shallows function as nursery grounds for a number of species. With growth and changing dietary needs, the young fishes make what are known as ontogenetic shifts to intermediate habitats or to their final home ranges. For example, larval grunts settle in seagrass beds where they hide among the blades to avoid predators. After a few weeks they move to a reef and form mixed schools near sea urchins or protective pockets. Later they join schools of larger juveniles above coral structures. Finally, as adults, the grunts become permanent members of aggregations, resting by day and moving in shoals each evening to feed in distant seagrass beds.

Nearly all reef fishes are strongly site-attached throughout their benthic life cycle. The few exceptions are species making daily migrations to feeding grounds, sleeping shelters or spawning aggregations. Territorial species may remain in one area to protect food resources, or because of strong social and reproductive bonds with small groups, known as harems, but the basic reason reef fishes flock to reefs is to find security from predators in the form of hiding holes. Few small fish dare venture far from their familiar set of hiding-holes. For most fishes, even a short excursion across open bottom, much less to a neighboring reef, would be extremely dangerous.

Determining the maximum age and growth rates of wild fishes is difficult. Fisheries researchers making stock assessments of groupers and snappers generally employ one of three methods, each with interpretation problems. The most widely used technique, the length-frequency analysis, is possibly the least reliable. It measures the growth of fishes by extrapolating a large population sample with mathematical formulas. The tag-and-recapture method provides information on the rate of growth between two dates. Analysis of scales, spines, vertebrae and otoliths presents the troubling problem of distinguishing time-markers from incidental marks. Otolith evaluations work extremely well as markers during the first five to six months of life for many reef fishes, but later their usefulness varies among species.

Growth and age studies show a marked difference among fish populations in the Caribbean, where water temperatures remain relatively constant, and in the southeastern United States, where temperatures drop each winter. The data indicate that species from cooler climes typically live longer and grow larger and more slowly. However, these results might be skewed by different sampling techniques. The Caribbean specimens usually were taken from shallow reefs by trawls and traps that capture fishes only to a certain size. Fishes sampled in the southeastern United States are primarily caught by hook and line at various depths including the deeper offshore banks where the larger and older specimens typically live. Gray Snapper, *Lutjanus griseus,* from Cuba were found to live seven to nine years and grow to 22 inches; in Florida the oldest individuals live 21 years and reach a length of 35 inches. Basically, the larger a species is, the longer it lives. Little gobies and blennies may live only a season; small wrasses, such as Blueheads, three years; angelfishes and other medium-sized fishes average 10 years, while large groupers survive 15 to 20 years. Goliath Grouper, *Epinephelus itajara,* the largest of all reef fishes, have been aged to 37 years.

Feeding

A seemingly uninterested Great Barracuda swims casually among potential prey.

After visiting a coral reef packed with schools of fish lazing in the shadows and notorious predators swimming past potential prey with no apparent interest, divers often leave the water with the impression that fishes seldom eat. Nothing could be further from the truth. Most reef fishes spend the majority of their time locating, stalking, capturing and digesting prey. In fact, the driving force behind every fish is to maximize its reproductive capacity by obtaining as much energy as possible from food. Because of the endless evolutionary arms race it is nearly impossible for a predator to capture prey without taking advantage of unusual circumstances or employing unorthodox hunting strategies. The inactive schools lingering close to protective structures are taking well-deserved rests after long nights scouring the seagrass flats for food, and the seemingly unenthusiastic predators are biding their time until twilight,

A mixed group of grunts and snappers rest in defensive schools during the day.

when the odds of making successful strikes turn in their favor. As divers become more familiar with the feeding behaviors of fishes, they will begin to understand the subtle rhythms of life and death on the reef.

Fishes are placed in broad categories according to what they eat: herbivores primarily feed on plants; carnivores prey on animals; omnivores consume both animals and plants, and detritivores ingest decaying organic matter that accumulates on the sea floor. Some carnivores are further categorized as planktivores that pick zooplankton (tiny pelagic animals) from open water, and piscivores that primarily eat fishes. The shape, size and position of a species' mouth generally indicates its feeding category. The mouths of bottom-feeding

Prominent incisors of a carnivore.

Tooth-lined mouth of a piscivore.

Protrusible mouth of a planktivore.

Beak of an algae-eating parrotfish.

carnivores are downturned, and typically surrounded by strong blubbery lips or decked with prominent incisors. Piscivores have large mouths either expandable, like groupers, or lined with sharp teeth, like barracudas, sharks and needlefishes. The small terminal mouths of planktivores shoot forward to secure

Feeding

An algae-eating Blue Tang switches from its typical diet to take advantage of a temporary bloom of Sea Thimbles.

prey. The teeth of parrotfishes are fused into a bone-hard beak to scrape algae from rocks, and angelfishes use overlapping rows of tightly packed needlelike teeth to cut through the tough tissue of sponges.

The diets of reef fishes are normally quite varied and customarily change as they grow. Most species begin life as larval planktivores feeding on tiny animals in the open sea. After settling to shallow-water habitats, they transform into carnivorous juveniles foraging for small crustaceans before finally switching to their adult diets. Even after maturity, nearly all fishes readily abandon their daily diets to take advantage of temporarily abundant food supplies. For example, in the early summer small jellyfish, known as Sea Thimble, *Linuche unguiculata,* bloom in such profusion that they form rafts that wash over reefs. The sudden influx of jelly-plankton provides easy pickings for numerous species, including Blue Tang, *Acanthurus coeruleus,* which normally eat algae. Only a species' mouth size and structure and digestive capabilities limit the foods eaten. A fish's flexible eating habits and tendency to consume the most accessible prey maximizes energy gains with minimum energy expenditure. This strategy, known as the optimal foraging theory, is the guiding principle of fish-feeding ecology.

The Great Barracuda uses a primitive mouth structure lined with sharp teeth to sever the bodies of fishes with a single strike.

Because all reef fishes are to some extent opportunistic feeders, the frequently used terms specialists, generalists and opportunists are somewhat misleading and should be used only to describe temporary feeding

The vast majority of reef fishes use suction to draw prey into their mouths with the force created by rapidly opened jaws and flaring gill covers. It is common for predators to slowly expand their mouths to full size. This action is believed to keep the mechanism limber for the fast action required to inhale prey.
Longlure Frogfish (upper left), Nassau Grouper (upper right), Spotted Trunkfish (lower left), Shortnose Batfish (lower right)

behaviors. Specialist refers to species that typically concentrate on a single food source. Generalists eat a variety of foods without a specific preference for one type. Opportunists willingly eat foods not included on their normal menus. The ability to convert from one food to another is made possible in great part by suction feeding, a nearly universal characteristic of fishes. Even though mouth structures vary, food, whether algae, zooplankton or crustaceans, is ingested by suction when the jaws rapidly open and the gill covers flare. A few piscivores, such as the Great Barracuda, *Sphyraena barracuda,* are exceptions: using a primitive mouth structure lined with sharp teeth, the swift hunters sever fishes with a single strike, then circle back to pick up the pieces.

Even specialists may permanently alter their diets if their preferred food

A pair of Spotted Eagle Rays scour the sand for gastropods.

becomes scarce. Two species of western Caribbean toadfishes provide an illustration of a specialist's ability to adapt. Both reef-dwelling species dined almost exclusively on Long-spined Urchin, *Diadema antillarum,* before 1983, when an unidentified pathogen wiped out more than 95 percent of the abundant echinoderms throughout the tropical western Atlantic. After the loss of their long-time food supply, both species quickly became generalists. One switched to a diet of fishes and hermit crabs, and the other began hunting mobile, bottom-dwelling invertebrates. Neither species suffered a population decline.

Some specialists have difficulty adjusting to a rapid decline in their customary food. Spotted Eagle Ray, *Aetobatus narinari,* forage over sand flats for hard-shelled clams, oysters and large gastropods which are crushed by specialized dental plates. The spectacular fish suffered a marked decline in the Cayman Islands after commercial interests overharvested their primary food, Queen Conch, *Lobatus gigas.* But after the government mandated no-take reserves for Queen Conchs, the tasty mollusks made a dramatic rebound. With their primary food supply reestablished, the islands' population of Spotted Eagle Rays also made a strong recovery.

Individual members of many fish species stake claims to and vigorously defend areas of the bottom, known as territories. This strategy conserves energy by transferring the cost of searching for food to the less demanding duties of defense. Territories also provide for the protection of mates, nests and reproductive sites and offer residents the advantage of knowing where every hiding hole is located. Once a territory is established, most fishes remain almost exclusively within its boundaries. A few, however, commute to and from alternate sleeping locations, and others migrate to distant spawning sites. Home range, a term frequently confused with territory, refers to the entire area an individual frequents but does not defend.

The courtship and spawning pattern of a male Longfin Damselfish is indicative of his heightened belligerence toward intruders during reproductive periods.

Typically fishes occupy territories in relation to their size with small fishes, blennies, gobies, jawfishes and juveniles, defending a few inches or feet of turf, while large species, such as angelfishes and Hogfish, *Lachnolaimus maximus,* partition tracts the size of football fields. The territories of some fishes lie side-by-side, covering a large area of the sea floor; those of other species are distributed randomly. Territorial boundaries are defined by the extent of the occupants' defensive efforts. The size of an individual's territory appears to be more closely related to intra-species competition than to protecting food supplies. When the defender of an adjoining territory disappears, the neighbors compete for the vacant turf. However, even after acquiring additional feeding sources, the victors' consumption of food seldom increases.

A Longfin Damselfish challenges a Whitespotted Filefish encroaching on its territory.

Territories are customarily established and defended by a single inhabitant, as in the case of algae-eating damselfishes. These spunky herbivores single-handedly challenge intruders—including divers—who approach their personal algal patches adorned with tasty epiphytes. They become even more pugnacious when tending a fresh clutch of eggs. Parrotfishes, wrasses and several smaller sea basses form harems with a single dominant male overseeing the territories of several females. Each group member participates in defense. Intruders,

A Great Barracuda picks up the head portion of a Schoolmaster severed with an initial strike.

usually food competitors or members of the same species, are challenged by lateral displays, chases, sounds, parallel-swimming, and occasional nips. These showdowns, generally lasting less than a minute, seldom cause injuries.

Mated pairs of Gray Angelfish, *Pomacanthus arcuatus,* French Angelfish, *P. paru,* and Harlequin Bass, *Serranus tigrinus,* share territories and the responsibility of protecting their mutually held feeding grounds. In a more complex arrangement, the boundaries of Yellowtail Damselfish, *Microspathodon chrysurus,* incorporate the smaller territories of several Dusky Damselfish, *Stegastes adustus*, and Redlip Blenny, *Ophioblennius macclurei.* The three herbivores feed from the same field of filamentous algae without conflict, but the smaller Duskies are expected to expend the greatest amount of energy defending the communal algal crop.

Approximately one-half to two-thirds of the species in a reef fish community forage by day, and one-quarter to one-third feed during the night. The hunting activities of many piscivores peak during the low-light hours of dawn and dusk. For daytime feeders, vision and vigilance are key to survival. Every reef fish is well aware of just how far it can wander away from a shelter hole and expect not to be eaten. This sphere of relative safety greatly limits the size of a species' feeding grounds.

By the time the sun rises the profiles of feeding planktivores fill the water column. But only the streamlined species with long tail bases and forked tails designed for speed, such as Brown Chromis, *Chromis multilineata,* and Atlantic Creolefish, *Cephalopholis furcifer,* dare to venture high above the reef where the richest plankton streams flow; the smaller, deep-bodied plankton-pickers must be satisfied to dine in the less fertile water closer to hiding holes. Sand-dwelling planktivores such as Yellowhead Jawfish, *Opistognathus aurifrons,* and Hovering Dartfish, *Ptereleotris helenae,* nab passing specks just above the safety of their sand burrows. Crustacean-eating carnivores, including wrasses, goatfishes and porgies, spend their daylight hours rooting in sand only a short sprint away from the safety of the reefs' fringes. Cleanerfishes, primarily juvenile wrasses and gobies pick parasitic crustaceans from client fishes at cleaning stations. Because of their utilitarian behavior, the little carnivores are graced with a degree of immunity from predators. Large parrotfishes, emboldened by size, range widely scraping algae from coral rock, while smaller, more vulnerable family members feed in loose groups near bottom cover. Deep bodies, tough skin and a formidable, stilettolike foredorsal spine provide triggerfishes and filefishes a substantial degree of protection while they nibble algae and search for bottom-dwelling invertebrates. Likewise, the flat, deep bodies of larger angelfishes and butterflyfishes, are difficult for piscivores to seize, providing ample protection for the fishes to roam away from structures with secure hiding places.

Barracudas, groupers and other piscivores hang around the reef during the day, but their ability to make successful strikes is severely limited by the excellent vision of their prey. If reef fishes remain alert, do not stray too far from cover or allow themselves to become distracted or confused, the odds of escaping an attack are decidedly in their favor. However, when opportunities present themselves, large piscivores will not hesitate to attack in broad daylight.

During their safety stop, a group of divers were casually watching a five-foot Great Barracuda hovering just under the surface. The large predator suddenly angled downward aiming its elongate body at two initial phase Stoplight Parrotfish, *Sparisoma viride,* moving across a wide stretch of open sand 40 feet below. The barracuda bent into an S-shape, and, as if shot from a bow, streaked toward the bottom in a blur completely severing the body of one of the Stoplights in the attack. Quickly circling, the predator grabbed the head portion off the sand, bit down on the fresh kill repeatedly, and casually swam away with its bleeding prey.

Piscivores often hunt as a team to gain an advantage. Mated pairs of Harlequin Bass hunt together within their small territories. One partner momentarily distracts a shrimp or other small crustacean while its companion sneaks in from the opposite direction. An underwater photographer diving in the Los Roques Islands, Venezuela spent 20 minutes observing a group of 10 Western Comb Grouper, *Mycteroperca acutirostris,* working in concert to keep schooling silverside minnows corralled above a knoll where the grouper fed. From a position under the school the grouper raised their bodies upward at a 45-degree angle and shot up through the mass in open-mouth rushes. Each time the minnows dispersed the predators herded them back into a concentrated school above the killing-zone and continued to feed.

When daylight begins to fade and vision becomes less reliable, feeding fishes grow restless and even more cautious, and with good reason, for this is when the odds begin to swing in favor of the piscivores. The smaller, more vulnerable species are first to retire for the night, larger fishes follow, and the big parrotfishes bed down last. During the last 20 minutes before complete darkness the low-light conditions are so favorable for the big predators that the water above the reef is significantly devoid of fish life. This precarious interval, known as the quiet period, precedes the emergence of nocturnal foragers.

Just after sunset grunts, such as these Tomtates, stream away from the reef and begin feeding as they migrate to customary nightly hunting grounds.

Night-feeders, such as cardinalfishes, bigeyes and squirrelfishes, are easily recognized by their large eyes facilitating better night vision and reddish hues that make them difficult to see in the dark. With fewer piscivores around after the sun goes down, nocturnal foragers have little need to aggregate or remain near shelters. Many species range freely over open sand flats where their daytime counterparts wouldn't dare feed. Small cardinalfishes ascend by the thousands from their shadowy daytime haunts into the water column to feed on resident zooplankton that, like themselves, hide during the day often buried in sand and rubble. As the quiet period ends grunts, snappers, squirrelfishes and sweepers make nightly pilgrimages to distant feeding grounds. Grunts, wearing muted night colors, leave the reef in streaming schools and feed randomly as they travel over open terrain toward seagrass flats. Some of these migration routes take the fishes a mile or more from their home reefs. After arriving the fishes fan out and forage for crustaceans until just before dawn, when they gather once again for the return trip to their home reefs. Several species of squirrelfishes leave the reef just after the grunts, traveling individually to the same seagrass meadows. Glassy Sweeper, *Pempheris schomburgkii,* journey to seaward reef banks to feed on zooplankton. Unless the sky is heavily overcast, these migrations, synchronized by light intensity, occur at a predictable time each evening.

Detritus is a potpourri of decomposing plant and animal matter mixed loosely with sand. Much of the content is made up of single-celled plants called diatoms and the organic waste of fishes. Pockets of the matter harbor a variety of living organisms, ranging from bacteria and fungi to tiny worms. Only a few reef associated fishes, primarily mullets and bottom-dwelling gobies, consume sufficient amounts to be considered detritivores, but many secondary consumers, especially bottom-grazing herbivores, ingest varying amounts of detritus when foraging.

Opportunistic carnivores are extremely fond of human food. Yellowtail Snapper, *Ocyurus chrysurus,* Sergeant Major, *Abudefduf saxatilis,* and Bermuda

Chub, *Kyphosus sectatrix,* hang out in the shade of dive boats and piers waiting for handouts or garbage. To entertain customers, divemasters and tour guides often attract swarms of fishes by chumming. At sites such as shallow snorkeling reefs where fish feeding is routine, fishes mob divers whether they are carrying food or not. At times the fishes can be intimidating, nipping hair, scalp and hands. On the other hand, being close to so many wild creatures is exhilarating. Reef visitors generally enjoy the experience, and the fishes certainly aren't complaining. However, some ecologists argue that these free lunches might permanently alter the fishes' ability to search for more traditional foods. This seems unlikely. Remember, these are opportunistic feeders; if their handouts suddenly come to a halt they are certainly not going to starve.

Our favorite fish-feeding story occurred in a palm-lined lagoon on the Caribbean coast of the Yucatán Peninsula in Mexico. As usual, a tour group loaded with stale bread for the fishes arrived from a cruise ship. One young tourist, tiring of his frequent swims to shore for bread, stuffed half a loaf inside his trunks and headed back into the boil of feeding fishes. It wasn't pretty, but the vacationer did survive more or less intact.

Carnivores and Omnivores

The first step in understanding the nature of fishes is to know the foods they eat. And there is no better source for such information than *Food Habits of Reef Fishes of the West Indies* by John E. Randall. During the seven-year study (1958 to 1965), Randall and his colleagues analyzed and cataloged the stomach contents of 5,526 specimens representing 212 species of fishes. To insure a complete picture of a species' food habits, specimens were collected from a variety of reef-related habitats in Puerto Rico and the Virgin Islands. More than half a century later, the unprecedented work remains a standard of reef fish ecology.

The majority of reef fishes are generalized carnivores which, as a group, eat just about every animal on the reef, including such unsavory fare as sponges, fire corals, and hydroids. The Sergeant Major provides a good example of just how diverse a carnivore's diet can be. The species' indiscriminate feeding habits allow the little damselfish to proliferate in a variety of habitats. They are equally at home picking zooplankton high in the water column above the reef, nipping algae from rocks or plucking marine animals off the reef. Randall's analysis of 35

Sergeant Majors feed in a variety of habitats.

specimens taken at 13 stations showed that the Sergeant Major's diet consisted of 43 percent zoanthid polyps; 14 percent pelagic copepods; 9 percent algae; 7 percent tunicates; 5 percent Lettuce Sea Slug, *Elysia crispata;* 5 percent fish eggs; 4 percent minnow-sized herring; 3 percent shrimp larvae; 3 percent barnacle appendages, and about 1 percent each of winged ants, feather duster plumes and jelly plankton.

Generalized carnivores can be categorized by the types of foods they typically eat. The more specialized feeders—piscivores and planktivores—are discussed later in this chapter (pages 74 and 92), and cleanerfishes are reviewed separately in the Symbiosis chapter (page 110). The remaining carnivores eat primarily from one of three invertebrate groups: mobile invertebrates, attached (sessile) invertebrates, or hard-shelled mollusks. The species eating mobile invertebrates typically have the most extensive diets, often including a substantial portion of fishes.

Shrimp, crabs and lobsters are the most frequently eaten invertebrates. Fishes commonly consuming crustaceans include snappers in the genus *Lutjanus* (except the Cubera Snapper, *L. cyanopterus,* a piscivore), sea basses, *Epinephelus* spp., hamlets, *Hypoplectrus* spp., mojarras, goatfishes, Sand Tilefish, *Malacanthus plumieri,* Slippery Dick, *Halichoeres bivittatus,* the larger jawfishes, *Opistognathus* spp., night-feeding squirrelfishes, grunts and the Spotted Drum, *Equetus punctatus.*

A Chain Moray hunts the sun-splashed shallows for crabs, its primary food.

The Chain Moray, *Echidna catenata,* is the king of the crab-eaters. The golden-streaked eels match the dark, sun-splashed tidal rocks where crabs congregate. The eels, accustomed to hunting in a world of both water and air, think nothing of being temporarily stranded while waves go out. During low tide they often wedge into pockets of pools well inland from the water's edge. Crabs are also the favorite food of snake eels and Flying Gurnard, *Dactylopterus volitans,* which use modified ventral fins to turn rocks when hunting. At night, foraging squirrelfishes, grunts and snappers plunder crab-rich seagrass beds. Shrimp top the food list of most scorpionfishes, jawfishes and smaller sea basses, including hamlets and Harlequin Basses. Mantis shrimp have been

The sharp incisors of the Slippery Dick indicate its preference for mobile prey.

found in the stomachs of over 40 fishes but never in large numbers. Few fishes eat lobsters because of their size, but they are the main food of the largest fish on the reef, the Goliath Grouper, *Epinephelus itajara.* Dog Snapper, *Lutjanus jocu,* and Nassau Grouper, *E. striatus,* occasionally eat smaller lobsters.

Sea urchins and brittle stars have good reason to hide deep within crevices: the brittle echinoderms are regularly wrestled from the rocks by wrasses and other carnivores armed with conspicuous incisors. However, few fishes go to the trouble of eating the urchins' tough-skinned relatives, sea stars and sea cucumbers. Only the entrails of a sea cucumber were found inside a Spotted Trunkfish, *Lactophrys bicaudalis,* leading to speculation that the echinoderm might have been harassed by the trunkfish until it expelled its intestines, a last-resort defensive tactic. In times of plenty, opportunistic carnivores can change their preferences quickly. Spearfishermen discovered dozens of inch-long juvenile sea cucumbers packed inside a Sheepshead, *Archosargus probatocephalus,* a member of the porgy family inhabiting inshore temperate zones.

The diet of the Slippery Dick is representative of carnivores feeding on small, mobile prey. Randall found that the fast-moving, cigar-shaped wrasse consumed 22 percent crabs, 18 percent echinoids, 17 percent polychaete worms, 12 percent gastropods, including more than 20 species, 7 percent brittle stars, 4 percent bivalves, 4 percent shrimp, 3 percent chitons, 3 percent fish remains, 3 percent mantis shrimp, and 1 percent hermit crabs.

Omnivores primarily eat attached invertebrates and a substantial amount of plant material. Commonly sighted omnivores include Black Durgon, *Melichthys niger,* Scrawled Filefish, *Aluterus scriptus,* Orangespotted Filefish, *Cantherhines pullus,* Whitespotted Filefish, *C. macrocerus,* French Angelfish, *Pomacanthus paru,* and the Sharpnose Puffer, *Canthigaster rostrata.* Although animals in the phylum Cnidaria, which includes stony and soft corals and hydroids, are the most abundant attached invertebrates on the reef, they are not important sources of food in the Caribbean. Only random fragments of stony coral, believed to have been accidentally ingested, have been found in a few fishes. However, Stoplight Parrotfish, *Sparisoma viride,* and Queen Parrotfish, *Scarus vetula,* have been documented to take just over nine percent of their

Feeding

A group of Sergeant Majors dine on a Moon Jellyfish.

grazing bites from living Boulder Star Coral, *Orbicella annularis,* and, to a lesser extent, two species of brain corals. The damage, which leaves open white lesions, often requires several weeks to heal, while the tissue removed by smaller bites can regenerate in less than five days. Although grazing seldom causes the death of an entire colony, it is responsible for the largest amount of chronic tissue loss of the Boulder Star Coral. Gorgonians avoid heavy predation by giving off pungent odors. Only Scrawled Filefish were found with significant amounts of gorgonian polyps (12 percent) in their stomachs. The same species also dines on the calcareous, nematocyst-filled skeletons of fire corals, *Millepora* spp. Hydroids and anemones are nibbled only sparingly; but the polyps of zoanthids are important in the diets of Sergeant Major, Foureye Butterflyfish, *Chaetodon capistratus,* and Banded Butterflyfish, *C. striatus.* Nutrient-poor jellyfishes, siphonophores and comb jellies, collectively known as jelly plankton, are picked at from time to time by Sergeant Majors, Scrawled Filefish, Creole Wrasse, *Clepticus parrae,* Boga, *Haemulon vittatum,* Ocean Surgeonfish, *Acanthurus tractus,* Yellowtail Snapper, *Ocyurus chrysurus,* and other water-column feeders.

Sponges, covered with tough fibrous exteriors and filled with tiny indigestible mineralized supports and repellent chemicals, are distasteful food. Only 11 fish species regularly eat the primitive multicellular animals. It is interesting to note that all of the sponge-eating species, including the Atlantic Spadefish, *Chaetodipterus faber,* four angelfishes, boxfishes and two filefishes, evolved in recent geological time, possibly as a response to increased competition for food on coral reefs. Numerous attached worms including feather dusters and Christmas Tree Worm, *Spirobranchus giganteus,* are popular food items for butterflyfishes, grunts and boxfishes. The protruding mouths of butterflyfishes are ideal for nipping off the featherlike crowns of worms before they can withdraw into their protective tubes.

Atlantic Spadefish, offer an example of the diverse diets of bottom-picking carnivores. The 29 specimens dissected by Randall contained 33 percent sponges; 19 percent zoanthid polyps; 14 percent Magnificent Feather Duster,

Sabellastarte magnifica; 13 percent pelagic tunicates; 6 percent gorgonian polyps; 5 percent algae; 3 percent gastropod eggs and 2 percent seagrasses. Less than 1 percent each of crabs, lobsters, and shrimp larvae and a half-inch fragment of live stony coral, *Oculina diffusa,* were also found inside the specimens.

The blubbery lips of a Permit—a gastropod eater.

Fishes that feed on thick-shelled gastropods have powerful mouths with big lips or beaklike jaws for crushing. Notable shell-crackers include Spotted Eagle Ray, *Aetobatus narinari,* Black Margate, *Anisotremus surinamensis,* Spanish Grunt, *Haemulon macrostomum,* Caesar Grunt, *H. carbonarium,* Jolthead Porgy, *Calamus bajonado,* and, to a lesser extent, other members of the porgy family. Other gastropod-eaters include the Queen Triggerfish, *Balistes vetula,* the Trunkfish, *Lactophrys trigonus,* the Hogfish, *Lachnolaimus maximus,* and several members in the spiny puffer family. Permit, *Trachinotus falcatus,* also have blubbery lips and strong jaws indicative of their preference for gastropods. The stomachs of eight specimens contained 48 percent gastropods; 25 percent echinoids; 17 percent bivalves; 6 percent hermit crabs and 4 percent true crabs.

Although most large carnivores generally feed alone, many smaller species regularly forage in roving bands of mixed species in an attempt to take

Only a few species, including the French Angelfish, eat tough, distasteful sponges.

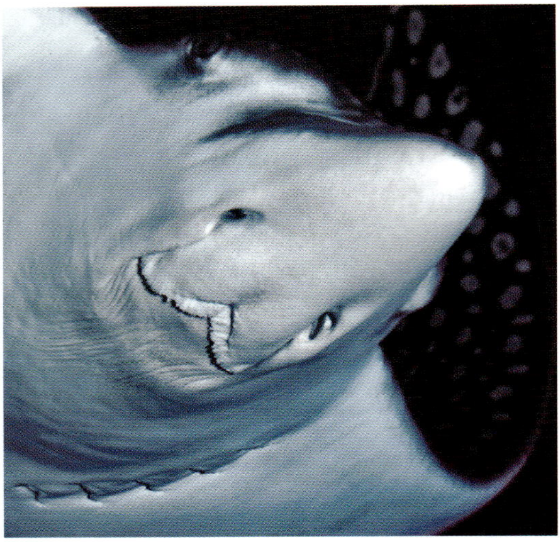

The shell-crushing mouth of a Spotted Eagle Ray.

advantage of the other members' expertise and luck as well as creating confusion, which works to the predators' advantage. Bar Jack, *Caranx ruber,* Yellowhead Wrasse, *Halichoeres garnoti,* Yellowtail Snapper, and porgies often follow goatfishes as they probe the sand for hidden prey in a behavior, known as nuclear hunting. The goatfish is the nucleus of the hunting parties, acting as the disturber, to locate buried prey in the form of polychaete worms, shrimp and crabs. When detected, the band of opportunistic followers attempt to be first to snatch the exposed morsel. The goatfishes locate the prey with two elongate appendages packed with chemosensory organs that extend from their lower jaws. The semi-rigid probes, known as barbels, are

Bottom-feeding carnivores improve their capture rate when feeding with goatfishes.

Blueheads devouring damselfish eggs.

Fishes enthusiastically picking gametes from a spawning star coral.

also used to excavate soft bottoms. In the process goatfishes can dig depressions half the length of their bodies. Their frenetic pursuits often create billowing clouds of suspended debris. Not surprisingly up to 25 percent of a goatfish's stomach content can be made up of sand. Other species with bottom disturbing habits, such as stingrays, eagle rays, sea turtles and even box crabs, also attract followers. We watched one piggybacking Bar Jack follow a Spanish Hogfish for over an hour. In this case of nuclear hunting, the hogfish was busily probing nooks and crannies, which periodically flushed concealed prey into the open. Whether foraging in pairs or groups, the rate of strikes increases substantially for participants.

Every carnivore delights in a meal of fat-filled fish eggs. When a Sergeant Major's nest is full, as they often are, packs of young Bluehead, *Thalassoma bifasciatum,* periodically test the male's guardianship abilities. If defenses are overwhelmed, the ever alert wrasse packs swarm the nest nipping off egg capsules until the male can regain control. Divers inadvertently assist the wrasses' efforts by approaching nesting sites too closely. After an initial charge at oncoming divers males retreat a short distance leaving their nests defenseless. Nearby carnivores immediately join the rout. At times Sergeant Majors turn the tables becoming egg predators themselves. Alone with Yellowtail Snapper and occasionally Black Durgons the damselfish hover in the wings as daytime pelagic spawning parrotfishes prepare to release their gametes in open water high above outer reef slopes. The instant eggs are released Sergeant Majors rush in and gobble down as many fertilized eggs as possible before the gamete cloud disperses. During the annual mass coral spawns of summer, many carnivores stay up past their bedtimes to grab packets of coral gametes released just after dusk. We have watched greedy little Foureye Butterflyfish use their protracted mouths to pull gametes from the polyps of Star Coral, *Orbicella* spp., before they were even released.

A huge school of silversides attracts piscivores from far and wide.

Piscivores

Even though prey and predators share reefs, it is uncommon to see a fish eat another fish. Barracudas and groupers seldom waste energy making random strikes in broad daylight. Their prey are too accomplished at surviving in a world of big mouths and sharp teeth for the hunters' conspicuous attacks to be effective. The grunts hovering together next to the reef instinctively understand, like all schooling fishes, that there is security in numbers. Their close parallel formation and bold stripes make it difficult for predators to single out targets. Every eye tracks the movement of passing piscivores; before a split-second

Right: *Fishes have evolved numerous defenses against piscivores.*

Flat bodies offer protection.

French Grunts schooling for safety.

A defensive foredorsal spine.

A burrfish inflates its body.

A Lined Seahorse relies on camouflage.

Razorfish escape by diving into the sand.

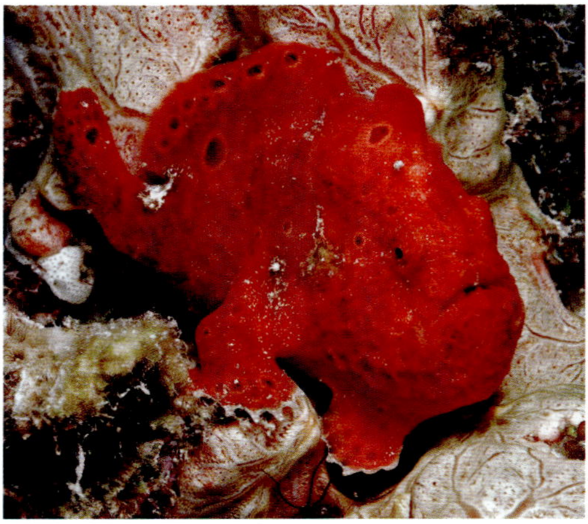

Longlure Frogfish

Sand Diver—a lizardfish

Spotted Scorpionfish

strike can meet its mark, the prey vanish inside the reef. Small solitary fishes, constantly mindful of their vulnerability, rarely wander far from a hiding hole.

Predation not only affects behavior but also influences the shape, color and markings of reef fishes. To discourage attacks, the bodies of reef fishes have evolved into tall, flat and difficult-to-bite shapes or become streamlined for speedy escapes. Others protect themselves by displaying spines, inflating their bodies beyond bite size, displaying confusing markings, or by quickly blending into their backgrounds. Others survive on open spaces by building burrows or diving beneath the sand.

Fish-eating species are divided into two groups according to their hunting technique. Roving predators such as barracudas, sea basses in the genus *Mycteroperca*, sharks, mackerels, jacks, morays in the genus *Gymnothorax,* Tarpon, *Megalops atlanticus,* needlefishes, Atlantic Trumpetfish, *Aulostomus maculatus,* Tobaccofish, *Serranus tabacarius,* and several snappers, stalk or chase prey. The lie-in-wait predators—frogfishes, lizardfishes and scorpionfishes—perch stationary in concealed positions for extended periods while waiting to ambush unsuspecting prey.

Under normal circumstances prey fishes are virtually impossible to capture. However, a predator's capture rate increases dramatically when it takes advantage of abnormal situations. Concentrating hunting efforts during the brief, low-light period of twilight has proven to be a productive strategy. During the remainder of the day,

A hunting pack of Bar Jacks attempting to disrupt the schooling behavior of grunts.

roving predators have two basic options for acquiring food: they can wait patiently for disturbances to occur naturally or they can create distracting situations themselves.

Raiding parties of large Yellow Jack, *Caranx bartholomaei,* and Bar Jack, *C. ruber,* are masters of pandemonium. We have watched packs ranging from 6 to 20 large jacks terrorize reefs on numerous afternoons. Seconds before the hunters appear, schooling reef fishes, sensing their approach, tighten ranks and begin to swirl while solitary fishes inch toward shelters. At times the jacks make only a few cursory passes, but on other afternoons they come flying in toward the reef, cutting fast and with authority.

The packs circle polarized aggregations of grunts and snappers, attempting to splinter the groups before slashing through the escaping fishes. In the confusion, the panicked prey plunge into overcrowded reef pockets, where the big predators come crashing in behind. In the swirls of sand and fleeing fishes, it is difficult to determine success; but an occasional jack rises out of the free-for-all compulsively working its jaws. The more aggressive the attack, the longer it lasts. Some forays are over after one or two strikes. Others continue for 10 minutes or more as the hunters work up and down a reef line, continually disrupting schools before they are able to regroup.

On two occasions we have watched jacks flush Spotted Moray, *Gymnothorax moringa,* out of the reef. The frenzied pack pounded the terrified eels as they dashed for cover. Even after the morays disappeared back into the reef, the jacks continued to lunge at the openings where the eels disappeared. One hunting party chased an eel out of five different shelters during a 10-minute period before losing interest. It appears that the jacks were taking pleasure from the chase rather than attempting to kill or eat the eels. If they had wanted, it seems the jacks could have torn the eels to shreds—but maybe not; a moray's skin is as tough as rawhide.

Other common reef predators, particularly Graysby, *Cephalopholis cruentata,* and Coney, *C. fulva,* have a much more favorable attitude toward eels. Similar to carnivores nuclear hunting with goatfishes, rays and other fishes that regularly disturb the soft substrate, the two small sea basses, along with a suite of other perceptive piscivores, form nuclear hunting confederations with eels and octopuses to increase their odds of capturing prey.

A close gathering of predators is a sure sign of nuclear hunting.

Coral reefs appear solid, but surprisingly their interiors are typically cavernous and riddled with unseen passages and chambers where multitudes of small fishes hide from all major reef predators except eels and octopuses. Eels, with long, slender, malleable bodies and one of the keenest senses of smell in the animal kingdom, are well suited for home invasions. When a prowling eel slips inside a coral head, small resident reef fishes, such as cardinalfishes, know exactly where the exits are located; but if escape is blocked by a waiting grouper, the fishes panic and bounce around inside the dark like pinballs making easy targets for hungry eels—an abnormal situation and a predator's dream.

We first became aware of this form of nuclear hunting late one afternoon in the mid-1990s, when our attention was drawn to three Graysbys and a Coney resting on the bottom, staring intently at a small opening in a limestone ledge. Graysbys are usually wary, but this bunch was so intent on their business they didn't appear concerned with our close approach. Moments later the head of a

Eleven Coneys nuclear hunt with a foraging octopus.

Spotted Moray appeared in the entrance. To our amazement one of the Graysbys moved forward and calmly reclined its beefy 10-inch body against the eel's head. The odd couple remained cheek-to-cheek for nearly a minute. The moray didn't appear to be concerned in the least by the close proximity of the grouper. Without warning the moray shot forward and raced across the open sand, with the Graysby shadowing its head and the others following close behind. Ten feet away, the eel slithered inside an opening at the base of a brain coral. Two fishes poked their snouts into the same crevice while the others moved anxiously around the structure, peering first in one opening and then another. Presently, the speckled head of the moray reappeared. A Graysby moved forward and

An aggregation of predators attempts to stay close to a foraging Sharptail Eel.

An Atlantic Trumpetfish and a Coney, both piscivores, hunt with a Goldentail Moray—a behavior known as nuclear hunting.

nudged the moray with its snout, while the others hovered just off the bottom. The eel dodged back into the dark pocket but almost immediately reappeared and streaked off to another crevice, with its entourage close behind. After several runs the moray finally settled under a ledge with only its head exposed. While the minutes passed and light faded, the two grouper remained nearby for several minutes before losing interest and slipping away.

After observing the behavior for the first time, we began to notice fish hunting with eels on a regular basis and soon a pattern began to emerge. The eel nearly always waits for a grouper to position itself next to its body before making a break across the bottom. In every instance, the fish in contact at the beginning of a run shadows the moray's head. It appears that the animals cooperate to insure that they arrive at the next coral head simultaneously so the grouper can cover an escape route while the eel invades the dark interior.

Although we watched groupers make frequent strikes into the dark holes occupied by foraging eels, it was a long time before we finally confirmed a capture. Three Coneys and an Atlantic Trumpetfish were following a very active Goldentail Moray, *G. miliaris.* On a short run across the sand, a Coney crowded the eel's head as usual, while the other fishes trailed close behind. As soon as the moray disappeared inside a coral pocket, one of the larger Coneys started jumping eagerly from one opening to another. Suddenly the Coney made a strike that sent sand flying. The

A Coney perches next to a Goldentail Moray waiting for it to make a run.

piscivore appeared from the cloud with a small fish protruding from its mouth. When the Coney tried to reposition its catch, the prey, a three-inch Orangespotted Filefish, *Cantherhines pullus,* popped free and rocketed away. The grouper pounced, recapturing its victim. But nothing comes easy in the sea. The little filefish locked its piercing foredorsal lancet firmly into the roof of the predator's mouth. The Coney, unable to swallow, swam sluggishly over to a sea fan, settled hard on the bottom, and with its mouth awkwardly bulging and eyes glazed over, rolled to one side and slowly changed into its blotched night-colors. When we left the area 10 minutes later, the Coney, still in the same position, had still not swallowed the filefish.

Every eel hunt we've observed has involved either a Spotted Moray, a Goldentail Moray (the only two morays that regularly hunt on the reef during

A Coney shadows a Goldentail Moray as it dashes to a new hunting location.

A pair of Coneys and Schoolmasters follow a roving octopus.

daylight hours), a Sharptail Eel, *Myrichthys breviceps,* a slender snake eel common in the eastern and southern Caribbean, or on one occasion a Green Moray, *Gymnothorax funebris,* trailed by a Tiger Grouper, *Mycteroperca tigris.* Over the years we have observed a galaxy of fish trailing the eels including Atlantic Trumpetfish, Schoolmaster, *Lutjanus apodus,* Mutton Snapper, *L. analis,* Yellowtail Snapper, *Ocyurus chrysurus,* young Black Grouper, *Mycteroperca bonaci,* Nassau Grouper, *Epinephelus striatus,* Spotted Soapfish, *Rypticus subbifrenatus,* Bar Jack and even goatfish. Occasionally the affairs last the better part of an hour. In one instance we counted 11 predators accompanying a single snake eel.

Opportunistic piscivores also nuclear hunt with large octopus during the day. In the Caribbean, only the Common Octopus, *Octopus vulgaris,* has been documented leading packs of predators similar to those hunting with eels. Twice a day, the Common Octopus leaves its den to forage for gastropods, primarily bivalves, and crustaceans. The octopus pokes around inside crevices or throws its pliable body over sponge and coral formations while tickling out concealed prey with its arm tips. Either way its disruptive behavior unsettles the routine of prey, leaving them vulnerable to a ring of waiting mouths.

Accounts of piscivores nuclear hunting with eels have been reported from the Gulf of California, the Red Sea, and the Pacific as well as the Caribbean. Each study described small reef predators following day-feeding morays. One interesting facet of the relationship yet to be sorted out is the occasional rubbing interaction that takes place between Spotted Morays and members of their entourage. As mentioned above the behavior appears congenial and often involves the grouper rubbing its side against the eel's upper body for as much as a minute before the moray slips away. Even more puzzling, from time to time we have watched algae-eating Ocean Surgeonfish, *Acanthurus tractus,* non-participants in group hunts, also lean and rub their bodies again Spotted Morays in the same fashion. A research paper also documented an Ocean Surgeonfish as well as a Princess Parrotfish, *Scarus taeniopterus,* another herbivore, involved in "rubbing and posing" similar to what we have seen, but the author didn't venture an explanation of the curious behavior.

Of course morays routinely hunt alone, and sometimes they hunt each other. We had the opportunity to view an extraordinary piece of video showing a Spotted Moray hunting a shallow reef crest of Finger Coral, *Porites porites,* in St. Lucia. It was apparent that the eel was following the scent of prey by the way it tracked across the coral, periodically thrusting its head into openings. Suddenly the hunter plunged into a fissure, struggled, and pulled out a Chain Moray, *Echidna catenata,* half its length, tightly gripped in its jaws. The desperate moray reached around and bit its captor across the eyes. Intertwined, the two eels spun violently. The Spotted Moray thrust its tail into the reef to gain traction, then raked its head through coral branches to break the blinding grip. After pulling free, it swallowed the prey in less than 10 seconds.

Day in and day out, trumpetfish are the most entertaining piscivores on the reef. Stealth is their game, and persistence is their strength. Their long, thin bodies scarcely throw shadows as the solitary hunters cautiously weave their way through a maze of bottom structures in search of prey. Disguising their positions by aligning with branching gorgonians and sponges and shadow-feeding with foraging fishes are favorite tricks. Unlike nuclear-hunting where a piscivore closely follows another species with the hope of grabbing prey flushed out of restrictions, shadow-feeding involves a close association with a non-threatening herbivore, typically parrotfishes, occasionally sponge-eating angelfish, in one curious case a traveling lobster, and from time to time divers. Trumpetfish, as is their nature, are attempting to disguise their presence in order to get closer to unsuspecting prey.

An Atlantic Trumpetfish—one of the reef's most prevalent piscivores.

The elongate hunters occur in three basic color patterns: brown to reddish brown, tan to gray with blue to purple snouts, and yellow. Each color scheme is inclined to shadow-feed with fishes of similar hues. Brownish trumpetfish associate with groupers and parrotfishes, and yellow individuals prefer Spanish Hogfish, *Bodianus rufus.* Those with blue snouts join swarming feeding aggregations of Blue Tang, *Acanthurus coeruleus,* to take advantage of the general confusion caused when the herbivores mob damselfish gardens. Trumpetfish are seldom welcomed as hitchhikers. On occasions, parrotfishes and groupers attempt to lose the predators by swimming erratically and brushing against bottom structures.

Trumpetfish hunt throughout the day, only occasionally taking time out to visit cleaning stations. While on the move, they generally swim horizontally just above the bottom but quickly assume a vertical pose to conceal their presence or position for a strike. They often position their elongate bodies vertically with their mouths just inches from the bottom for minutes at a time while waiting for a holed-up fish to reappear. Downward strikes generated from bowed bodies are short and lightning-quick. Their mouths flare like trumpets to engulf fishes, which generally measure less than two inches, although a five-inch squirrelfish was once found inside a two-foot trumpetfish. Bicolor Damselfish, *Stegastes partitus,* larval fishes, blennies, cardinalfishes and juvenile grunts are their most frequent victims. Trumpetfish also pursue groups of foraging fishes while horizontal to the bottom and occasionally lie on their sides

Shadowing a Spanish Hogfish.

Shadowing a Queen Angelfish.

Shadowing a Caribbean Spiny Lobster.

Even predators make mistakes.

The expanded mouth of a trumpetfish illustrates the source of the predator's common name.

Four trumpetfish stalk the same prey.

Postured for a strike.

Lying flat on the sand—an unusual stalking position.

or turn upside down to gain an advantage. A feeding study recorded 8 successful trumpetfish strikes in 45 attempts during 19 hours of observed stalking.

The same study analyzed the response of Threespot Damselfish, *Stegastes planifrons,* to models of trumpetfish manipulated by observers. The research found the Threespots (prey) most vigilant when the models (predators) were large, close or held vertically. When the Threespots sight horizontally swimming models moving in their direction, they stop feeding, rise slightly off the bottom and swim toward the threat. This behavior is possibly meant to better assess the danger or to let the predators know they have been seen. The damselfish then drop to the bottom and retreat. If the predators continue or assume vertical positions, the damselfish disappear inside hiding holes. After the mock predators pass, the damselfish often chase after the adversaries and nip at their tails. Large damselfishes frequently charge and occasionally nip trumpetfish simply swimming near their territories. These attacks are quite effective, causing the trumpetfish to jump and quickly leave the area. However, once poised in a vertical strike position, the hunters seem impervious to damselfish assaults.

Coneys regularly stalk small fishes during daylight hours. Once in pursuit, the groupers' intensity and persistence are reminiscent of terriers after rats. The hunters often lie motionless on their sides for minutes at a time, with their eyes glued on their intended victims. During one encounter, a determined Coney worked a targeted goby for 10 minutes before moving eight feet away and settling to the sea floor as if no longer interested. Moments later, when the goby made a dash for a nearby rock, the Coney sprang into action, nabbing the fleeing prey well before it reached safety. One afternoon we watched a lengthy standoff between a Bicolor Damselfish and a Coney. There was no obvious reason why the predator singled out its prey; the damselfish didn't appear injured, and there were numerous larger and smaller Bicolors around the same isolated coral mound. The Coney stared intently at a hole where the prey hid for long stretches, then moved several feet away. Each time the one-inch fish attempted to rise into the water column the Coney charged, driving the damselfish back into the hole. One hour and nine unsuccessful lunges later, the stalemate continued.

A piscivore's fortunes are not always what they appear. When a fish makes a strike, competitors rush to the scene. To avoid injury and sharing their prey, most piscivores stalk fishes small enough to fit inside their mouths. However, if an opportunity arises to capture prey too large to swallow easily few predators can resist. While descending to a 30-foot patch reef we noticed a 20-inch Red Hind, *Epinephelus guttatus,* rapidly cutting across a sand channel with the back half of a bright red Bigeye, *Priacanthus arenatus,* projecting awkwardly from its mouth. Even before we could settle to the sea floor, the successful hunter was joined by a smaller Red Hind that made a grab for the prey. With a toss of its head that sent sand swirling, the hunter defended its catch and accelerated toward the cover of a ledge.

At that instant, the first of several Queen Triggerfish, *Balistes vetula,* appeared. It struck with the speed and intent of a dive-bomber, burying its beak deep into the Bigeye's flesh. The unexpected jolt broke the grouper's hold.

Feeding

A Red Hind dashes across the sand with a freshly captured Bigeye.

The smaller Red Hind pounced but was knocked clear by the enraged hunter. To secure a better grip, the larger Red Hind unhinged its jaws and flared its gills, drawing in the Bigeye with such powerful suction that the outgoing blast sent sand flying. Just as the prey was secured, the water darkened with the swooping silhouettes of a large strike force of triggerfish intent on a free lunch. Their usual pastel shading of purple, blue, turquoise and yellow darkened to a menacing black. In this belligerent state, the triggerfish relentlessly attacked the Bigeye's exposed tail until the hapless prey, again wrenched from the grouper's mouth, lay on the sand, pale and eyeless, in two tattered pieces.

We counted 38 fishes attempting to hijack the Red Hind's freshly caught prey.

The Red Hind, now joined by three Coneys and a Graysby, formed a circle around the kill just out of reach of the flying circus of snapping triggerfish beaks. While the grouper waited for an opportunity to make a move, a school of Blue Tangs (the normally algae-eating members of the surgeonfish family taking advantage of the chaotic situation to nab bits of flesh to accelerate growth) added their numbers to the ever-growing feeding frenzy that swirled in a chaotic cloud 10 feet off the bottom. Kneeling just outside the vortex of flying fishes and sand, we quickly counted 38 Queen Triggerfish, Blue Tangs, Yellowtail Snapper, groupers and Black Durgons. Suddenly, the big Red Hind dashed forward, grabbed the head segment and took off in the direction of a ledge with a streaming column of irate fishes in pursuit.

The Red Hind disappeared under a limestone overhang. Just as it appeared that the big fellow was finally going to reap a portion of his just reward, several pursuing triggerfish, without so much as breaking stride, turned on their sides and dived under the ledge. A sandstorm rose from the narrow opening as three lobsters wisely fled. It took 20 minutes before the sand settled, the last of the fishes dispersed, and the lobsters marched back across the flat to their lair.

Even though piscivores must know better, they occasionally succumb to temptation and grab small fishes protected by piercing dorsal spines. We have seen both a frogfish and a scorpionfish engulf heavily spined butterflyfishes and immediately blow them back out. An Indigo Hamlet, *Hypoplectrus indigo,* made the mistake of plucking a dainty little Slender Filefish, *Monacanthus tuckeri,* from the branches of a gorgonian where the species traditionally hides. The half-swallowed prey locked its elongate foredorsal spine inside the predator's widely stretched mouth. A Yellowhead Wrasse, *Halichoeres garnoti,* swooped down and, in a single bite, clipped off the filefish's exposed tail then returned and chased the beleaguered hamlet in circles attempting to steal the remainder of its meal. An underwater photographer reported a Nassau Grouper dashing six feet off the bottom to take a Black Durgon in midwater. The

An Indigo Hamlet having trouble swallowing a Slender Filefish.

A Yellowhead Wrasse exploits the hamlet's dilemma.

Sand Diver with a captured flounder that eventually escaped. Prey fishes taken tail-first are difficult to swallow because of their backward pointing fin spines.

Sand Diver with Blackbar Soldierfish.

triggerfish extended its single foredorsal spine on impact momentarily preventing it from being taken into the grouper's mouth. The grouper repositioned and shoved the prey against the bottom for five minutes before it was finally swallowed. In another instance a Peacock Flounder, *Bothus lunatus,* grabbed a Sharpnose Puffer, *Canthigaster rostrata.* Like all members of the puffer and porcupinefish families, the two-inch puffer inflated its body with water to the point it couldn't be swallowed and was eventually released. On two separate occasions, we have seen Sand Diver, *Synodus intermedius,* capture a Sharpnose Puffer; in one instance the puffer, after swelling its body to the size of a ping-pong ball, eventually escaped, but the practical adaptation didn't work for a second inflated puffer that was successfully swallowed after several minutes.

Few piscivores, other than sharks, a few eels, Tarpon and soapfish are believed to feed regularly after dark. However, under favorable circumstances such as bright moonlit nights, or when an unusual abundance of nocturnal prey is present, the opportunistic predators freely extend their daytime feeding hours. Fishermen report taking Nassau Grouper on live bait well after dark, even though the species is commonly thought of as a daytime feeder. Night-stalking piscivores can be quite resourceful. One evening a Greater Soapfish, *Rypticus saponaceus,* grabbed five cardinalfishes in succession before we realized that the predator was using our light beams to locate prey. We also have heard accounts of a Great Barracuda, Cubera Snapper, Schoolmasters, Mutton Snapper and Tarpon using divers' lights to locate food. Interestingly, a few researchers theorize that nighttime prey can anticipate the oncoming rushes of piscivores from flashes of bioluminescence activated by an attack.

A Black Durgon's erect foredorsal spine makes it difficult for the Nassau Grouper to swallow.

Hiding behind large mouths and camouflage coats, lie-in-wait predators patiently wait for small fishes to mistakenly swim too close. Prey is captured with explosive lunges and suction created by a rapidly opening mouth. Frogfishes, not satisfied with waiting for prey to happen by, entice small fishes with a dangling lure attached to the end of a modified foredorsal spine. Peacock Flounder not only use their remarkable camouflaging ability and bury their flat bodies just beneath the sand's surface to ambush prey, but also actively stalk and capture small fish by raising their forebodies and striking like a cobra. Lizardfishes are one of the most aggressive bottom predators. They prowl the sea floor, stopping to rest on coral, sponges or sand for minutes at a time before resettling in a new location a short distance away. Instead of changing colors to match their background, the slender piscivores often bury in the sand with only their eyes and large mouths exposed. Their high-speed, upward angled attacks make short work of juvenile grunts and other small fishes passing overhead. Although the feeding strategies of Caribbean lizardfishes have not been studied in detail, the Variegated Lizardfish, *Synodus variegatus,* a family member from the Indo-Pacific, was found to average one strike every 35 minutes with an 11 percent success rate.

Many night-feeding fishes generally rest in schools for protection during the day. By far, the largest and most spectacular of these gatherings are the teeming schools of herrings and anchovies, collectively known as silversides, that swirl together in the entrances of coral caves and open wrecks in polarized schools. These minnow-sized silvery fishes have good reason to congregate; they are the most frequently eaten fishes on the reef. By joining together, their odds of survival improve dramatically. The continually moving and flashing schools make difficult targets for hunters. An attack from any direction is signaled through a school at twice the speed of the oncoming rush, setting into motion one of several synchronized evasion tactics. Using vision, otoliths and lateral lines to maintain position, the fishes can open a hole around an advancing predator, contract into balls or explode into a thousand shimmers. Attacks become more profitable at dusk, when the schools begin to disperse and the silversides no longer glint and gleam at every turn.

Planktivores

One or two species from nearly every major fish family have evolved the physical characteristics necessary to feed on the endless stream of zooplankton swept past the reef from the open sea. Zooplankters are primarily swimming crustaceans so small they cannot be detected easily by the unaided eye. Most are clear, shrimplike copepods less than two millimeters in size that hop erratically as they swim. There are two basic types of planktivores: particulate feeders, usually reef associated species that use binocular vision to focus on oncoming specks which are ingested one at time, and filter feeders that swim with opened mouths, engulfing large volumes of water containing prey.

Particulate feeders, also known as plankton-pickers, have small, upturned, protrusible, fine-toothed mouths that shoot forward to reach and draw in prey with suction. Long, closely spaced gill rakers keep the minute particles, which are typically swallowed whole, from passing through the gill openings. A growing number of researchers believe that mucus coating the gill rakers plays an important role in entrapping particles. Even with these specialized modifications, plankton-pickers are able to capture only the larger zooplankters with conspicuous pigmented eyes or body parts. Surprisingly, daytime planktivores seldom feast on the swarms of tiny Mysid Shrimp, *Mysidium* spp., associated with reef structures. It seems that planktivores are unable to focus on an individual within the swarms.

Brown Chromis feeds on plankton with its protrusible mouth.

Reef-related planktivores feeding high in the water column developed sleek bodies and deeply forked tails necessary to evade the strikes of openwater predators such as jacks and mackerels. Their trim forms also allow them to maintain stationary positions in brisk currents. The most noticeable planktivores, Brown Chromis, *Chromis multilineata,* Creole Wrasse, *Clepticus parrae,* and, in the southern Caribbean, Atlantic Creolefish, *Cephalopholis furcifer,* gather in large aggregations in the current-swept waters over outer reefs, where the greatest number of transient zooplankters occur.

When threatened by roving predators, the clouds of feeding planktivores close ranks and ease toward the bottom. If attacked, they dive bomb into the reef below. Water-column foragers such as Striped Grunt, *Haemulon striatum,* and Boga, *H. vittatum,* which feed in polarized schools, flee en masse rather than seeking protection in the reef. A few plankton feeders, including Black Durgons, and Sergeant Majors, have taken an alternate route to survival by relying on deep flat bodies and sharp spines to discourage predation. On overcast days and during the low light of late afternoon, all planktivores feed lower in the water column than normal.

Those smaller planktivores without streamlined bodies, such as Bicolor Damselfish, *Stegastes partitus,* Fairy Basslet, *Gramma loreto,* Masked Goby, *Coryphopterus personatus,* and Purple Reeffish, *Chromis scotti,* feed alone or in small groups within restricted areas close to the reef's hiding holes. Planktivores inhabiting open sand, including Yellowhead Jawfish, *Opistognathus aurifrons,* Hovering Dartfish, *Ptereleotris helenae,* and Brown Garden Eel, *Heteroconger longissimus,* rise just above their burrows to feed on passing zooplankters.

As darkness falls and the daytime crew heads for night shelter, an entirely new set of planktivores enters the water column to dine. The most common

species include cardinalfishes, bigeyes, Blackbar Soldierfish, *Myripristis jacobus,* Glassy Sweeper, *Pempheris schomburgkii,* and silversides. These large-eyed night feeders, unable to see the small transient zooplankters in low light, consume the larger resident copepods and mysids that ascend from the reef after dark. With their food source distributed randomly and predators limited, there is little need to aggregate. Most feed alone or in small groups near their daytime hiding places, except for Glassy Sweepers that make migrations from inshore ledges and caves to outer reefs each night.

The filter-feeding planktivores vary in size from small, minnowlike herrings to Whale Shark, *Rhincodon typus,* the largest fish in the sea reported to reach a length of 55 feet. While maintaining a stationary position, herrings employ pump filtration, using rapid bouts of suction to draw in water and prey. Oceanic Manta Ray, *Mobula birostris,* the largest ray, feed by ram filtration. The great fish swim through clouds of microscopic plankton with their oval-shaped mouths opened wide. Gill rakers sieve out food particles as the water exits through flared gill covers. Extensions of the pectoral fins, known as cephalic lobes, funnel food toward their mouths. The lobes fold out of the way when not in use. While feeding in concentrated clouds of plankton, the fish often perform graceful loops. Whale Sharks use both filtration methods to sustain their massive size. They typically swim near the surface, engulfing plankton and small fishes as they go, but, when feeding in thick patches of plankton, they hover and pump food into their mouths.

Terminal phase male Creole Wrasse feeding on plankton.

Left: *Creole Wrasse feeding in the rich plankton stream above the outer reef.*

Oceanic Manta Ray filter feeding.

Whale Shark filter feeding.

Herbivores

Herbivores transfer energy from plants to animals higher in the food chain. This pattern holds true for coral reefs, where roughly 25 percent of the fishes eat plants. Large populations of plant-eating marine fishes are essentially restricted to tropical reef and seagrass environments. Why herbivores have not become better represented in subtropical or temperate waters is not clearly understood. In the Caribbean, parrotfishes and surgeonfishes, the two major families of plant consumers, feed from an abundant supply of fast-growing filamentous algae. Using powerful beaklike mouths, parrotfishes scrape the short (1–2 mm) algal turf from dead coral and rock with flurries of bites. During the process, significant amounts of calcium carbonate are ingested. In contrast, surgeonfishes browse selectively, taking quick, deliberate bites. Because algae provide few nutrients, herbivores must harvest and process large quantities of plant material to meet their growth and energy demands. When opportunities arise (such as the Red Hind unable to quickly devour Bigeye too large to swallow, described on page 87) parrotfishes and surgeonfishes will take advantage of the situation to grab bites of animal tissue to supplement nutritional needs for growth.

The rapid growth of turf algae makes shallow forereefs extremely productive habitats. Marine herbivores are estimated to consume between 50 and 100 percent of the primary production—the highest grazing rate on earth. Without herbivores to harvest the rapidly growing algae, coral reefs could not exist in their present form. If left ungrazed, the competitively superior algae would quickly smother existing corals and blanket reef rock, leaving nowhere for coral larvae to settle and establish new colonies. When plastic-mesh cages were placed over porous clay tiles during herbivore exclusion studies in the Florida Keys, tiny filaments of red algae, typically found in mats of algal turf, changed structure, growing into fleshy upright plants. After the cages were removed months later, parrotfishes and surgeonfishes picked the lush growth clean in less than two hours.

Parrotfishes and surgeonfishes are not the only plant-eaters on the reef. Algae dominate the diets of the Goldspot Goby, *Gnatholepis thompsoni,* Bridled Goby, *Coryphopterus glaucofraenum,* Redlip Blenny, *Ophioblennius macclurei,* Cherubfish, *Centropyge argi,* and several species of bottom-dwelling damselfishes. Numerous invertebrates also consume algae. Microherbivores, limpets, amphipods and small gastropods keep tiny foraging areas clear of growth. Clinging crabs, *Mithrax* spp., also depend on plants for food. But by far the most important algae-eating invertebrate in the tropical western Atlantic is the Long-spined Urchin, *Diadema antillarum.* A single urchin, using its complicated five-part mouth structure, known as an Aristotle's Lantern, keeps a square yard of rock or reef clear of algae each night. The abundant, shallow-water night-feeders consumed so much algae that they were a major food competitor of herbivorous fishes before 1983–84, when a Caribbean-wide epidemic almost eliminated the Long-spined Urchin. Following the die-off, parrotfish and surgeonfish populations monitored in St. Croix increased two- to four-fold in response to the additional food supply. However, in Jamaica, where parrotfishes and surgeonfishes had been severely overfished for decades, the loss of *Diadema* had a devastating effect on the island's northern reefs. Without the urchins' help, the depleted fish stocks were unable to keep the burgeoning algal crop in check. Corals, which had already experienced severe damage from Hurricane Allen in 1980, were overwhelmed. Living coral coverage plummeted in three short years from 45 to 55 percent to 5 to 15 percent. Although they may be dense in a few isolated locations, urchins are unfortunately making a slow recovery in Jamaica, as well as across most of the Caribbean Basin. Because of the scarcity of *Diadema,* coupled with continued

Terminal phase Queen Parrotfish scraping filamentous algae from reef rock.

overfishing and nutrient-rich runoff from coastal development and agriculture, algae continue to dominate the reefs in many areas.

Instead of foraging freely over the bottom, several species of damselfishes spend considerable energy tending and defending private algal patches. The little farmers weed out unwanted plants, leaving a luxuriant turf that grows one and one-half to three and one-half times taller than the unprotected algae on surrounding rocks. Although individual plots measure only a few feet across, in places up to 70 percent of a reef's surface can be covered with contiguous territories, greatly increasing the habitat's primary production. The thick mats harbor thriving colonies of tiny invertebrates and support secondary gardens of appetizing diatoms and other epiphytes. It is not known if, or to what extent,

Ocean Surgeonfish and other family members regularly form large feeding aggregations.

the damselfishes' diet depends on the hidden fauna for protein, but encroaching carnivores and herbivores are chased with equal vigor.

Grazing is so intense in shallow water that 30-foot, plant-free sand zones, known as halos, are created around inshore patch reefs. Their margins represent the limit to which herbivores dare to venture from the reef's protection. As algal biomass decreases with depth, the number of foraging herbivores also declines. However, there is a tendency toward a surplus of ungrazed plants on the deeper reefs. This disparity might be brought about by a greater threat of predation in the lower light. The density and diversity of herbivore populations remains stable throughout the year, even though grazing increases significantly during the warm-water periods of summer and early fall. In the Florida Keys, where there is a marked difference in seasonal water temperatures, grazing rates are six times greater during summer than winter.

Digesting plant material is not easy. Chemicals break down the rather selective diets of damselfishes and surgeonfishes as the food passes quickly through their thin-walled, elastic stomachs and winding guts. In contrast, the digestive process of parrotfishes has been characterized as a concrete mixer, grinding plants and calcareous sediment with flattened pharyngeal teeth, known as the pharyngeal mill. The chalky mixture moves rapidly through a stomachless gut before being discharged in long, white sand plumes. In fact, it has been estimated that a large parrotfish may produce a ton of sand annually! How the large herbivores obtain sufficient protein for growth from their diets is still under study. Adults could possibly acquire the necessary nutrients from organic waste, microorganisms or small invertebrates that accumulate in algal turf, or as mentioned above, from occasional opportunities to consume animal tissue. Whatever the source, it is effective; parrotfishes grow quickly, reaching maturity in 3 to 5 years and live for perhaps 8 to 12 years.

It has long been assumed that parrotfishes are strict herbivores. However, a recent study reported that three members of the genus *Sparisoma,* Redtail Parrotfish, *S. chrysopterum*, Redband Parrotfish, *S. aurofrenatum,* and Stoplight Parrotfish, *S. viride,* consume several species of crevice-dwelling sponges. Each time researchers removed the sponges from the crevices the parrotfishes eagerly devoured the colonies. They even chased one another from the sponges, although they never defend their customary food. It was obvious from the parrotfishes' enthusiastic response that the cryptic sponges would be exterminated if they didn't inhabit inaccessible crevices.

Reef algae are divided into four groups based primarily on their morphological structure. Crustose coralline are unpalatable, highly calcified encrustations that form important protective coverings and function as cement for the reef's structure. Calcareous species are green, leafy upright plants that avoid predation by calcification and chemical deterrents. Turf or filamentous species form the primary food source of herbivores although some species are chemically defended. Frondose or fleshy algae, typically vulnerable to grazing, grow primarily on reef flats and in back-reef areas.

Calcareous leaf algae in the genus *Halimeda* provide an interesting example of how one group of marine plants has evolved defenses to survive on the reef. These rich green clumps and vines not only reduce their nutritional value by calcification, but also produce chemical compounds repugnant to many herbivores. The plants use their deterrents judiciously. On reefs with an abundance of herbivores, the chemicals produced by *Halimeda* are much more potent than those manufactured by the same species in areas with few grazers. The plants have also adapted to grow their tender, uncalcified leaf tips at night, when herbivorous fishes sleep. The new growth, produced without sunlight for photosynthesis, is packed with high concentrations of toxins until the tips develop calcification. When researchers coated blades of Turtle Grass, *Thalassia testudinum,* with *Halimeda* toxin, Bucktooth Parrotfish, *Sparisoma radians,* avoided the treated blades but readily ate unmanipulated control grasses. Chemical deterrents are far from foolproof, however. The toxic compounds that repel one species have limited, or no, effect on others.

Colors & Camouflage

The bright "poster colors" of a Queen Angelfish.

Reef fishes are known for their exquisite colors and markings. Adding even more variety to the reef's catalog of wonders, many species display strikingly different, non-reversible colors and patterns during various life phases; others have the ability to quickly alter their typical hues and markings with mood swings and for intra-species communication, or lighten or darken their bodies within seconds to blend with new surroundings. Generally the brightest, most conspicuous patterns are used to communicate reproductive readiness or aggression, whereas dull, broken markings provide a protective cloak of camouflage.

When stimulated by nerves or, in some cases hormones, pigments in flat, irregularly shaped skin cells known as chromatophores either concentrate in the core, causing the fish to pale, or expand through radiating branches, intensifying colors. Layers of chromatophores containing different pigment hues overlap in the skin producing colors in the longer wavelengths—red, yellow, orange, black and brown. Those species able to switch between the brightest and most complex color patterns have higher densities of chromatophores. Many fishes display permanent structural colors of shorter wavelengths—greens, blues, pinks, silvers and iridescence—by reflecting light from stacked plates of protein crystals inside flattened skin cells called iridophores. Silvery fishes have several layers of iridophores that mirror nearly all the light striking their bodies.

Several of the large groupers in the genus *Mycteroperca,* including Black Grouper, *M. bonaci,* Yellowmouth Grouper, *M. interstitialis,* Tiger Grouper, *M. tigris,* and the Yellowfin Grouper, *M. venenosa,* can pale, darken, display various patterns or flush red, depending on their moods or activities. Some of the most skilled pattern-changers are lie-in-wait predators such as flatfishes, frogfishes and scorpionfishes. These masters of disguise, whose livelihoods depend on camouflage, are replete with chromatophores. Within seconds of settling in new locations flounders can vanish like mirages by using visual cues to expand or contract their chromatophores to match the colors and patterns of the seafloor where they rest. Although frogfishes sometimes take hours to alter their colors to correspond with new surroundings, any experienced fish watcher can attest to their genius at hiding in the open. The Sargassumfish, *Histrio histrio,* and Sargassum Pipefish, *Syngnathus pelagicus,* endemic inhabitants of the floating weed for which they are named, display color patterns to match their drifting homes. The remarkable little frogfish even acquire the shape of the surrounding stems and leaves. The fleshy tabs and cirri on bottom-dwelling Spotted Scorpionfish, *Scorpaena plumieri,* and the Lined Seahorse, *Hippocampus erectus,* often grow to resemble the algae where these predators wait in ambush. The disguise is so realistic that many casual observers believe algae actually grow on the sedentary fishes.

Some small fishes use color-changing camouflage to hide from big mouths.

A Yellowfin Grouper displays its typical color pattern.

Yellowfin Grouper often change their color scheme radically, such as developing reddish blotches when near the bottom.

The cryptic pattern of Sargassum Pipefish blends well with its floating home.

Relying on camouflage, instead of fleeing when threatened, initial phase Redband Parrotfish, *Sparisoma aurofrenatum,* drop to the bottom and blend with the background and only retreat when closely approached. Slender Filefish, *Monacanthus tuckeri,* keep low profiles by hanging out within gorgonians and alter colors, patterns and skin texture to disappear among the branches. Some bottom-oriented species like Slippery Dick, *Halichoeres bivittatus,* and Rock Hind, *Epinephelus adscensionis,* spontaneously lighten or darken as they move between light sandy areas and darker sections of the reef.

Males of many species crank up their chromatophores during courtship. Unfortunately, their eye-catching patterns also attract predators. To strike a balance between reproductive success and untimely death, males of many small, vulnerable fishes flaunt their brightest hues only during brief courtship bouts. This behavior is notable among seagrass residents, which must be extremely cautious when away from bottom cover. Male Bluelip Parrotfish, *Cryptotomus roseus,* flash their neon-bright panels of gold, pink, blue and green

Sargassumfish, a member of the frogfish family, spend their entire lives as part of the Sargassum community.

Right: *Many small reef fishes use color-changing camouflage to hide from big mouths.*

Female Sailfin Blenny

Juvenile Atlantic Trumpetfish

Slender Filefish

Pipehorse

Juvenile Green Razorfish

Initial phase Bucktooth Parrotfish

A Hogfish with its normal color pattern.

A Hogfish with bottom feeding colors.

only when swooping down to court females grazing in or near the seagrass. As they move away their colors become partially muted. A few bottom-dwelling blennies also show off vibrant colors when reproductively active. Members of the Hairy Blenny complex, *Labrisomus* spp., change from their drab brown camouflage wardrobes to dazzling orange, yellow and white courtship costumes. The brown head and forebody of the one-inch male Red Banner Blenny, *Emblemariopsis ramirezi,* turns jet black and it pops its red-lined dorsal fin erect when attracting females to its hiding hole. If a rival male approaches the dark features instantly turn brown, but the head remains extended and the dorsal fin stays taut to emphasize their size.

While it seems logical to assume that the varied markings and colors of reef fishes benefit survival, advantages are not always readily apparent. For instance, there is little agreement on why so many fishes display such eye-catching colors and patterns. Famed Austrian naturalist Konrad Lorenz coined the term "poster colors" to describe brightly colored fishes, a phrase in common

A Bluelip Parrotfish displaying its normal pattern.

A Bluelip Parrotfish turns on its bright courtship colors.

The nocturnally active Bigeye normally displays a solid red coat.

A group of Bigeye temporarily blanches while hovering just above the bottom during the day.

usage today. In his 1960s best-selling book, *On Aggression,* Lorenz hypothesized that the flamboyant colors act as no-trespassing signs announcing territorial sovereignty. Further research has proven his idea too simple to explain fully the purpose of conspicuous colors, but no one has yet offered a more convincing argument. It has been suggested that the bright colors worn by spiny, deep-bodied species such as butterflyfishes and angelfishes deter predation by advertising the fishes as bony, low-quality, difficult-to-swallow meals. Others have proposed that the graphic colors might serve to help gregarious species remain in contact with one another.

Several recurring colors and patterns are more easily explained. Nocturnal species are customarily red, an obscure color in the low light of night or depth. However, for an unknown reason, reddish representatives of the bigeye, squirrelfish and cardinalfish families regularly pale to a pinkish white, possibly for intra-species communications. Many fishes, especially openwater species, use counter shading to conceal their movements. When observed from above, their dark backs blend with the dimness below, while their light bellies viewed from below tend to be lost in the surface light. The bold horizontal stripes of schooling species help individuals to orient visually with their neighbors and at the same time make it difficult for predators to single out individuals from a shifting panorama of lines. When Hogfish, *Lachnolaimus maximus,* porgies, snappers and many other bottom-foragers settle to the sea floor to feed, they often acquire vague bars that instantly vanish as they rise back into the water column. The bars help mask the bottom-feeders' profiles when they are viewed against a backdrop of vertical structures such as gorgonian branches. Many

Juvenile Banded Butterflyfish display eyespots and eyebars.

fishes also alter their colors while attending cleaning stations to have external parasites removed.

Predators routinely key on the eyes of intended prey to evaluate their size and likely direction of escape. Dark eyebars worn by various species help conceal the eye's location. False eyespots in the form of dark spots with a contrasting outer ring, typically located near the tail, may misdirect attacks away from the head or cause the predator to misjudge the prey's size or direction of escape. Residents of shallow habitats typically display more eye camouflage than deep-water species. All five Caribbean butterflyfish species display permanent eyebars. Adult and juvenile Foureye Butterflyfish, *Chaetodon capistratus,* as well as juvenile Banded Butterflyfish, *C. striatus,* and the juvenile Rock Beauty, *Holacanthus tricolor,* a common member of the angelfish family also display eyespots. While living among mangrove roots, young Gray Snapper, *Lutjanus griseus,* and Schoolmaster, *L. apodus,* display distinct eyebars that are lost when they mature and migrate to deeper water.

A Red Hind in its sleeping colors.

An initial phase Redband Parrotfish changes its colors and pattern dramatically when sleeping at night.

During the night most daytime species sleep inside the protection of the reef or bury in the sand, but a few bed down in the open and display gaudy night colors. The complexions of slumbering parrotfishes and groupers change so radically that it is often difficult to identify the sleeping species. Their blotched patterns, so striking when illuminated by light beams, possibly function as disruptive coloration in the low light of night. Several members of the butterflyfish family, which typically sleep just inside ledges at the edge of the reef, develop vague bars and eyespots during the night. Divers should be cautious when illuminating a sleeping fish. A direct beam can cause a slumbering fish to bolt blindly possibly causing a harmful collision with a nearby structure. Daytime predators frequently adapt to hunting after dark using divers' lights to locate sleeping prey. For some unknown reason Creole Wrasse, *Clepticus parrae,* intensify their already gaudy purple, yellow and red colors while bedded down at night inside the protection of tube sponges and tight crevices.

As they mature, most parrotfishes and wrasses change color patterns three or four times to signify a new sexual or social status. However, most reef-associated species either wear the same wardrobes throughout their benthic lifecycles or make only a single transition when changing from juveniles into adults. Plankton-feeders, nocturnal species and small bottom-foragers that have no need to protect limited food resources rarely exhibit distinct juvenile patterns, whereas larger bottom-foragers, especially territorial species, typically have juvenile patterns unmistakable from adults. It has been hypothesized that

A terminal phase Redband Parrotfish bedding down on the open sand changes into its nighttime wardrobe.

A Blue Tang displays night colors.

Solid adult pattern.

Bicolor adult pattern.

Golden adult pattern.

Transitional phase—juvenile to adult.

Juvenile golden pattern.

Juvenile mottled pattern.

Resting and night pattern.

Male courtship pattern.

the radically different color patterns of youngsters conceal their species identity which, in turn, protects them from attacks by their territorial adults. Because of their susceptibility to predation, small juveniles are often more cryptically patterned than their elders. Young Green Razorfish, *Xyrichtys splendens,* and Great Barracuda, *Sphyraena barracuda,* are well patterned for hiding in the vegetation where they customarily settle after their pelagic larval stage. However, the juveniles of a few species are far more colorful than the adults. This is especially true of territorial damselfishes. Many of these conspicuous juveniles, which live isolated from other members of their species, are extremely territorial. It is possible that, like Lorenz's "poster color" hypothesis, eye-catching colors serve to space individuals from one another.

The Coney, *Cephalopholis fulva,* a common medium-sized sea bass, has one of the most widely varied repertoires of patterns displayed by a Caribbean fish. Juveniles either display cryptic mottled reddish brown-and-white markings or, in direct contrast, a bright golden scheme. Adults are most frequently reddish brown with a scattering of blue or pale dots, although a few individuals are golden. This striking xanthic coloration, believed to result from a single recessive gene, occurs sporadically in a few other species, including Atlantic Trumpetfish, *Aulostomus maculatus,* and the Smooth Trunkfish, *Lactophrys triqueter.* It is not known if golden Coney juveniles mature only into golden adults, but it appears doubtful because there is a much higher ratio of golden juveniles than of golden adults in a given population. Adults regularly alternate between reddish brown and a bicolored pattern, with a reddish brown back and white belly and tail. When making the transition, which in some instances occurs quite rapidly, the fish pass through an intermediate pattern. While sleeping at night Coneys nestle on the bottom next to a structure and take on an irregular barred and blotched pattern. Finally, courting males pale and become blotched, with a tan streak extending from behind their pectoral fins to their tails.

Left & above: *Various color patterns displayed by Coneys.*

Symbiosis

Tiger Grouper with parasites and wounds attends a cleaning station.

Symbiosis, or "living together," refers to a close interrelationship between two species. There are three primary types of symbiosis: mutualism—two species forming an alliance for the well-being of both; commensalism—one species benefits while the other is unharmed by the association, and parasitism—one species benefits at the expense of a host species. In mimicry, a special category of symbiosis, a species imitates the colors, patterns and, in some cases, the behavior of another species to gain benefit. The advantages of cleaning symbiosis, a common relationship between parasite-picking fishes and shrimp and numerous reef fishes, are still under investigation even though great strides in understanding the relationship have been made in both the Pacific and the Caribbean since the turn of the century.

Cleaning Symbiosis

Many species of fishes are plagued by an assortment of external parasites (ectoparasites), often so small they can hardly be seen. The little bloodsuckers, mainly isopods and copepods, armed with razor-edged mandibles, burrow into the tissue around the eyes and nostrils, under scales and even invade the tender, blood-rich lining of gills and mouths. To keep their infestations under control, client fishes spend significant amounts of time at sites, known as cleaning stations, where they often assume stationary trancelike poses while parasite-eating fishes and shrimp dine on their pests. In the process, the cleaners also remove dead or injured skin and occasionally take bites of beneficial mucus and attached scales—a form of cheating that causes the client fish to jerk in apparent pain and swim away, or in some cases chase the offender.

Cleaning behavior is not exclusive to the marine world; some terrestrial animals also form similar associations. Galapagos finches pick parasites from marine iguanas; Oxpeckers feast off the backs of hoofed mammals on the African savanna, and Black-tailed Deer, responding to the calls of Scrub Jays, hold still and extend their ears while the birds remove ticks. However, most cleaning occurs underwater where animals have inherent difficulty removing their parasites. At first glance, the alliance appears to be a model of mutualism: the cleaners tap a renewable food source, and the clients obtain a dependable means of pest control. But as researchers regularly discover, the behavior of animals is not always what it seems.

Cleaning symbiosis between fishes was originally described during the early part of the previous century, but was not linked to the removal of parasites until the mid-1950s. The behavior gained widespread attention after the publication of an attention-grabbing account by Conrad Limbaugh in a 1961 issue of *Scientific American*. The author described a modest field experiment conducted in the Bahamas where he removed all the cleaning organisms from two small, isolated reefs. His reported results were dramatic. Limbaugh wrote, "Within a few days the number of fish was drastically reduced; within two weeks all except the territorial fishes had disappeared. The experiment also demonstrated the importance of cleaning symbiosis in maintaining the health of the marine populations. Many of the remaining fish developed fuzzy white

A Bar Jack extends its mouth while being cleaned by a juvenile French Angelfish.

blotches, swelling, ulcerated sores and frayed fins." Limbaugh concluded, "From the standpoint of the philosophy of biology, the extent of cleaning behavior in the ocean emphasizes the role of co-operation in nature as opposed to the tooth-and-claw struggle for existence." Almost overnight, the glowing portrayal of the cooperative behavior captured the hearts and minds of both the public and the scientific community. But at the same time, Limbaugh's article unwittingly planted some broad, untested generalizations about the remarkable behavior that remain with us today.

A decade later, marine biologists thousands of miles away in Hawaii began conducting their own cleaner-removal experiments on Pacific reefs. Their controlled studies revealed the relationship to be more complex than first believed. Far from duplicating Limbaugh's celebrated results, the new findings showed insignificant changes in the abundance of fishes and no increase in their rates of parasites or disease. Suddenly, everyone's favorite example of cooperation in the animal kingdom was open to reevaluation. And sure enough, the scattering of research papers published on the subject in the 1970s and 1980s laid siege to the conventional wisdom. One line of reasoning proposed that cleaners have only limited effect on parasite control and actually act as parasites themselves, removing bits of flesh and mucus. In his 1987 paper *Cleaning Symbiosis,* George Losey, a biologist at the University of Hawaii, presented an idea of interest, supported by intriguing evidence. Losey wrote, "Cleaners are nothing but very clever behavioral parasites ... [that] perhaps have taken advantage of the rewarding aspects of tactile stimulation, found in nearly all vertebrates. They may have parasitized this reward system to train hosts to visit them for rewarding stimulation and, at the same time, provide a dining table for the cleaners." During aquarium experiments, fishes, whether infested or not, reacted to objects rubbed against their sides in the same

An initial phase Redband Parrotfish assumes a classic head-up cleaning pose within the branches of a gorgonian—a puzzling behavior, known as pseudo-cleaning.

manner they responded to cleaners even though free of ectoparasites.

A puzzling behavior, termed pseudo-cleaning, also supports this concept. Four species of Caribbean parrotfishes—Stoplight Parrotfish, *Sparisoma viride*, Redband Parrotfish, *S. aurofrenatum*, Redtail Parrotfish, *S. chrysopterum*, and Blue Parrotfish, *Scarus coeruleus*—often assume classic, head-up cleaning poses within the upright branches of two species of sea plumes, *Pseudopterogorgia americana*, and *P. rigida*. The fishes remain in zombielike states for seconds or even minutes at a time, often repeating the behavior in the same gorgonian bushes throughout the day. Close inspections did not find cleaners in the vicinity. Could the caress of waving branches prompt the peculiar behavior? In

Creole Wrasse are regular clients of cleaning stations manned by Spanish Hogfish.

the long run, it might turn out that tactile stimulation and parasite control are not mutually exclusive. The pleasure of touch could well be the primary incentive for clients to attend cleaning stations, which by having parasites removed, possibly improves their health and ultimately, reproductive success.

At the turn of the 21st Century many fundamental questions concerning cleaning symbiosis remain unanswered. It appears from past research that rates of parasitism and the benefits of cleaning vary considerably between regions, with studies from one area having little or no relationship to those made elsewhere. For instance, more external parasites were found on reef fishes in Puerto Rico than in Hawaii.

A wave of new cleaning symbiosis research began at the turn of the century. Much of the effort has been conducted by the Coral Reef Ecology Laboratory under the direction of Dr. Alexandra Grutter, at The University of Queensland, Brisbane, Australia. A large segment of the lab's research concentrated on the behavior of Bluestreak Cleaner Wrasse, *Labroides dimidiatus*, the most common cleanerfish in the Indo-Pacific. In turn, their efforts have triggered a flurry of investigations into cleaning behavior in the Caribbean.

Symbiosis

The most prevalent cleaning organisms inhabiting tropical western Atlantic waters include: 11 of the 17 gobies in the genus *Elacatinus*, primarily the Sharknose Goby, *E. evelynae,* the Neon Goby, *E. oceanops,* the Cleaning Goby, *E. genie,* and the Broadstripe Goby, *E. prochilos,* juvenile Spanish Hogfish, *Bodianus rufus,* initial phase Bluehead, *Thalassoma bifasciatum,* and Pederson Cleaner Shrimp, *Ancylomenes pedersoni.* Other fishes and shrimp also participate in cleaning activities, but are considered secondary or incidental cleaners because of the limited scope of their interaction. In addition, the juveniles of numerous carnivorous fishes act as incidental cleaners taking occasional nips at stationary fish resting nearby.

Generally a single mated pair to a dozen of small one- to two-inch cleaner gobies, *Elacatinus* spp., live as a group on large brain, star and plate corals where they maintain a cleaning station. The gobies inhabit depths as shallow as five feet, but tend to be more common at 45 feet and are infrequent at 90

Nine species of gobies in genus Elacatinus *act as cleaners in the Caribbean.*

Juvenile Spanish Hogfish are prominent cleaners in the Caribbean.

Initial phase Blueheads account for 10 percent of cleaning activity on Caribbean Reefs.

A Pederson Cleaner Shrimp (with an abdomen full of eggs) is the most common cleaner shrimp in the Caribbean.

feet and below. Although Sharknose Gobies are dedicated cleaners that most frequently live on coral, a field study in St. Croix produced some intriguing results. There the Sharknose live on both coral, where they act as cleaners, as well as sponge where they rarely clean. Instead the sponge-dwellers feed on non-parasitic, sponge-dwelling copepods. Surprisingly, the study found that immature Sharknoses living on sponge not only grow faster than their coral-dwelling counterparts, but also live longer. At the same time the Sharknose population in Barbados does not inhabit sponges at all. The result not only questions the long held belief that cleaning behavior provides a more beneficial lifestyle than other strategies, but also indicates that the cleaner/client relationship is situation dependent.

Research from the past decade has also shown that the extensively examined cleaning behavior of the Indo-Pacific Bluestreak Cleaner Wrasse cannot be used as a textbook model of how cleaning symbiosis functions elsewhere. For

Neon Goby—common cleaner in Florida.

Smooth Trunkfish cleaned by a goby.

Lizardfish cleaned by gobies.

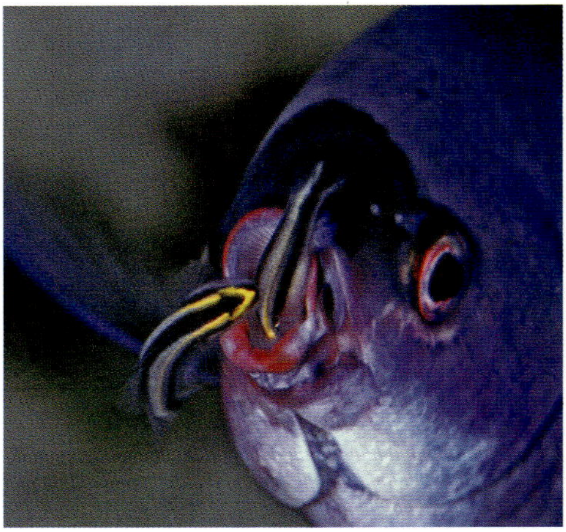

Creole Wrasse cleaned by gobies.

Atlantic Creolefish cleaned by a goby.

A goby darts inside the mouth.

Gobies stream from a coral head to clean a Yellowmouth Grouper.

instance, when offered a choice the Bluestreak Cleaner Wrasse not only prefer to eat mucus from client fishes rather than their parasites, but they also perform advertising dances to attract clients, provide tactile stimulation before and while feeding, and chase cheaters that take nips of flesh rather than pick pests. The opposite appears to be true of Caribbean cleaning fishes. Where removal studies of cleaner wrasse from reefs in the Pacific show an increase in parasite loads and eventual decrease in clients' numbers these results have yet to be duplicated to a satisfactory extent in the Caribbean.

Gobies scamper over the head of a Spotted Moray.

Juvenile Spanish Hogfish, a common Caribbean cleaner while young, prefer to clean hovering fishes. Up to seven of the purple-and-gold wrasse, typically measuring an inch or two, form cleaning groups in the general vicinity of tall coral structures, and one or two regularly associate with cleaning gobies and initial phase Blueheads at mixed-species cleaning stations. The agile cleaners pick here and there along the sides and fins and, now and then, explore beneath gill covers or dart inside opened mouths. The young hogfish tend to be more active at depth, around 40 feet. Some of their nips, usually directed at the tail, are evidently painful, causing the client to flinch and abruptly swim away or, in a few cases, turn and chase the offending cleaners. It is not known whether these unappreciated bites are necessary to dislodge well-attached parasites or to scavenge mucus, scales or other healthy tissue. But it is interesting to note that the nips nearly always occur well away from a client's mouth. At times cleaners are much in demand. Tightly packed schools of hovering grunts are regularly serviced by patrolling squadrons of young Spanish Hogfish. Frequently the grunts gang around the two-toned cleaners simultaneously opening their mouths as wide as choirboys in the hope of having parasites plucked from their throats.

Initial phase Blueheads, one of the most common parasite-pickers on Caribbean reefs, account for only 10 percent of cleaning activities. The bulk of their diet is obtained by plucking plankton from currents or foraging for crustaceans on the sea floor. Many cleaning stations established by other species have a few initial phase wrasse buzzing about, but young Blueheads

Tomtates gather around a juvenile Spanish Hogfish in the hopes of having their parasites picked from inside their mouths and gills.

Small cleaners enter the mouths of larger client fishes without fear.

also form independent cleaning clusters above gorgonian bushes and coral formations. The bright yellow juveniles work quickly, momentarily swarming clients suspended in open water before moving on to their next client. Unlike the other primary cleaning species, which have immunity from predation,

Stoplight Parrotfish poses with snout up and fins spread for initial phase Bluehead cleaners.

Blueheads are regularly eaten. This is probably a strong factor in the species' reluctance to clean piscivores. Although Blueheads in the terminal phase do not tend to function as cleaners because of their well-developed mouth structure, we once observed a solitary terminal phase male cleaning in association with initial phase cleaners.

Pederson Cleaner Shrimp are the most abundant and active cleaner shrimp found in the tropical western Atlantic. As many as a dozen, but commonly less than four of the wispy little transparent shrimp with lavender highlights live within the protective tentacles of a single Corkscrew Anemone, *Bartholomea annulata,* and on rare occasions the Giant Anemone, *Condylactis gigantea.* To solicit prospective clients, the dainty shrimp prance about near the ends of the spiraling translucent tentacles, waving conspicuous white antennae twice the length of their bodies. Clients must take a position on the bottom within a few inches of an anemone before the shrimp venture out to pick parasites, including larval gnathiid isopods, monogenean flatworms and juvenile cymothoid isopods. The shrimp are believed to provide tactile stimulation during cleaning bouts, which last from seconds to more than 15 minutes, but most interactions end in about 20 seconds. Parrotfishes make some of the shortest visits, lizardfishes and groupers some of the longest. The soothing behavior is believed to help mitigate stress hormones, which in turn encourages clients to return to the anemone throughout the day. The anemones also gain from hosting shrimp. While the anemones' stinging tentacles provide protective shelter for shrimp, inorganic nutrients excreted by shrimp fertilize the hosts' beneficial

Pederson Cleaner Shrimp inspects the open gill cover of a Brown Chromis.

Red Snapping Shrimp share a home with Pederson Cleaner Shrimp within the stinging tentacles of Corkscrew Anemones. Although not documented to be a cleaner species, Red Snapping Shrimp readily pick at extended fingers.

zooxanthellae algae living within the anemone tentacles. Although it has not been reported, we suspect that Red Snapping Shrimp, *Alpheus armatus,* are cleaning organisms. Usually one, but in some cases a pair, of the stocky, one- to two-inch crustaceans with large, single snapping claws, share Corkscrew Anemones with Pederson Cleaner Shrimp. Whenever we rest a hand within a half inch of an anemone, the snapping shrimp will pick at our fingertips for several minutes. Even after repeated attempts with numerous individuals, we were never able to coax them away from the anemone's protective tentacles. Unfortunately, the fishes that we believe are being cleaned by the shrimp lie flat over the opening of the anemone, blocking our view of the activity.

Several additional species of Caribbean fishes and shrimp act as secondary or incidental cleaners. Juvenile Gray Angelfish, *Pomacanthus arcuatus,* and French Angelfish, *P. paru,* obtain 20 percent of their diet from ectoparasite plucking. We have on rare occasions observed cleaning behavior by juvenile Rock Beauty, *Holacanthus tricolor,* Spotted Drum, *Equetus punctatus,* and Spotfin Hogfish, *Bodianus pulchellus.* Once, in the Gulf of Mexico, we watched an adult Spotted Soapfish, *Rypticus subbifrenatus,* picking parasites from Goliath Grouper, *Epinephelus itajara.* Published accounts describe cleaning by juvenile Porkfish, *Anisotremus virginicus,* and Black Brotula, *Stygnobrotula latebricola.* Caribbean shrimp mentioned in cleaning literature include Scarlet-striped Cleaner Shrimp, *Lysmata grabhami,* Banded Coral Shrimp, *Stenopus hispidus,* a circumtropical species, Golden Coral Shrimp, *S. scutellatus,* Spotted Cleaner Shrimp, *Periclimenes yucatanicus,* and Twoclaw Shrimp, *Brachycarpus biunguiculatus,* a nocturnal species that inhabits crevices and tube sponges. Of these, we have only seen the Scarlet-striped Cleaner Shrimp, cleaning on a regular basis, and have never observed the Spotted Cleaner Shrimp or the Twoclaw Shrimp make contact with a client.

The eight members of the family Echeneidae, commonly known as

Juvenile French Angelfish

Juvenile Spotfin Hogfish

Juvenile Porkfish

Juvenile Spotted Drum

Black Brotula

Juvenile Rock Beauty

Above: *Fishes known to be secondary cleaners.*

Scarlet-striped Cleaner Shrimp

Banded Coral Shrimp

Golden Coral Shrimp

Spotted Cleaner Shrimp

Twoclaw Shrimp

Red Snapping Shrimp

Above: *Shrimp known to act as cleaners.*

A fleshy oval ridge transversed by riblike lamellae with the ability to move up and down to adjust suction, makes the hitchhikers difficult to shake.

sharksuckers or remoras, are moderate-sized pelagic fishes known for their habit of hitching long-term rides on large marine animals, such as sharks, mantas, rays, sea turtles, manatees, sizeable bony fishes, and occasionally, when preferred hosts aren't available, divers and boats. Family members tend to favor particular hosts: the Sharksucker, *Echeneis naucrates,* a common, circumtropical species, opts for sharks and turtles; the Remora, *Remora remora,* specializes in larger openwater hosts, like manta; while Whalesucker, *R. australis,* associates with whales. As often happens when dealing with symbiosis, relationships between hosts and associates are not always clear cut, frequently blurring the lines between commensalism, mutualism or parasitism. The manta-riding Remora are believed to be parasitic because they consume the same food as their host. Sharksuckers appear to be commensal, dining on dead skin, scales and ectoparasites, as well as scavenging leftovers from their hosts. Young individuals act as cleaner fish, not only picking parasites but also removing dead tissue and bacteria from wounds. However, the bulk of the family members' diets turns out to be fecal matter provided by their hosts. Although not particularly nutritious, the food supply is plentiful.

Poor swimmers, lacking swim bladders, family members rely on their ability to grip onto mobile hosts with an ingenious suction disc located on the top of their flattened heads. The unique structure, derivative of dorsal fin spines, begins to develop on tiny pelagic larvae and is fully functional by the time a juvenile reaches a little over an inch in length. The structure, a fleshy oval ridge transversed by numerous riblike lamellae with the ability to move up and down to adjust suction, makes the hitchhikers difficult to shake once they latch onto a host. All a sharksucker has to do is pull backwards to tighten its grip or move forward to release its hold.

It is assumed that sharksuckers act as cleaners when small. We watched a seven- to eight-inch juvenile Sharksucker slip inside the gill cavity of a resting Nurse Shark, *Ginglymostoma cirratum,* each time the shark opened its gill covers to take in water. With each opportunity, the spry little Sharksucker darted in, grabbed a bite and backed out in less than two seconds. After a series of successful forays, luck ran out when the gill cover slammed shut on the cleaner's head. The instant the flap reopened the Sharksucker made a beeline for the back of the shark, where it huddled next to the rear dorsal fin for the remainder of our observation.

A paper from Brazil describes a six-inch juvenile Sharksucker inhabiting what appeared to be an established cleaning station on a wreck. The cleaner was observed swimming a short distance away from the wreck to clean the flanks of a terminal phase Stoplight Parrotfish, *Sparisoma viride,* which posed near the site in a classic head-up cleaning posture. The following day the small sharksucker once again was seen cleaning the same parrotfish in the same location. Ichthyologist John Randall also noted a young Sharksucker with its head inside the gill chamber of a large parrotfish.

If you have dived for long you've more than likely already had a run in with a sneakerhead—at least that's what we call Sharksuckers in Florida because of their suction discs resemble gym shoe soles. If you haven't experienced the misfortune yet, odds are you will one of these days. Be ready, sneakerheads appear out of nowhere, come in close, nosing around, circling and darting. The fish can be annoying and quite persistent. If uncontested the fish usually end up attached to your scuba tank, from where they like to nibble a diver's hair. Otherwise the fish are harmless unless they attach to exposed flesh, which can leave a nasty welt. To break the grip, simply push the fish forward.

So what to do when confronted by a free-swimming sneakerhead? You'll probably do what most divers do. Lash out, which only makes things worse. After a miss or two you start to get mad. At this point three options remain:

A juvenile Sharksucker ventures inside the mouth of a Goliath Grouper to remove parasites.

get out of the water, tolerate the creepy fish, or foist the nuisance off on a buddy.

Bluestriped Grunts open their mouths to be cleaned by an adult Porkfish even though only juveniles of the species act as cleaners.

Perhaps half of the fish species on a Caribbean reef spend part of their day at cleaning stations. The cleaners' best customers are typically groupers, parrotfishes, surgeonfishes, Brown Chromis, *Chromis multilineata,* and Creole Wrasse, *Clepticus parrae;* however such lists vary between regions, often closely representing the local population structure. Members of different families display classic posing styles to indicate their desire to be cleaned. Some fishes show their eagerness by resting on the bottom and listing to one side near a cleaning station, while others maintain stationary positions in mid-water with their heads angled up or down. Posing fishes regularly spread their fins exposing otherwise hidden parasites and also open their mouths and gill covers to allow cleaners access. Goatfishes hover with their heads and barbels pointed straight down; surgeonfishes and grunts hover horizontally. Studies have demonstrated that posing fish are customarily more likely to be cleaned.

Clients normally visit cleaning stations alone; however, two common planktivores, the Brown Chromis and the Creole Wrasse, regularly converge on cleaning stations in groups. On outer reef banks, where these plankton-pickers feed in large aggregations, Creole Wrasse numbering in the hundreds stream along reef slopes in late afternoon. Groups of up to a dozen dash down to cleaning stations where they hover in tight clusters in head-up or head-down poses, while cleaner gobies swim off coral heads, and juvenile Spanish Hogfish quickly shuttle from one to the other. Many of the wrasse pucker their small, white protruding mouths allowing the slim gobies to slip inside.

Cleaning can also take place at what would seem inopportune times. We have watched aggressive hunting packs of large Yellow Jack, *Caranx bartholomaei,* wildly working their way down a reef, slamming their bodies beneath ledge after ledge in pursuit of hiding fishes, when suddenly one of the big predators stopped cold in the midst of the frenzy and assumed a composed head-up cleaning stance allowing a two-inch Spanish Hogfish to calmly pick its side. In what must be an innate reaction to the typically bright colors of cleaners, we have seen Bluestriped Grunt, *Haemulon sciurus,* solicit cleaning by gaping in front of adult Porkfish over twice their size, even though only juvenile Porkfish act as cleaners. We have also seen a photograph of an adult Bluestriped Grunt opening its mouth wide to an adult Foureye Butterflyfish,

Left: *Scarlet-striped Cleaner Shrimp ventures inside the mouth of a diver to feed on food particles.*

Chaetodon capistratus, even though the species has never been reported to clean as juveniles or adults. In another observation young fish born in an aquarium and never exposed to adults would on occasion open their mouths to cleaners, indicating another possible innate reflex to cleaning.

Cleaning stations frequented by groupers tend to be regulated by dominant hierarchies dictated by size. Large groupers regularly chase smaller groupers away from cleaning stations, even if they are not interested in being cleaned at the moment. Clients settle to the sea floor or coral within inches of a station and lean in the direction of the cleaners. Groupers often open their jaws and flare their gill covers so wide that daylight can be seen when looking into their mouths. A few species change colors when being cleaned. Tiger Grouper, typically turn deep red; Yellowmouth Grouper, become so dark their spotted pattern is obscured; and Hogfish, *Lachnolaimus maximus,* display rust-colored body bars. The purpose of the color shifts is not understood, perhaps the color changes are simply camouflage patterns employed when near the bottom. On the other hand the color change could possibly render parasites easier to see, or function as a signal to the cleaners indicating the need to be cleaned.

Once settled, the predators appear oblivious to their surroundings but remain quite alert. When divers approach, fishes being cleaned often become uneasy, shifting slightly or moving away. To get a closer view, pause at least 10 feet away for a minute or two before making a slow approach, gauging the clients' restlessness as a signal to proceed or stop. Occasionally, large groupers remain at cleaning stations for more than an hour, but most clients leave within minutes, some within seconds. When ending cleaning sessions, groupers and other fishes twitch, insuring the gobies and shrimp a safe passage back to their home base before the clients swim away.

A cleaner goby dares to enter the open mouth of a predator to pick parasites with little fear of being eaten.

A cleaner goby and three species of shrimp clean the outstretched hand of a diver.

Because of their beneficial relationship, cleaning organisms have a substantial degree of immunity from predation. This advantage allows them to dart inside the gaping, tooth-lined mouths of predators such as barracudas without fear of being eaten. Studies in the lab measuring the cleaning gobies' cortisol (a steroid hormone released in response to stress) levels after being introduced to predators indicate that the gobies probably perceive predators as threats. But instead of freezing in place or initiating the expected flight response, cleaning gobies become proactive approaching predators faster and servicing them longer on average than nonthreatening clients. However, immunity does not seem to apply to young Blueheads that are routinely found inside the stomachs of predators. This inequity might be due to limited number of initial phase Blueheads involved in cleaning on a regular basis, and also might indicate why the prolific little wrasse seldom if ever clean piscivores.

Patient divers can usually entice Pederson Cleaner Shrimp, Scarlet-striped Cleaner Shrimp and gobies to briefly clean their outstretched fingers. Your best bet is to look for Pederson Cleaner Shrimp advertising for business by waving their conspicuous antennae. Slowly extend the back of your open hand with the palm angled slightly away. The shrimp usually make tentative contact with their antennae before hopping aboard, where they seem particularly fond of picking around the nails or at rough broken skin. The gobies are quite cautious and only smaller individuals fall for the ruse, but on rare occasions we have had several of the tiny fish simultaneously picking at our hands. Once we had three species of shrimp and a cleaning goby scampering over an extended hand

A cardinalfish opens its gill cover offering access to a Scarlet-striped Cleaner Shrimp.

at the same time. Scarlet-striped Cleaner Shrimp, which tend to live inside crevices, can from time to time be enticed to clean a diver's hand. Occasionally, the dainty shrimp will even enter a diver's open mouth. But be aware, small groupers often make proprietary claims to cleaning stations defending their turf with aggression.

In Bonaire we spent two dives unsuccessfully attempting to photograph a Scarlet-striped Cleaner Shrimp picking inside the gill opening of a cardinalfish, one of several dozen of the small, nocturnal species taking refuge inside a hollow coral formation. The problem was too many fish. No sooner did I get a clear view of the image I was after than a skittish companion would block the scene. Returning the following morning I found the cardinalfish clustered toward the back of the recess away from the cleaners, so I settled in for a wait. Having nothing better to do, I casually slipped my hand toward the shrimp—a mistake. A 10-inch resident Graysby, *Cephalopholis cruentata,* bolted for my bare hand. I yanked my fingers back in the nick of time and hugged them tightly to my chest. The grouper turned, glared then swam away sending up a huff of sand in its wake before slipping inside the recess through a side passage, resting its head on an incline and opening its mouth wide. A nearby shrimp literally galloped across the sand and scrambled onto the waiting jaws. No sooner was it settled than the grouper sent it flying with a shake and swam away. In previous visits to the coral head I had never seen the Graysby show interest in being cleaned by the shrimp until I offered my hand.

Three species of sea turtles—the Hawksbill Sea Turtle, *Eretmochelys imbricata,* Green Sea Turtle, *Chelonia mydas,* and the Loggerhead Sea Turtle, *Caretta caretta*—have been observed being groomed by surgeonfishes, a blenny, juvenile Spanish Hogfish and other species, in Brazil. It seems likely that the surgeonfishes and blennies, being herbivores, were removing algae from the turtles' shells. However, the carnivorous hogfish, known to be a dedicated cleaner in the Caribbean was probably removing ectoparasites. Once in Bonaire we encountered three Yellownose Goby, *E. randalli,* a common cleaner fish in the southern Caribbean, presumably picking parasites for more than 10 minutes near the head of a Green Sea Turtle resting next to a small coral head.

Another benefit from cleaning organisms is wound-healing. A researcher made extensive observations of the healing process of three severely wounded Blue Tang, *Acanthurus coeruleus,* and a small Yellow Jack, *Caranx bartholomaei.* During the first five days after injury, the Blue Tangs spent an average of 25 minutes each hour at cleaning stations with gobies, juvenile Spanish Hogfish and Blueheads. Two weeks later, after the wounds completely crusted over, the time was reduced to less than two minutes an hour. The cleaners picked at loose muscle threads, the edges and, to a lesser extent, the surfaces of the fresh wounds. Their bites often caused the injured fishes to jump and, at times, move away from the station. In four weeks the black scabs covering the injuries disappeared, leaving only slight indentations. The researcher also documented a small Yellow Jack, a non-reef resident that

Scarlet-striped Cleaner Shrimp inspects the open mouth of a Graysby.

Cleaner shrimp attend to an injured Viper Moray.

arrived at the patch reef missing one-third of the muscle tissue from its left side. It spent an average of just over 18 minutes an hour in a cleaning station of gobies and juvenile Spanish Hogfish. After 16 days, just four days after the massive wound sealed over, the jack left the area. A juvenile Gray Angelfish picked at an open wound on the leg of a researcher for as long as 20 minutes on three successive days. Over a three-day period, we observed Pederson Cleaner Shrimp and Banded Coral Shrimp picking at wounds on the head of a seldom-seen Viper Moray, *Enchelycore nigricans,* which repeatedly jerked, causing his cleaners to retreat temporarily. By the time the moray disappeared from the cleaning station three days later the wound appeared to have healed.

Larval gnathiid isopods tend to be the most common ectoparasite of reef fishes in both the Pacific and Caribbean. These tiny larvae of the crustaceans, measuring only 1.0 to 1.5 mm, emerge from the bottom during the day and night in three growth stages before finally maturing into non-parasitic breeding adults. At each transition the active swimming larvae rapidly seek out and latch onto fishes with specialized mouthparts modified for piercing and attachment. Once in place they begin to feed immediately on the prey's blood and body fluids. After engorging, which in most instances takes less than an hour, the larvae return to the bottom to molt and metamorphose into the next of their four life stages. The large number of the predatory larvae in the water column during the day and night coupled with their rapid feeding regime explain why so many fishes return to cleaning stations periodically throughout the day. Studies indicate that the population size and behavior of gnathiid isopods vary significantly among locations.

A field study conducted in the Exuma Cays Land and Sea Park in the Bahamas during the 1990s removed external parasites from 63 groupers representing four species (44 were Nassau Grouper, *Epinephelus striatus*). The fishes were captured in mesh traps, dipped in an anesthetic, immersed in a freshwater bath for five minutes to remove the parasites, injected with an antibiotic, tagged and released where captured. Four species of crustaceans, three from Order Isopoda and one species from Subclass Copepoda, and a single flatworm from Phylum Platyhelminthes, were recovered and examined. As many as 60 ectoparasites were removed from a single fish, although a few specimens appeared to be free of infestation.

Typically, isopods are flat on top and bottom, with no definition between body sections. Four hundred fifty parasitic forms infest fishes and invertebrates. All three isopods attached to the infested groupers were free-living species that spend only a portion of their life cycle as parasites. Two species attach to hosts as egg-laying females, temporarily feeding on blood before returning to the bottom. Isopods in the genus *Excorallana*, live in intertidal areas of rubble, seagrasses, sponges and corals; *Alcirona krebsii* inhabit sponges and dead corals.

Copepods display diverse body shapes but are generally cylindrical with a division between the head and thorax. More than 1,800 species of parasites are known to infest marine and freshwater fishes. Nearly 100 species have been described in the genus *Lepeophtheirus,* to which the specimens collected in the Bahamas belong. If there is a typical shape for copepods, these species certainly don't fit the template. Flattened bodies offer ease of movement; great

Ectoparasites of Reef Fishes

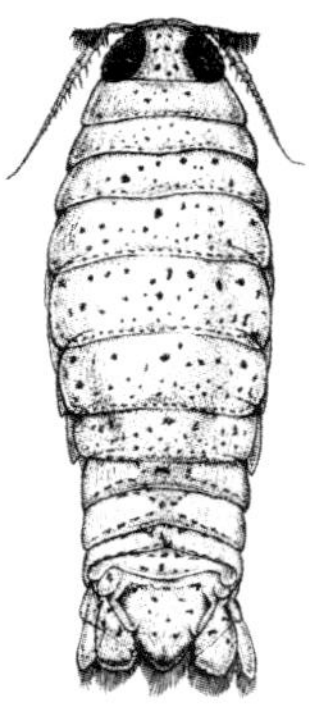

Isopod, Excorallana *sp., 6 mm.*

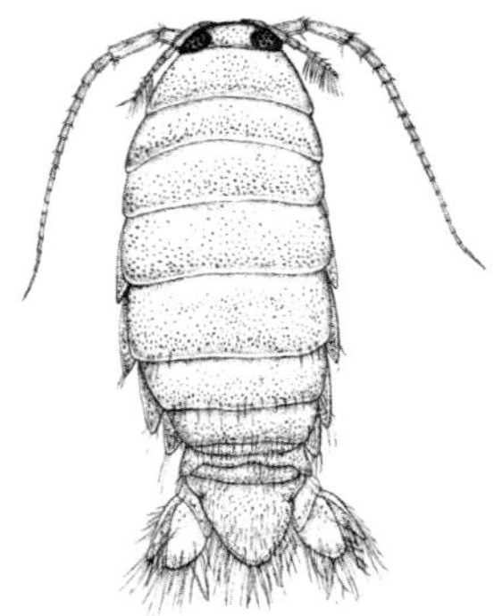

Isopod, Alcirona krebsii, *6 mm.*

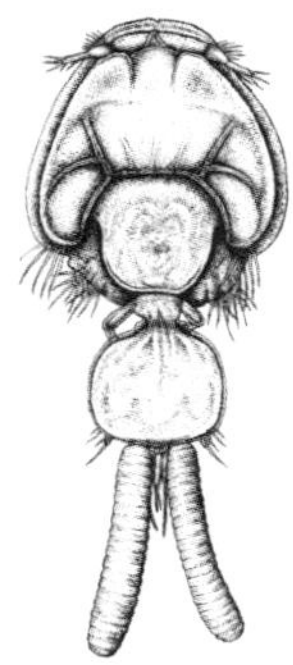

Copepod, Lepeophtheirus *sp., 4 mm.*

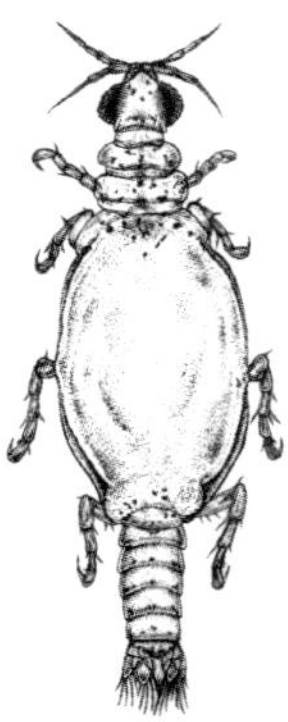

Isopod, Gnathia *sp., 3 mm.*

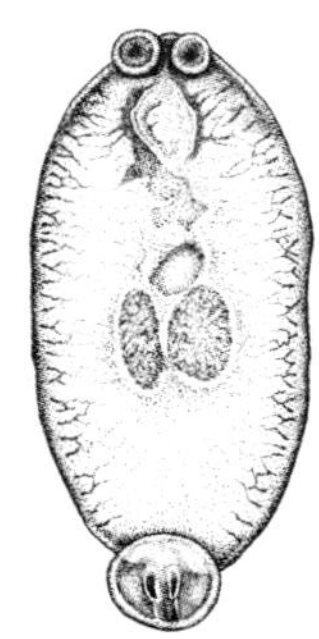

Platyhelminth, Neobenedenia *sp., 2 mm.*

head-shields ringed with suckerlike organs provide attachment; and sharp mandibles penetrate into the flesh below. Two sausage-shaped egg sacs trail from the abdomens of females. These four-millimeter brutes were the most prevalent parasites found during the study.

A Coney scraps against the bottom—a behavior, known as flashing, believed to ease the aggravation of parasites.

The single specimen from Phylum Platyhelminthes belonged to genus *Neobenedenia.* The free-swimming ciliated larvae of the tiny two-millimeter worms attach permanently to the body or gills, where they feed on mucus and blood. Another genus member, *N. melleni,* causes considerable problems for the tropical aquarium fish and cultured fish trades. It is believed, but as yet unconfirmed, that hosts employ different cleaning organisms to relieve infestations of specific parasites.

Some fishes attempt to eliminate or relieve the irritation of pests themselves. Flashing occurs with a fish swimming to the bottom where it turns on its side, arches laterally and scrapes against the sea floor. The spectacular water-clearing leaps of Oceanic Manta Ray, *Mobula birostris,* and Spotted Eagle Ray, *Aetobatus narinari,* culminating in great belly flops possibly serve the same purpose.

Mutualism

A seemingly straightforward mutualistic relationship exists between Orangespotted Goby, *Nes longus,* and a snapping shrimp from genus *Alpheus.*

An Orangespotted Goby lives in a mutualistic relationship with a near-blind snapping shrimp. The goby keeps guard for the shrimp which, in turn, keeps a burrow in good repair for both animals.

The Orangespotted Goby is the only species known to associate with snapping shrimp in the Caribbean, although dozens of similar goby/shrimp partnerships occur throughout the tropical Indo-Pacific region. Wherever the relationship takes place the behavioral interaction remains the same. The nearly blind shrimp excavate burrows in shallow mud or sandy bottoms near mangroves, along the sloping edges of canals and calm seashores. The sediment-filtering crustaceans have little to worry about while underground, but once above the surface they become easy pickings for hungry carnivores hunting the sand. Gobies also find life dangerous on the open sand. They have no problem seeing potential predators, but finding a suitable place to hide is another matter. To adapt, the blurry-eyed shrimp and the homeless gobies join forces for their mutual benefit.

From dawn to dusk, the tiny, nearsighted shrimp maintain winding burrow tunnels connecting two or three enlarged chambers by hauling material up from below. The gobies perch on furrows leading from burrow openings, except now and again, when they dart a few inches away to grab a mouthful of sand filtered for food and discarded through the gill openings. When above ground, the shrimp nearly always keep at least one antenna in contact with the goby, usually near the tail. Signals range from a slight tail twitch, indicating caution, to a thrash for alarm. Once a warning is given, the time it takes for both animals to disappear can be measured in tenths of a second. But even in panic there are proprieties—the shrimp always enters the burrow first. The subterranean passages where they hide are typically one inch in diameter, two to four feet long and run a few inches beneath the surface. The instability of the upper sections creates constant maintenance, and new entrances must be dug frequently.

Because of their vigilant nature, the duos are difficult, but not impossible, to approach closely. Once a goby is sighted, remain still for a minute or so before moving slowly in its direction. The reward of watching the shrimp bulldozing their entranceways is worth the wait. The brawny crustaceans work like power robots heaving and hoisting, jamming and cramming, poking and plowing and at time lifting shell fragments twice their weight with their enlarged snapping claw in a never-ending battle against shifting sand.

Commensalism

Large, segmented isopods from the family Cymothoidae attach primarily to the heads and gill covers of several common reef fish species. Unlike their flesh-burrowing brethren, Cymothoids attach to the skin with seven pairs of hooklike legs and benignly scavenge specks of floating food. Far from killing the hosts, their long-term associations cause only minor skin discoloration, scale erosion or, at worst, slight bone deformities. For this reason a long-running debate continues as to whether the association is considered commensal or parasitic.

These monster isopods, sometimes growing to three inches, begin life as tiny, one-eighth-inch, free-swimming males. Once associated with a fish, they lose their ability to swim and remain in place for life. Typically only a single

Cymothoid isopods are quite particular about which species they infest and the exact location on the host where they attach.

female or a mated pair colonizes a host. To increase their odds of finding mates, male Cymothoids have the ability to transform into females. When a male settles on a host with a female already in place, it mates with the larger female. Later, after the female dies, the male changes sex and awaits the arrival of a young male. If a mate is not present at settlement, a male accelerates growth and prematurely changes into a female. The largest and darkest individuals are invariably brooding females, sometimes incubating more than 100 juveniles

A pair of isopods fastens to the gill covers of a Coney.

Cymothoid isopods only attach under the eyes of Foureye Butterflyfish.

inside their bulky ventral pouches. When fully developed, the offspring emerge one at a time. While collecting Cymothoids, researchers discovered that each time they speared a fish carrying a pregnant female, the isopod instantly discharged her entire brood. This behavior, termed a burst release, is evidently intended to save the offspring during attacks on hosts. The vulnerable youngsters are fair game while searching for a compatible host, and even after settling they continue to suffer predation from cleaner fishes and shrimp until they begin to grow. Like all crustaceans, isopods must molt to grow; however, if their entire exoskeleton were shed at one time they would lose their life-sustaining grip. The parasites resolve their dilemma by first molting the front half of their bodies, and later, after their legs harden, slough off the remaining shell.

Large cymothoid isopods, which can easily be seen, inhabit tropical seas around the world. To date, 12 species from two genera have been classified from the Caribbean. The nine members of genus *Anilocra* are broadly distributed, but the different species tend to inhabit specific regions. While diving in the southern Caribbean, you might find them attached to Atlantic Creolefish, *Cephalopholis furcifer;* in the Bahamas they are most frequently sighted on Coney, *C. fulva,* and Blackbar Soldierfish, *Myripristis jacobus.* The three species in genus *Renocila* appear to be restricted to isolated sites off Puerto Rico and the Dominican Republic.

The large isopods are quite choosy about their hosts. Of the nine *Anilocra*, eight infest a specific fish species or members of a single genus. One associates with three genera in the grunt and sea bass families. The isopods' scientific names are derived from the host. For example, *Anilocra abudefdufi* associate with Sergeant Major, *Abudefduf saxatilis; A. myripristis* with Blackbar Soldierfish and *A. chromis* with Brown and Blue Chromis, *Chromis multilineata* and *C. cyanea.* Other infested fishes include four species of butterflyfishes, four species of sea basses, nine species of grunts, Squirrelfish, *Holocentrus adscensionis,* Doctorfish, *Acanthurus chirurgus,* Ocean Surgeonfish, *A. tractus,*

An isopod fastens beneath the eye of a Blue Chromis.

A damselfish carries an isopod behind its gill cover.

Glass Goby with parasitic isopod attached.

Bicolor Damselfish, *Stegastes partitus,* and the Rock Beauty, *Holacanthus tricolor.* In addition, *Renocila* infest two cardinalfishes and the Harlequin Bass, *Serranus tigrinus. Anilocra* also are quite particular about where they attach. *A. abudefdufi* always settles beneath the eyes of Sergeant Majors; *A. acanthuri* fastens under the pectoral fins of surgeonfishes; and *A. myripristis* make their homes between the eyes of Blackbar Soldierfish. Of the thousands of *Anilocra* observed by researchers, only four females were located in abnormal positions.

The isopod *A. chromis,* which is known to infest both Blue Chromis and Brown Chromis, for a yet unexplained reason, attaches only to Blue Chromis in a range extending from south Florida through the Bahamas to the Dominican Republic. In the U.S. Virgin Islands and Puerto Rico the same species avoids the Blue Chromis and attaches exclusively to the Brown Chromis, even though both fishes are well represented in each of the two regions. To test regional compatibility, researchers performed an experimental transfer of parasites between the two species of *Chromis.* Isopods were removed from Brown Chromis captured in the U.S. Virgin Islands and reattached to previously uninfested Brown Chromis and to Blue Chromis, a non-compatible host in the area. Then both sets were released back to their home reefs. The Blue Chromis

A Redcheek Goby makes its home beneath the protective spines of an urchin.

reacted violently to the isopods, and most of the 22 experimental fish had ditched their pests by the end of the first day. All were lost after day two. However, the Brown Chromis seemed undisturbed by the freshly attached females, and all but three still carried the isopods at the end of the seven-day experiment. The study indicates that host species appear to be predisposed to accept the presence of the hulking pest. This might explain why fishes don't go berserk attempting to dislodge the brawny freeloaders.

For protection in the open sea, juvenile jacks cluster within the draping tentacles of jellyfishes.

Numerous commensal relationships occur on the reef and in the open ocean. The Greenbanded Goby, *Tigrigobius multifasciatus,* Redcheek Goby, *T. rubrigenis,* and Ninelined Goby, *Ginsburgellus novemlineatus,* shelter beneath the protective spines of sea urchins, and the Diamond Blenny, *Malacoctenus boehlkei,* often takes refuge among the tentacles of the Giant Anemone, *Condylactis gigantea,* apparently unaffected by the stinging cells. In open water, small fishes often travel in the company of large fishes and invertebrates. Juvenile jacks regularly convoy near the heads of Great Barracuda, *Sphyraena barracuda,* and sharks. They also cluster within the draping tentacles of jellyfishes. Schools of Man-of-war Fish, *Nomeus gronovii,* live beneath the purplish, gas-filled floats of the Portuguese Man-of-

Diamond Blennies often seek sanctuary beneath the tentacles of Giant Anemones. Apparently the blennies are unaffected by the anemones' stinging cells.

A cardinalfish, known as the Conchfish, inhabits the mantle cavities of living Queen Conch.

war, *Physalia physalis,* scavenging bits of food captured by their hosts' toxic tentacles. If food is scarce the fish are occasionally devoured by the deadly siphonophores. Boldly banded Pilot Fish, *Naucrates ductor,* accompany sharks, rays and whales.

The Conchfish, *Astrapogon stellatus,* a member of the cardinalfish family, inhabits the mantle cavities of living Queen Conch, *Lobatus gigas,* during the day. It is not known if the mollusk benefits from the association. The one- to two-inch fish only venture out after dark to feed on planktonic crustaceans just above the bottom or in the water column, often up to 25 feet away from the protection of their hosts. Before sunrise they return, normally entering their hosts through the anterior siphon canal, whether the mantle cavity is open or closed. Most populations occur on the leeward side of islands. Only a single Conchfish inhabits most hosts, but as many as 16 have been found crowded inside a single conch. The small fish have also been discovered inside penshells and helmet shells, but tests show they prefer living in the Queen Conch. Eggs have been found year-round inside the mouths of brooding males.

The letter-opener-shaped Pearlfish, *Carapus bermudensis,* and the sluggish, bottom-dwelling Five-toothed Sea Cucumber, *Actinopyga agassizi,* form one of the most peculiar commensal relationships in the Caribbean. The elongate, two- to five-inch fish wriggles, head or tail first, through the anus of the sea cucumber and spend its days inside the alimentary canal of its host. At night, it exits the same opening to feed on an assortment of shrimp, other small crustaceans and occasional fishes. While inside their hosts, Pearlfish cannibalize recently settled larvae or juveniles of their own species seeking sanctuary inside the same sea cucumber. Probably for this reason, only solitary individuals inhabit most hosts, although researchers have found as many as 10 Pearlfish infesting a single sea cucumber. Other than occasionally raising the rear portions of their bodies as the fish make their entries, the sea cucumbers seem unaffected by the association. *C. bermudensis,* the only shallow-water member of the pearlfish family, Carapidae, in the Caribbean, has also been found in limited numbers inside Donkey Dung Sea Cucumber, *Holothuria mexicana,* and Three Rowed Sea Cucumber, *Isostichopus badionotus,* in depths below 100 feet. Pearlfish produce jellylike rafts of elliptical eggs that hatch in open water. The elongate pelagic larvae with a streaming dorsal fin ray festooned with floats settle to the sea floor approximately three months later (photo page 52). Those finding vacant sea cucumbers shrink to one-third of their presettlement length before changing into their definitive shape.

Sargassum is the genus of two floating forms of brown seaweed, *S. natans* and *S. fluitans,* as well as 150 bottom-dwelling algal species that attach to the seafloor by holdfast and hold themselves upright by air bladders. Like seagrass and mangrove habitats, floating masses of *Sargassum,* known as mats, floats or rafts, act as safe havens for juvenile fishes and invertebrates. The seaweed's fame comes from the Sargasso Sea—a self-perpetuating universe of algae, roughly the size of the United States, thriving within the bounds of four clockwise currents in the North Atlantic. Storms regularly tear huge rafts free from the Sargasso Sea that sail east with the prevailing wind along with their cargo of animals and a hungry escort of predators lurking in the shadow.

A substantial amount of pelagic *Sargassum* also grows in the northwestern corner of the Gulf of Mexico. Gulf rafts harbor an equally diverse ensemble of hitchhikers that rides the Loop Current out of the Gulf and into the Florida

Sargassumfish—members of the Frogfish family are endemic to Sargassum *floats.*

Large rafts of Sargassum *can be home to thousands of marine animals.*

Juvenile Atlantic Flyingfish

Juvenile Almaco Jack

Juvenile Chub

Juvenile Atlantic Tripletail

A two-inch juvenile Dolphinfish patrols the fringes of a Sargassum *raft.*

Current that sweeps east around the tip of south Florida where it joins the Gulf Stream for a ride up the eastern seaboard to Cape Hatteras.

The predators patrolling the fringes of the rafts feed on juvenile fishes, shrimp, crabs, and many of the other 145 invertebrates hiding within the tangled mats. Samples of *Sargassum* collected off the Carolinas produced 80 species of fishes including the endemic Sargassumfish, *Histrio histro,* Sargassum Pipefish, *Syngnathus pelagicus,* and 10 invertebrate species that have evolved to live nowhere else. Sixty species of fishes traveling with the rafts were also recorded. Filefishes, jacks and flying fishes made up 90 percent of the passengers. The five most commonly collected fishes were juveniles of the Planehead Filefish, *Stephanolepis hispidus,* Blue Runner, *Caranx crysos,* Atlantic Flyingfish, *Cheilopogon melanurus,* Gray Triggerfish, *Balistes capriscus,* and Almaco Jack, *Seriola rivoliana.*

Even when the surface water is calm, finding small animals in *Sargassum* is tricky business. The easiest way to enjoy the little magicians is to dip up a net full of seaweed and shake it over a bucket of seawater. The freshest seaweed (yellowish tan) yields the most animals. Older patches, turning brown, have been too picked over by predators to be worthwhile.

Mimicry

Mimicry is employed in two primary ways: aggressive mimicry, where the mimic imitates the appearance and behavior of a harmless species to gain closer access to unsuspecting prey, and Batesian mimicry, where a harmless species imitates a predator for protection. Several criteria must be fulfilled to distinguish between true mimicry and the chance resemblance of two species. The mimic and model must share the same habitat and range; the mimic must be less abundant than the model; it must alter its behavior to enhance the resemblance; and it must gain an advantage from the resemblance. When considering the possibility of mimicry it is important to keep in mind that there are so many similar-appearing fishes living on the reef there are bound to be numerous cases of coincidental resemblance.

The Wrasse Blenny, *Hemiemblemaria simula,* is an excellent imitator of the initial phase Bluehead, *Thalassoma bifasciatum.* Not only does the Wrasse Blenny bear a close visual resemblance to Blueheads, but they are also the only blenny in the Caribbean that, like wrasses, swim with their pectoral fins. These solitary little impersonators hang out in abandoned wormholes in coral heads with only their heads protruding. When herds of Blueheads swim close by, the blennies join the feeding packs. It is still uncertain whether this mimicry is aggressive, Batesian, both, or neither. Wrasse Blenny mimic Blueheads possibly because they are part-time cleaners, which may grant a degree of immunity from predation. On the other hand, their mimicry may be aggressive, enabling them to approach unsuspecting prey more easily. The Wrasse Blenny's disguise is so effective that only sharp-eyed fish watchers can pick them out of Bluehead herds. Subtle clues—a red rather than yellow iris, and a larger mouth—help distinguish the impostors.

A second possible mimic of initial phase Blueheads, juvenile Tiger Grouper, *Mycteroperca tigris,* closely resemble their models' yellow bodies and broad, dusky midbody stripes. When mixing with foraging Blueheads or when small fishes approach, juvenile Tiger Grouper begin swimming with their pectoral fins and collapse their medial fins for a more wrasselike appearance. The disguise works. A researcher watched a juvenile Tiger Grouper slip into a group of one-inch juvenile Redband Parrotfish, *Sparisoma aurofrenatum,* without causing a flight reaction and capture one of the fish. The young grouper change into adult colors between four and five inches, at about the same size that initial phase Blueheads make their transition to the terminal phase.

One study proposes that Yellowtail Snapper, *Ocyurus chrysurus,* are aggressive mimics of the Yellow Goatfish, *Mulloidichthys martinicus.* These two species are the only reef fishes in the region with bold, bright yellow lateral stripes joining forked yellow tail fins. Although Yellowtail Snapper grow larger than goatfishes, nearly 90 percent are approximately the same size as the goatfish with which they associate. Goatfish scour open sand in groups of up to 20 individuals for buried invertebrates. By mixing with the similar-appearing goatfish, snappers were found to be nearly twice as successful in capturing small fishes such as juvenile Slippery Dicks, *Halichoeres bivittatus,* that do not flee from passing goatfish or other carnivores attached to the groups' sand clouds.

Initial phase Blueheads are believed to be models for mimicry because of their role as cleaners.

Juvenile Tiger Grouper

Wrasse Blennies often reside in holes.

Wrasse Blennies are believed to mimic initial phase Blueheads.

Senses & Sound Communication

Schooling fish use vision and lateral lines to maintain their position within a group.

In the clearwater world of tropical reef fishes, life and death revolve around good vision. Correspondingly, most species have excellent eyesight. Their eyes are much like those of humans, with external corneas and lenses that focus light on retinas packed with photoreceptor cells. However, there are a number of important differences. Because water rapidly absorbs sunlight, fishes in general, and nocturnal-feeders in particular, have large eyes with extremely clear spherical lenses to gather the scarce light. Like our eyes, the retinas of fishes contain rod-shaped and cone-shaped photoreceptors. Species with more rods can see better in low-light situations, whereas fishes with more cones can see details better in bright daylight. Cones provide color vision in humans, but their presence does not mean that fishes distinguish colors as we do. There is a tendency for the cones of reef fishes to be blue and green sensitive, a good combination to contrast targets against the blue backlighting of tropical reefs but not for discriminating among a wide spectrum of colors.

Fish eyes bulge from the head, providing an arching panoramic field of vision that extends almost completely behind the fish. Round lenses positioned in the center of oval-shaped eyeballs allow nearsightedness directly ahead and farsightedness to the side. Fishes such as plankton-pickers, with eyes set close together, have overlapping binocular vision with good depth perception to focus

Surgeonfish eye

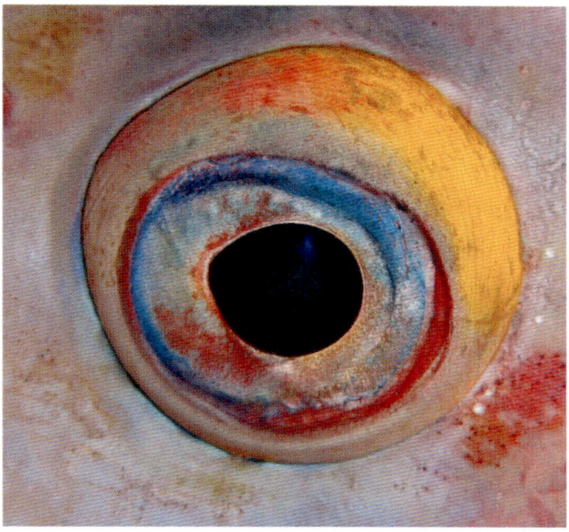

Spanish Hogfish eye

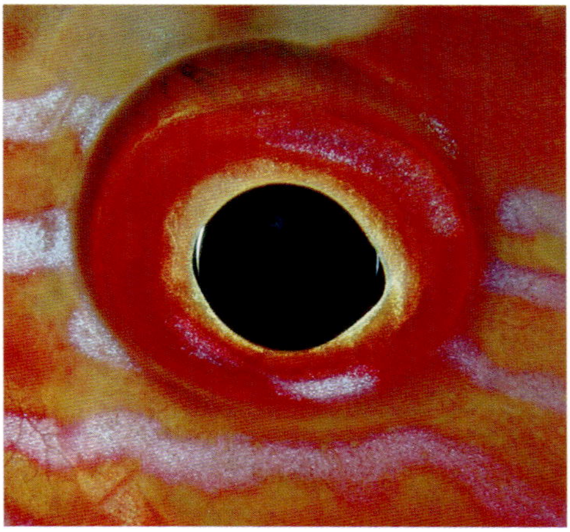

Goatfish eye

Mushroom Scorpionfish eye

Stalked flounder eye

Yellow Stingray eye

on incoming specks. Those species with eyes farther apart have a wider range of vision. As fishes grow, vision improves. Bluehead, *Thalassoma bifasciatum,* a plankton-feeding species with high cone density for better resolution, were found to detect particles at greater distances as they matured.

Several nocturnal species, including reef sharks, have a reflecting layer of tissue behind the retina, known as the *tapetum lucidum.* The *tapetum* reflects light back through the photoreceptors greatly enhancing low-light vision. During the day the light-sensitive reflective layer is covered to prevent overload and requires nearly an hour after the sun sets to regain its full potential. It is this layer that creates the dazzling eyeshine in members of the bigeye and sweeper families when their eyes are illuminated by light beams at night.

The Atlantic Flashlightfish, *Kryptophanaron alfredi,* is the only Caribbean representative of family Anomalopidae. The three- to four-inch, primarily deep-dwelling, nocturnal fish have a crescent-shaped light organ beneath each eye. On dark nights individuals occasionally ascend from the depths and temporarily inhabit caves and overhangs along steep slopes and walls as shallow as 40 feet. The organs' blue-green glow, which is believed to either attract or illuminate prey, is generated by a colony of symbiotic bacteria. The fish turn the light on and off by rotating and simultaneously covering the organs with skin folds.

Although eyesight is used primarily to capture prey, smell, taste and electroreception also are used in varying degrees by different species to locate food. Reef fishes have two paired sets of nasal openings (nares), one each side of their heads. Damselfishes, with only a pair of single nares, and members of the pufferfish family, with no nasal openings, are exceptions. Depending on the species, ciliated cells, a pumping action or the force of water pressure when swimming carries water into one of the paired openings. Incoming odors pass along a canal and over a series of epithelial lobes lined with olfactory receptor cells (olfactory bulbs) before exiting the second nare. All animals, including reef fishes, exude chemical compounds, known as pheromones, from their skin and excretions. Fishes use their keen sense of smell to trace these chemical markers

Deepwater Atlantic Flashlightfish, which invade shallow water on moonless nights, attract prey with luminous patches beneath their eyes.

to recognize members of their own species as well as for locating prey and mates, and avoiding predators. During courtship, males of most species repeatedly approach and often nudge the vents of egg-laden females to assess their readiness to spawn. It has been shown in laboratory experiments that male Bicolor Damselfish, *Stegastes partitus,* do not respond to nearby spawn-ready females if the pairs are kept chemically isolated by glass partitions. Recent studies have proven that fishes only a few days old have the ability to locate their birth reefs with smell, during their pelagic larval stage.

Morays and sharks, both night-feeders with highly developed senses of smell, have much longer olfactory bulbs than day-feeding fishes. Morays, with virtually no eyesight, have large, obvious incurrent nares on their snouts, often in the form of tubes, and excurrent openings beneath their eyes. Nurse sharks and hammerhead sharks, with widely separated nares, home in on a distant food source by angling toward the side receiving the strongest olfactory clues. Sharks have been found to respond to fish extracts diluted to as much as 1 part per 10 billion parts of water. Taste buds located in the mouth and on the lips also detect chemical stimuli.

To locate invertebrate prey living beneath the sand goatfishes, in the family Mullidae, have evolved pairs of fleshy, elongate appendages beneath their chins covered with thousands of taste buds. When not in use each tapered, somewhat ridged barbel is housed inside a groove below the jaw. Extended barbels provide a means of "distant taste," allowing the fish not only to sense hidden prey, but to also determine its depth. A single internal nerve trunk running the length of each barbel branches repeatedly into numerous taste bud clusters. Each of a cluster's nerve terminals is capped with a large taste pore opening at the skin's surface allowing sensory clues to enter the system. Swimming over the bottom with the head tilted slightly downward the barbels flick rapidly over the sand and regularly probe below the surface. When their preferred prey, in the form of tiny crabs, shrimp, and segmented polychaete worms, are located goatfish strike rapidly often penetrating several inches below the surface. At times their vigorous excavations leave behind lingering clouds of debris.

Goatfishes use paired chin barbels to scour the sand for hidden invertebrates.

Senses & Sound Communication

Sharks and rays use electroreception to detect prey at close range. Numerous pores located on the snout, head and lower jaw open into canals filled with electrically conductive gel that lead to clusters of hair-lined chambers, known as ampullae of Lorenzini. When a hunting shark or ray passes its head within a foot of prey, even if buried, it is able to sense bioelectric energy generated primarily by muscle tissue. The system also is believed to function in some way as an electromagnetic compass to aid navigation.

Because fishes do not have external ear openings, it is often assumed that they are unable to hear. In fact, bony fishes have an elaborate inner ear system consisting of a series of interconnected, semicircular, fluid-filled canals. These structures, located next to the brain, typically detect sounds of lower frequency than can be perceived by humans. The same system also controls equilibrium. Each of two lower "hearing chambers" widens to house a crystalline calcium carbonate otolith, or ear stone, nestled in a bed of hairlike sensory cells. The density of water carries sound in the form of vibrations nearly five times faster and farther than air. These sound waves, when produced at close range, cause the fish to shift ever so slightly. The ear stones, three times denser than the fish, do not move as readily, forcing them to contact surrounding hairs which, in turn, transmit signals to the auditory center of the brain. In many fishes, gas-filled swim bladders behind the inner ear amplify many distant or weak sounds sufficiently to move the otoliths. Three upper canals, each angling in a different direction from a common base, and housing a third ear stone, are used to maintain balance and to monitor acceleration. They function like the auditory chamber: movement from side to side, up and down or forward and back shifts fluids creating pressure on sensory hair receptors.

Fishes also have the ability to detect movement in the water with a system of hairlike sensory receptors, similar to those used in the internal ear, referred to as the lateral system or lateral line. The system's central component, tightly

The lateral lines of Common Snook are quite obvious.

packed bundles of sensory hair cells and their supporting cells, known as neuromasts, are each encased inside a flexible, upward protruding, jellylike, dome-shaped cupula. Neuromasts collectively provide a sense of "distance touch" assisting in the detection of predators, locating prey, perceiving current flow, and, to a limited extent, orientating individuals when schooling. The system can also sense stationary objects, a significant advantage when navigating in low-light conditions.

There are two types of sensory bundles in the lateral line system: free neuromasts and canal neuromasts. Free neuromasts are singular units scattered over the surface of the head and body. Their cupulas extend through the skin where they are directly affected by water movement. Canal neuromasts are enclosed within short subcutaneous canals that form networks on the head, and more notably inside a pair of much longer lateral line canals that begin behind the gill cover and extend to the base of the tail on each side of the body. A series of pores opening the lateral line to the environment generally appear as distinct lines. Depending on species lateral lines can be straight, curve upward or downward adjacent to the pectoral fin (limiting disturbance from the fins' movement), or in the case of parrotfishes, be broken in two parts. Clusters of the neuromasts' sensory hairs originating in the base of the canals reach into the enclosed gel-filled channels. Each is positioned approximately midway between adjacent pores. External water movement creates pressure at the pore openings causing the canal fluid to bend the receptors in a way that allows a fish to sense nearby disturbances, objects and their general locations. Because the sensory cells are housed internally, the canal neuromasts only function when a fish is on the move, and tend to be more prominent on actively swimming species. Sedentary species, such as lie-in-wait predators, typically have a larger number of free neuromasts.

When on the move, each species of fish, including predators, generates a unique signature of water disturbance and pressure fluctuations. It has been proposed that the internal ear system and lateral lines work together to predict the size, speed and distance of approaching predators. Researchers have observed Slippery Dick, *Halichoeres bivittatus,* diving into the sand at the same instant that hydrophones intercepted the sound of an advancing school of Blue Runner, *Caranx crysos.* We have often witnessed a similar defensive response from schooling grunts. Seconds before hunting packs of jacks come into view, the fish draw closer together and begin to swirl. To help confuse signals picked up by a prey's lateral line receptors, predators have evolved the ability to disguise their approaches with hydrodynamic camouflage, by randomly altering their movements and making intermittent stops. The periodic pauses not only allow predators to disguise their presence, but also enable them to pick up additional signals from their potential prey.

Perceiving a current's direction and force, believed also to be a process of free neuromasts, enables species feeding in open water to face directly into the flow in order to more efficiently decipher olfactory information and pinpoint food particles coming their way. Although schooling fishes primarily employ vision to maintain their precise positioning, the lateral line is also believed to assist in the process, but exactly how or to what extent has yet to be determined.

Sound Communication

A few fishes use a limited number of sounds to communicate with members of their own and other species. Hard body parts rubbing together (stridulation) produce most sounds. The swim bladder, located next to, or connected by taut muscles to the sound-producing structure, serves as an amplifier. Vibrating muscles surrounding the swim bladder also generate sound in a few species. The deep rumblings of grunts, drums and croakers are so distinctive they became the origin of the fishes' common family names. Grunts produce sounds by grinding their flat pharyngeal teeth together. Drums and croakers communicate during the spawning season by vibrating muscles. Because the majority of fish sounds are transmitted at low frequencies, divers seldom hear them, but there are a few noteworthy exceptions. Large Nassau Grouper, *Epinephelus striatus,* Black Grouper, *Mycteroperca bonaci,* and Goliath Grouper, *E. itajara*, produce loud "booms" when startled or cornered. A first-time encounter with a thundering grouper can be quite unsettling. The loud sounds could be intended to frighten predators, but this idea has never been substantiated. Goliath Grouper also produce similar sounds during courtship, and researchers have heard juveniles in mangrove shallows boom before taking bait, conceivably to stun small prey. No one will forget a night dive during the spawning season of toadfishes. The amorous males generate a cacophony of whistling calls, measured to be 30 to 40 decibels louder than any other known sound produced by fishes.

By using hydrophones accommodating lower frequencies, researchers have substantiated that Longjaw Squirrelfish, *Holocentrus marianus,* Longspine Squirrelfish, *H. rufus,* and possibly other members of the family are major sound-producers. The calls, resonating from air bladders, reach a peak during dawn and dusk. Night-foraging squirrelfishes, intensely territorial by day, routinely challenge fishes approaching their hiding holes. Neighboring squirrelfishes are met with a grunt or two, fin displays, head-shaking and nipping lunges. Single calls are directed at small or neighboring fishes, whereas

Territorial conflicts between Longspine Squirrelfish includes grunts, shakes and fin displays. The encounter ends with the fish pressing their rear bodies and tails together.

The mating calls of toadfishes are one of the loudest sounds produced by bony fish.

bursts of staccato grunts greet larger intruders. An end to the bickering is generally signaled when the antagonists assume parallel positions and briefly press their rear bodies together.

Without question the most sophisticated vocalists on Caribbean reefs are damselfishes, particularly the highly social Bicolor Damselfish, *Stegastes partitus.* The one- to two-inch plankton-pickers use a combination of displays, color changes, chemical clues and a repertoire of calls to communicate with other members of their size-related hierarchies. Territorial males constantly challenge similar-sized rivals for social position. Hovering head down an inch or so off the bottom, the contestants face one another and emit a series of combative pops. During courtship, inspired males wearing black masks perform a series of eye-catching dips, each accompanied by a chirp. Females respond to these acoustical courtship overtures from as far as 20 feet away. Interested females are not only able to distinguish the locations of the callers but also to differentiate between suitors and to assess their relative sizes.

Marine Wildlife Management

Nassau Grouper trapped during their spawning aggregation.

Commercial and recreational fishermen have long enjoyed a tradition of virtually unrestricted access to the sea that many of them believe to be their inherent right. Their vocal opposition to conservation management policies, including the establishment of areas restricting fishing, has closely paralleled the rapid decline of inshore fisheries stocks, especially members of the grouper/snapper complex, in the tropical western Atlantic during the last half-century.

Overexploitation of marine fisheries is not just a local problem. According to the World Wide Fund, fisheries are in trouble around the world. The Fund estimates that 60 percent of the 200 most valuable species, including many reef fishes, are presently overfished or fished to the limit. It is increasingly obvious that the current level of global fishing pressure is unsustainable, leaving fisheries management with few options but to impose fishing bans on threatened species for extended periods. Even after repeated calls for the establishment of additional no-take marine reserves from conservation groups, at present less than two percent of the world's fishing grounds are protected in any meaningful way.

As conservation tools Marine Protected Areas (MPAs) provide a high level of long-term protection to marine resources while Marine Reserves, a subset of MPAs also commonly referred to as no-take reserves, prevent extraction of resources—typically banning hunting, fishing or collecting in order to preserve sensitive habitats, protect threatened species and help sustain fisheries.

Overfishing, habitat degradation, destructive fishing practices, prey attrition (rapid reduction of baitfish and shrimp, principal food of large, overexploited food fishes such as groupers and snappers) together with "fishing down the food chain" are primary matters of concern around the world as well as across Florida and the wider Caribbean. Even so, the threat of losing limited access to fishing grounds of any size has fueled a long-festering dispute about the implementation of no-take reserves between the recreational and commercial fishing lobbies and fishery management. Overwhelming evidence compiled from stock assessments and historical population surveys confirm a continuing downward spiral of grouper and snapper populations. Often declines have resulted from the targeted exploitation of annual spawning aggregations when certain grouper and snapper species congregate by the thousands at traditional sites during specific times and put their entire reproductive payload into a single yearly effort. In some cases, 90 percent of a region's annual catch of aggregate-spawning species are taken during these brief events.

One of the most persistent arguments against establishing marine reserves revolves around the long-held belief that reef fish communities are open systems, reliant on larval recruits spawned on distant reefs. Fishers have argued for decades that limiting fishing access in home water or at spawning grounds is foolish if the offspring simply float off to populate someone else's reefs. However, research during the past two decades has demonstrated that the system is not as open as once believed. In fact, as much as 60 percent of surviving fish larvae have been shown to return from their short, early-stage metamorphosis in the open ocean to reefs near where they were spawned.

Even though marine biologists and conservation groups determinedly and consistently call for increasing the size and number of no-take reserves as the only realistic solution for managing multispecies fisheries, fishing lobbies continue to couch their arguments by claiming that the scientific community does not have conclusive evidence supporting the use of marine reserves for increasing fishing stocks and reestablishing overexploited inshore food fishes.

A Case for Marine Reserves

Current ecological data supporting implementation of marine reserves may not be conclusive, but is compelling. In the few areas where no-take reserves have been established and monitored, local fisheries stocks have increased, in some instances dramatically. Stocks of overexploited fishes within reserves have been found on average to increase in size and numbers far faster than once suspected. No-take reserves also provide much needed sanctuary for traditional spawning sites, and the larger, genetically superior members of a species' breeding population. These "ol' mamas and papas" have been documented to be hundreds of times more prolific than smaller individuals just reaching sexual maturity. For instance, a single 25-pound snapper can produce nine million eggs per spawn; it would take 212 smaller snappers weighing 2.2 pounds each to generate an equivalent number. Yet these same great fishes, the favored target of fishers, have few sanctuaries where they can escape hooks and nets.

It has become obvious that traditional management strategies are insufficient to cope with the accelerating declines in fisheries stocks. Imposed size-limits, based on the minimum reproductive length of a species, have proven ineffective partly because of the high mortality caused by releasing undersized or oversized fishes. Evaluation of minimum size regulations implemented in Florida during the early 1980s found no obvious increase in the size of Nassau Grouper, *Epinephelus striatus,* and Black Grouper, *Mycteroperca bonaci.* Catch quotas, which are difficult to determine and monitor, also have proven problematic.

By taking human consumption out of the equation, reefs will manage themselves. Protection from fishing pressure in more than 100 marine reserves from around the world where compliance and protection has been maintained have consistently increased biomass of breeding stock, with far more large individuals and elevated fertilization rates. Research also shows that stocks of exploited species within reserves typically increase three- to five-fold within 5 to 10 years of protection. What has been more difficult to substantiate is the "spillover effect," much touted by proponents of marine reserves. Spillover is purported to occur following an elevation in egg production when larval recruits as well as adults and juveniles immigrate into surrounding fishing grounds after protected areas reach their carrying capacity. Carrying capacity is defined as the number of individuals that can thrive in a given area without exhausting resources. One reason fishers resist reserves is the threat of reducing their economic yield. Indeed studies substantiating the benefits of spillover have been mixed. In most cases it can be confirmed, but not always to the degree necessary to produce a net benefit sufficient to compensate for closing fishing grounds. The trend in spillover research indicates a positive spillover effect for reserves bordered by poorly managed fisheries with severely depleted stocks, however it has been less conclusive when reserves were established in well-managed fisheries. Two dissimilar marine reserves discussed in the journal *Science* in 2001 present convincing evidence supporting the benefits of no-take reserves to adjacent fishing grounds. In 1995 the Soufriere Marine Management Area set aside five coastal marine reserves encompassing 35 percent of a reef-lined cove on the eastern Caribbean island of St. Lucia. The project, intended to restore a dwindling subsistence fishery, has become a hit with both tourists and a majority of local fishermen whose catches soon began to increase in the waters outside the reserves' boundaries. Visual surveys three years after establishment indicated a three-fold increase in the collective biomass of five commercially important fish families within the reserves, while at the time doubling the number in the surrounding waters.

The second example focuses on the Merritt Island National Wildlife Refuge, which happens to encompass the two oldest marine reserves in the United States. Both areas were established in 1962 as a security zone to protect rockets at the Kennedy Space Center rather than fish. The de facto protection from fishing has turned out to be a boon for Florida anglers fishing along the adjacent 120-mile coast, especially for those after record-sized trophy fish.

Comparisons of fish inside the protected areas to neighboring fishing grounds revealed a nearly 13 times greater relative abundance for Black Drum, *Pogonias cromis,* 6.3 times greater for Red Drum, *Sciaenops ocellatus,* and 2.3 times greater for Spotted Seatrout, *Cynoscion nebulosus.* Even though the region encompasses just 13 percent of the Florida coastline it accounted for 62 percent of the state's 39 records for Black Drum, and 54 percent of the 67 records for Red Drum between 1939 and 1999. Because the reserves were heavily fished before closure it was predicted that record catches bordering the reserves would increase for long-lived species with time. Sure enough, a record-setting growth trend that began in the mid-1980s was substantiated in the mid-1990s for Red Drum, with a long life span of 35 years, and for Black Drum, with a maximum life expectancy of 70 years. From 1985 to 1999, when the paper was published, every Florida record for Red Drum and nearly all the records for Black Drum have been awarded to fish taken just outside the Merritt Island refuge.

Even with the growing evidence demonstrating the conservation and fisheries benefits of marine reserves, their implementation has increased at a snail's pace. The United States lost a banner opportunity to set the standard for marine conservation when Congress created the 2,896 square nautical mile Florida Keys National Marine Sanctuary in 1990. The concerted efforts of marine scientists, conservation groups and Florida and federal agencies to implement a network of no-take reserves within the Sanctuary turned out to be no match for the fishing lobby. The original management plan, allocating 10 percent of the Sanctuary for three no-take reserves was, after an initial campaign, whittled down to six percent. Still dissatisfied, the fishermen rallied local residents and their crusade became nasty and personal. Sanctuary managers and fisheries personnel were embarrassed, ridiculed and hanged in effigy. During six years of controversy the number and size of the reserves steadily dwindled until the final draft contained only the single nine-square-nautical-mile Western Sambo Ecological Reserve, an area encompassing less than one-half of one percent of the Sanctuary.

Three decades later, of the Sanctuary's 2,896 square nautical miles only six percent has been designated as no-take marine reserves, including the 151 square nautical mile Tortugas Ecological Reserve establish inside the Dry Tortugas National Park in the far western boundary of the Sanctuary in 2001. In 2007 a 46-square-statute-mile Research Natural Area was added. Regrettably, these two small areas in the Tortugas together form the second largest no-take marine reserve in the continental United States. At the end of the second decade of the new century there are 1,600 marine protected areas in the United States, of these only 223 are no-take marine reserves. This might sound impressive until one considers that no-take marine reserves make up only seven percent of the country's total area of marine protected areas, and cover just over three percent of U.S. waters including estuaries and the Great Lakes. Painting a clearer picture, 94 percent of the country's marine reserves are encompassed inside the 583,000 square mile Paphānaumokuākea Marine National Monument created in Northwestern a Hawaiian Islands in 2006.

Information about marine reserves in the Caribbean and Bahamas is difficult to come by. A 2003 study identified 55 reserves in 21 countries. In most of the countries funds for fisheries, much less conservation, is minimal and enforcement in most cases lacking. Many are marine reserves "in name only," a mixed bag of entities often created for funding and tourism opportunities rather than fisheries management. The authors estimated that 20 percent of the Caribbean reserves are compliant and about 50 percent provided some measure of protection. The first reserves in the Caribbean were two small areas established in the U.S. Virgin Islands: Buck Island on St. Croix in 1961 and Trunk Bay on St. John the following year. The largest cluster of reserves extends along the Mesoamerica coastline primarily in Belize, a small coastal country with a short but impressive history of marine conservation. The largest marine reserve, Cuba's Jardines de la Reina, Gardens of the Queen, became a national park in 2002 protecting 849 square miles of island-dotted inshore water and 93 miles of south central Cuba's coastline. The second largest, and one of the oldest, the Bahamas' 176 square mile Exuma Cays Land and Sea Park, established in 1958, became a no-take reserve in 1986. Both of these large reserves have done an excellent job protecting the environment, supporting fisheries and providing exciting venues for divers and snorkelers replete with large groupers and sharks absent from much of the Caribbean and Bahamas.

Reserve opponents often point to the perceived difficulties of policing no-take zones. Fortunately, enforcement, which can be a substantial financial burden to reserve management, has not always posed a significant problem. In New Zealand, with 33 marine reserves covering nearly 5,000 square miles or approximately seven percent of the country's territorial seas, commercial fishermen became the unauthorized enforcers of boundaries—their attitude is that if the area is off limits to them it is certainly not going to be fished by their competitors. Without official monitoring of any form, there was 68 percent compliance by fishers just after the establishment of reserves, and compliance increased with time. Also, in the Pacific Islands, coastal communities traditionally have authority over their neighboring reefs. The model of provincial ownership has proven extremely valuable in the enforcement of no-take regulations.

In attempts to temper public support, fishing lobbies have often called for the exclusion of divers and snorkelers from no-take reserves, citing the damage inflicted on living corals by underwater sightseers. While careless or uncaring visitors do occasionally break branching corals and sponges or mar the surface of boulder-type corals, the damage is mostly aesthetic and rarely jeopardizes the health of the affected colony. This is not to say that contact with coral is acceptable or should be ignored. We certainly would not think of walking across a flowerbed in a public park. Environmental etiquette and buoyancy control skills should be encouraged and practiced. However, restricting divers from protected marine reserves makes as much sense as limiting hiking in Yellowstone Park, or not allowing anyone access to appreciate the natural resources. If it were not for the economic impact and the environmental concerns of underwater sightseers, there would be far fewer protected reefs in the Caribbean. Those fortunate enough to visit a coral garden become the reefs' most ardent advocates. If there were twice as many divers and snorkelers coral reefs would be twice as safe, not the other way around.

Opposition to marine reserves often makes the point that conventional fisheries methods, such as quotas, size limits and gear restrictions, are sufficient to regulate fisheries. Regrettably, nearly every management measure implemented can be and has been legally circumvented. In turn rules are amended and new restrictions put in place creating a complex matrix of ever changing regulations. Instead of building a sustainable fisheries as hoped these efforts have proven frustrating for everyone and only slowed the ongoing decline of overexploited fishes. At the same time no-take reserves are not the sole answer. If sufficient coastal areas were set aside as a means of sustaining fisheries it would be unrealistic in size and scope. A combination of management tools and marine reserves with their promise of biodiversity conservation, normalizing age structure, protecting genetic variability and increasing fish stock, is seen by many as the most realistic strategy for maintaining viable fisheries. Such actions would also provide protection for non-targeted species with the elimination of bycatch and destructive fishing practices such as trawling. Of course, this approach would entail a radical shift in attitude toward the implementation no-take fishing areas. Where we are today is a giant leap from where we would need to be in light of the World Parks Congress recommendation that 20 to 30 percent of every habitat in the sea should be given full protection from fishing by 2030. Still many fishers as well as fisheries managers remain unconvinced that establishing marine reserves is a meaningful approach to the problem. They point to a lack of experience administering marine reserves in general much less the implementation of the idea on a grand scale for long periods.

Since the publication of the first edition of this book, at the end of the last century, the greatest paradigm shift in marine biology has revolved around the geographical nature of the oceanic transport of larval reef fishes and invertebrates during the early pelagic phase of their bipartite lives. As the century began, mounting evidence indicated that the distribution of the pelagic larvae is not as random and wide spread as previously believed. Instead of being an open system with passive organisms at the mercy of oceanic processes, the tiny offspring are now known to develop quickly into proficient swimmers equipped with sophisticated senses capable of entraining themselves in near-shore waters. While the vast majority of fishes fail to survive their larval metamorphoses a significant number, overwhelmed by the pull of currents, eventually settle miles from their natal reefs. The few survivors of those unable to make their way home find safe havens within 60 miles from where they were spawned, while a scattering eventually locate suitable settlement habitat as far as a few hundred miles away. From genetic studies matching recruits with parentage it is now reasoned that settlers from 120 miles or more away are ecologically insignificant to a reef's population structure. Even so this interconnective process of give-and-take, as insignificant as it might seem, is critical in sustaining biodiversity and promoting genetic diversity on coral reefs across widespread regions.

An understanding of this widespread dispersal system makes it easy to grasp how the continued overfishing of local breeding stocks adversely affects the fisheries of down-current neighbors. A practical solution, purposed in the

1990s, would be the establishment of extensive networks of strategically placed no-take marine reserves interconnected by larval flow. However, doing the right thing is not always easy. A map of the region's prevailing surface currents—the probable routes of larval transport—paints a clear picture of the geo-political implications of managing a widespread interdependent resource. To sell such a politically sensitive idea, conservationists, fisheries managers and politicians will require a simple, easy-to-understand master plan adapted from working models of reserves with proven track records as well as the cooperation of partner nations.

The first geographic guideline for establishing a Caribbean-wide network of reserves was proposed by Callum Roberts in 1997 in the journal *Science*. To work effectively the reserves must be situated close enough to one another to receive an adequate larvae supply from up-current sources and, at the same time, deliver larvae effectively to down-current reserves and intervening reefs. Using the limited data at the time Roberts calculated that the average minimum and maximum inter-reserve distances should be 90 and 125 miles respectively. However, these distances were computed by multiplying the local rate of current flow by 30- and 60-day envelopes—at the time believed to be the average periods larvae remain offshore before settlement. His approach also made the assumption that larvae are basically passive planktonic drifters, and also disregarded the possible relevance of near-shore countercurrents and seasonal gyres on the distribution process. Based on his early assumptions Roberts produced a series of maps, indicating the probable larval dispersal routes for 18 locations. At the low end of the scale are isolated reef systems, such as Barbados in the far eastern Caribbean and the Flower Gardens in the Gulf of Mexico, which are totally dependent on self-recruitment. Andros Island in the Bahamas and the middle Florida Keys, both bathed by currents from numerous upstream reefs, represent the opposite end of the spectrum.

The majority of the 18 larval transport chains detailed by Roberts spans the waters of multiple nations. Reserve networks represented by the 30-day envelope involve from one to six national management partners, with an average of just over two. Expanding to the 60-day envelope, the average escalates to three and a half nations per network. For example, using the 30-day model, partners of the U.S. Virgin Islands include Puerto Rico and the British Virgin Islands; the 60-day envelope reaches east, encompassing St. Martin, Anguilla and the Netherlands Antilles. Convincing so many partner nations to cooperate in such a political complex proposition might appear impractical were it not for one common bond—rapidly declining fish populations.

In the late 1990s when Roberts recommended networks of marine reserves, the plan represented a much-needed beginning for managing a long-neglected resource. At that time it would require traveling all the way to New Zealand, which has 12 no-take marine reserves scattered around the

Right: ***(A)*** *Major surface current patterns in the wider Caribbean region. The thickness of arrows are approximately proportional to a current's strength. Few of the weak nearshore countercurrents, common to most coastlines, are shown.* ***(B)*** *One- and two-month envelopes of potential larval transport indicating the probable upstream larval sources for 6 of the 18 locations studied.* ***(C)*** *From the same six locations, one- and two-month envelopes of potential larval transport showing downstream areas where larvae could settle. Reprinted with the pemission of Callum Roberts.*

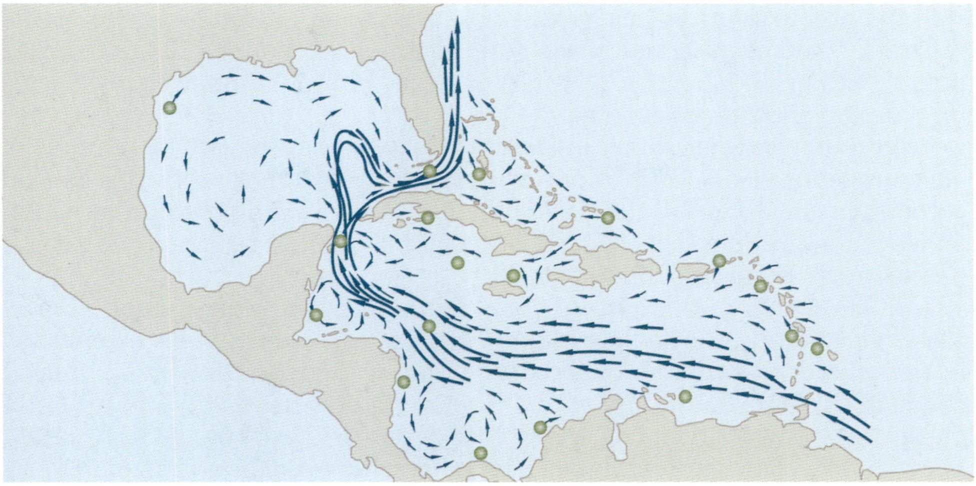

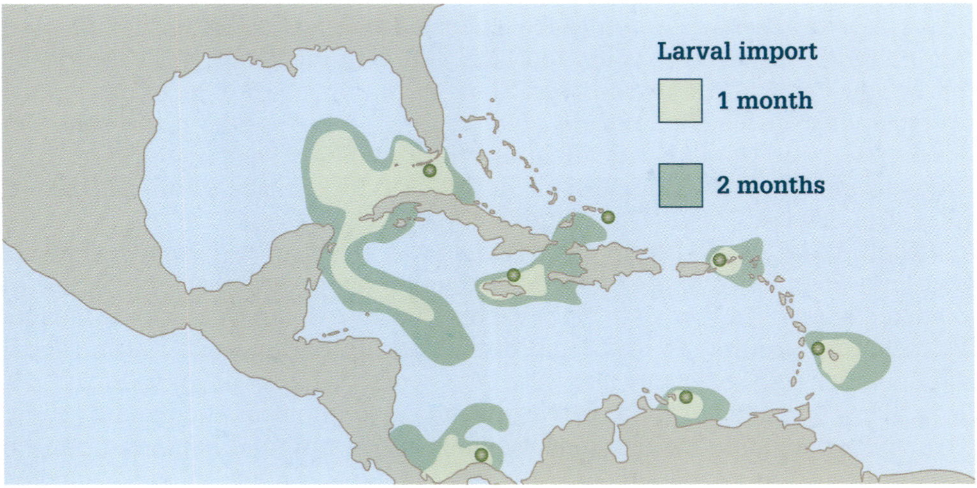
Larval import
1 month
2 months

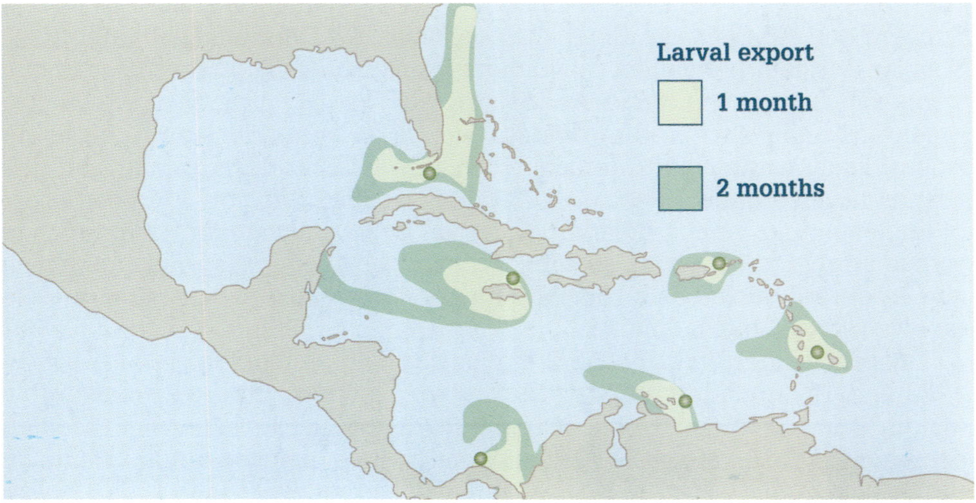
Larval export
1 month
2 months

country, and to South Africa, with eight, to find even a semblance of such a "system." And even though these nations were trendsetters, their protected areas were established for a variety of reasons and without clear policies for creating additional reserves.

With a growing realization in the 21st Century that the largest portions of surviving offspring tend to settle back to their home reefs the spatial component of Roberts' 1997 recommendations no longer appears valid. However his proposal is still relevant and a guiding principle of marine conservation groups around the globe.

A 2009 article published in *Coral Reefs* entitled *Management Under Uncertainty* offers an overview of three marine reserves networks established in the western Pacific as well as providing a guideline for incorporating connectivity into the protection of coral reefs and fisheries. The group of researchers presents their advice as "rules of thumb" based on the best current knowledge available. Readily acknowledging that much more needs to be learned about the implementation of marine reserve networks they state: "It is better to act with incomplete knowledge than to wait for detailed understanding that may come too late."—a sentiment currently shared by many advocates of marine reserves. Their three case studies, all located in the tropical western Pacific, include networks in the Bohal Sea in the southern Philippines, the Great Barrier Reef in the Coral Sea adjacent to the coast of Queensland in northeast Australia, and Kimbe Bay on the northern coast of West New Britain, Papua New Guinea in the Bismarck Sea.

When Angel Alcala began championing the idea of marine reserves in the southern Philippines in the early 1970s his primary goal was the protection of coral reefs from destructive fishing practices such as bomb and cyanide fishing. With help and support of a local university he began encouraging local communities to take destiny into their own hands by personally managing their own fishing grounds to enhance their livelihood for the near and distant future. Over time his sustained message explaining the potential benefits from the spillover effect and the idea of additional revenues from tourism won the day. His first success was the establishment of two small marine reserves at Sumilon and Apo Islands under the oversight of local municipal authorities. Sumilon, without permanent inhabitants, was historically under the edict of a village on nearby Cebu Island. As part of the community directive the protected areas and their surrounding waters would be regularly monitored using scientifically established protocol. Over the ensuing decades the two tiny islands have contributed some of the best evidence available for the concept of implementing marine reserves managed by local communities in order to safeguard reef habitats and sustain economic well-being. As a result, Sumilon and Apo have served as a template for the remarkable expansion of marine reserves across the Philippines.

The conclusions of a 10-year fish population study, implemented between 1983 and 1993 inside and outside the two reserves, provide considerable insight into the effects of subsistence fishing on local reefs. The research estimated that 15 percent of the fish biomass was removed annually outside the reserve at one island and 25 percent at the other. Even with such a heavy catch, species diversity remained steady with no loss in abundance noted. The study also showed a positive correlation with the concept of the spillover effect—an important factor in convincing other fishing communities in the wisdom of locating marine reserves in their home waters. The researchers concluded that the resilience of the local reef fish population was primarily due to a healthy supply of downstream recruits, the elimination of destructive fishing practices, and the communities' non-selective bias toward what species they ate.

Over the course of their existence Apo has consistently maintained its reserve status. However, Sumilon was another matter. In 1983, after nine years of protection, the no-take status of the Sumilon reserve lapsed for political reasons prompting the reopening of the reserve to fishing. More than 100 fishermen from nearby Cebu began fishing the reefs with hook and line, spears, bamboo fish traps, gill nets and blast fishing. A gallon canister of explosives dropped with a delay fuse flattens corals for 15 feet in every direction and kills fishes and invertebrates across a much broader area. Drive-netting involving a chain of humans, each equipped with a weighted scare-line dropped repeatedly herded fishes toward a central bag net. By the time the reserve regained its protective status a year-and-a half-later, 40 percent of the area's coral gardens lay in rubble. Six years later the reserve once again supported an abundant fish population, but the destruction of coral did not allow the reef to regain its previous community structure.

Nearly four decades after the humble beginnings of the two fledgling marine reserves at the edge of the Bohal Sea a vast network of 60 reserves stretches across the nation's southern region incorporating from 10 to 25 percent of traditional coastal fish grounds. Overall, since the 1970s more than 600 marine protected areas have been established in Philippine waters with many incorporating no-take reserves. By national law, areas up to nine miles from shore are co-managed by local communities and the national government. In 2010 a study of 156 reserves indicated an increasing trend of improvement, rating nearly 45 percent with good to excellent management. Those reserves with established revenue support streams, such as user-fees from dive tourism to support enforcement, have proven most effective generating local compliance. Backing from fishing communities, primarily generated by an understanding of the benefits provided by reserve networks, as well as local-based management and enforcement have been important keys to success. Although larval connectivity wasn't an early component of the of the Philippines network when implemented the idea's principles are being incorporated in recent management initiatives.

The Great Barrier Reef Marine Park, created in 1975, under the governance of the Great Barrier Reef Marine Park Authority stretches along Queensland on the northeastern coast of Australia for 1,400 miles (140 miles at its widest point) from Cape York in the north to just north of Bundaberg in the south. The park's 133,000 square miles encompasses the largest reef system on Earth consisting of 2,900 reefs. After a hard found battle with commercial and recreational fishers a new zoning plan was implemented in 2004 dramatically increasing the total area and number of no-take marine reserves from 4.5 percent to a network of

208 reserves covering one third of the iconic Marine Park. The impressive expansion instantly thrust the park into the position as world leader in the management of multiple-use protected areas including no-take marine reserves to "conserve exploited fish stocks and biodiversity." Following a vociferous public discourse directed against the park's rezoning effort the necessity of documenting the plan's benefits became imperative.

Just such a study, published in 2015, highlights an increasing biomass of coral trout, *Plectropomus* spp.—a heavily targeted grouper complex of seven medium-sized species. The paper provided just the sort of easy-to-understand analysis required to help validate the controversial strategy. The research charted how the biomass of *Plectropomus* steadily declined across the Great Barrier Reef from the 1980s until 1996 when, following the expansion of no-take reserves from one to four-and-a-half percent, the groupers began making a slow but steady comeback. In 2004 the biomass finally reached 1980s levels once again. During the same period, even the reefs outside the reserve boundaries showed a slight improvement. As expected, the increased biomass in fished areas was far less significant than those inside the no-take reserves, which were on average 2.5 times greater. Additionally, the grouper residing inside the reserves were larger than those outside. Overall the higher density combined with the increase in size of groupers within the reserves' boundaries translated to an 82 percent greater biomass than in fished areas. The increase in protected area also provided additional sanctuary for the older, larger, more fecund individuals which in turn not only enhanced the flow of larval recruits back to the reserves where they were spawned, but also help escalate the supply of recruits to adjacent reefs. Significant for political reasons, even with the decrease in the size of fishing grounds there was not a corresponding decrease in biomass in fished areas. At the same time, the Great Barrier Reef's network of no-take reserves has failed to show an improvement in the biodiversity, or population structure of non-targeted fishes. However, these findings were anticipated since the local hook-and-line fishery concentrates nearly all their effort on the grouper complex and seldom targets other reef fish species.

In 2016, and again in 2017, the Great Barrier Reef was radically altered by two devastating bleaching events induced by a rise in global water temperatures. When inshore waters warm to unnatural levels the photosynthetic symbiotic algae, known as zooxanthellae, which supplies the coral with its primary source of food, are expelled from the corals' living tissue causing the animals to lose their color. If the water temperature remains extreme, corals begin to die in a matter of weeks from starvation. In the summer of 2016 the northern third of the park lost a third of its living coral in matter of weeks from rising water temperatures. The following year a second bleaching event struck the central section. The combined episodes killed a billion corals, roughly half the colonies in the entire reef system.

At the end of 2017 a research paper examined how the connectivity of coral larvae might help reseed the damaged areas. The researchers identified 100 of what they termed "robust reefs." These areas, representing only three percent of the total coral system were selected using three criteria: location in relatively cool regions less susceptible to coral bleaching; connection to downstream reefs by significant current flow to deliver larvae; and areas without recent infestations of coral-eating sea stars, known as the crown-of-thorns. In their estimation these small healthy reefs, have the potential of seeding 45 percent of the entire reef ecosystem. Of course the parent corals and their larvae will also need help. The researchers suggest protection and supervision, and even more critical, an immediate reduction in carbon emissions around the world.

As early as 1992 The Nature Conservancy had identified Kimbe Bay as a site of significant conservation interest. The roughly 5,000 square mile area, located on the northern coast on New Britain Island, Papua New Guinea, sits on the southern edge of the Bismarck Sea. Like the Bohal Sea in the southern Philippines Kimbe Bay is situated in the Coral Triangle—the epicenter of marine biodiversity in the western Pacific. The Coral Triangle's boundaries encompass much of the Philippines, eastern Indonesia, northern Papua New Guinea and the Solomon Islands. The area, home to 76 percent of the world's coral species, over 2,000 fishes and six species of turtles, also includes some of the richest commercial tuna fishing ground in the ocean. Well thought out and long-planned, the Kimbe Bay Reserve, with 14 marine protected areas and a strong emphasis on no-take reserves, is intended to be a model for eventual expansion throughout the Bismarck Sea. During implementation, a strong priority was placed on larval connectivity among the sites and community involvement with a mandate of developing ecotourism, the only non-extraction economy in the region. In 2013, the West New Britain Provincial Government took the leading role in overseeing what is today known as the Kimbe Bay Marine Management Area.

Two studies published in 2005 and 2007 used Kimbe Bay to demonstrate support for the idea of larval replenishment in marine protected areas through self-recruitment. Three study sites, each part of the area's network of marine reserves, used genetic parentage analysis to compare larval distribution patterns of the Vagabond Butterflyfish, *Chaetodon vagabundus,* an open-water pelagic spawning species with an extended larval duration of four to seven weeks with the Clown Anemonefish, *Amphiprion percula,* a benthic egg-layer with a short period of less than two weeks. The research detailed a first-time comparison between two species with contrasting reproductive life styles over two different time periods. As anticipated the egg-laying Clown Anemonefish, with a much reduced larval stage compared with the pelagic-spawning Vagabond Butterflyfish, demonstrated notably less capacity to populate the two test sites located five and four miles away from where they were spawned. It was also noted that besides a short larval stage the anemonefish also have well adapted sensory and swimming capabilities for remaining near their reefs of origin.

The results from Kimbe Bay indicated a high degree of self-recruitment over time (around 50 percent) for both species at Kimbe Island where they were spawned, and although variable, a much lower pattern of settlement at the two test islands located on the west side of the Bay. It was assumed that the variability was an artifact of fluctuating current patterns in the Bay.

Similar to the coral trout study in the Great Barrier Reef the Kimbe Bay results, although employing non-food fishes as test subjects, provide additional confirmation of a strong tendency for locally spawned larvae to self-recruit back to their home reefs. These findings, coupled with recent larval transport and settlement studies from the tropical western Atlantic, speak favorably for the implementation of larval connectivity strategies among networks of no-take marine reserves for sustaining fisheries.

Protection of Grouper and Snapper Aggregations

Populations of groupers and snappers, both popular food fish, have suffered a faster decline during the past decades than any other reef fishes. In the 1960s, when divers first began extensive exploration of Caribbean reefs, numerous large groupers were regularly sighted on every dive. Things were quick to change. Within only a few years, spearguns, fish traps, powered fishing boats and an ever-growing demand for market fish began to take a heavy toll on every large and moderate-sized grouper species in the region. The big predators, with few natural enemies, had little or no innate fear of divers carrying spears. Instead of hiding, the curious fishes often swam off the reef to get a closer look at the approaching oddities. Those surviving the initial slaughter quickly learned to flee or relocated to deeper reefs. At the same time, a proliferation of wire-mesh fish traps by small-scale island fisheries removed substantial numbers of juvenile groupers from inshore and mechanized fishing fleets began taking fish from deeper water. However, the deadliest blow dealt the beleaguered groupers and snappers has been the exploitation of their annual spawning aggregations. With thousands of large reproductive groupers or snappers gathering from miles around at traditional sites at predictable times, it was like catching fish in a barrel. The brief, annual economic boom for local fisheries spelled disaster for large aggregate spawning species.

Although a small number of inshore spawning aggregations had been regularly fished since the early years of the 20th Century, a sufficient number of remote locations remained unscathed, helping to sustain breeding stock. But by the close of the 1960s, even the distant breeding grounds began to suffer from advanced technology and growing fishing fleets. In just a few years, many heavily fished aggregations, which may have existed for centuries were no longer active.

All targeted grouper species have shown marked decreases in the average size of individuals and population numbers. Aggregate spawning groupers, including Red Hind, *Epinephelus guttatus,* Goliath Grouper, *E. itajara,* Gag, *Mycteroperca microlepis,* and Scamp, *M. phenax,* have suffered heavily; but none has declined as precipitously as Nassau Grouper. Its populations have been so devastated in the past decades that the species is considered commercially extinct in the continental United States, U. S. Virgin Islands, Puerto Rico and Bermuda. Since the 1970s numerous spawning aggregations, some estimated to attract between 30,000 and 100,000 fish before heavy fishing pressures began, have ceased to reform in the Cayman Islands, Dominican Republic, Jamaica, Mexico, Belize and the Bahamas. In Bermuda, Nassau Grouper landings declined by 95 percent between 1975 and 1981 and today have vanished from the local seascape. Comparable declines have been reported throughout much of the Caribbean Basin and the Bahamas. Of the 80 Nassau Grouper aggregation once known to exist throughout the fishes' range fewer than a dozen remained active in 2019.

In the 1980s the United States, armed with strong evidence of an impending population collapse, took a timid step by instituting a minimum size limit of 12 inches for Nassau Grouper, with plans to increase the limit by one inch each year until reaching a length of 24 inches in the mid-1990s. The increments were designed to ease economic hardships on commercial fishermen. It was not until 1990, however, when there were essentially no more Nassau Grouper of any size to protect, that a moratorium prohibiting their removal (or landing) was finally imposed. However, the species has failed to rebound in Florida waters and in 2019 it is still rare to see Nassau Grouper in south Florida or the Florida Keys. It wasn't until 2017 before Nassau Grouper were listed in the United States as threatened under the Endangered Species Act. If more far-reaching management strategies are not implemented throughout the Bahamas and Caribbean Basin, Nassau Grouper may soon become the first major reef fish to go the way of the Passenger Pigeon. And if further extraction continues at its present rate, several other grouper species could quickly follow their path to extinction.

There have been a few other modest regulatory measures implemented in the last decade. After documenting an 86 percent decline in Nassau Grouper landings in Bahamian waters over a twenty-year period from its peak of 514 tons in 1997, the country's Fisheries Act was amended in 2015 to prohibit fishing Nassau Grouper during their aggregation months extending from December through February, but enforcement of the ordinance has been lax.

Near the end of 2003 the Cayman Islands closed all of its Nassau Grouper spawning aggregations to fishing for an eight-year period. Of the six sites known at the time two were judged to be fished out while the other three were in serious decline. One of the viable aggregation sites located off the west end of Little Cayman has been monitored annually during the yearly gathering since 2002 by REEF (Reef Environmental Education Foundation) in partnership with the Cayman Islands Department of Environment. During the eight years following protection the numbers of Nassau Grouper congregating at the site during the spawning season increase from just over 2,000 to more than 4,000 fish. The conservation measure was renewed for an additional eight years in 2011. In the winter of 2018, the count increased to well over 6,000 fish. The government implemented additional regulations to safeguard Nassau Grouper in 2016. They included a countrywide ban on taking Nassau Grouper during the aggregation period from December through April; readjusted size and catch limits during the remainder of the year, and prohibited the use of spearguns for taking the species at any time. Belize incorporated varied measures in 2001 to help protect its vanishing population of Nassau Grouper by establishing a network of 13 no-take zones at known spawning sites. The country also incorporated a four-month fishing moratorium during spawning seasons, and designated minimum and maximum size limits. Even with a rudimentary system of enforcement, population numbers continued to fall. Sixteen years later only the Northeast

Point aggregation at Glover's Reef Atoll remains active. Even this spawning population has shown considerable fluctuation in numbers ranging from 3,000 individuals in 2001 to a low of 450 in 2015. Other countries have tiptoed around the issue only implementing half-measures and compromises following opposition from commercial fishing interests.

Marine biologists under joint sponsorship of The Florida Department of Environmental Protection and National Marine Fisheries Service conducted extensive research on Gag, Scamp and Red Grouper, *E. morio,* populations to better understand how the species' reproductive behavior affects their susceptibility to overexploitation. Each of these three commercially important grouper species in the Gulf of Mexico and the southeastern United States has been confirmed to be protogynous hermaphrodites (changing from females to males), with a naturally female-heavy sex ratio. However, while Gag and Scamp migrate in large numbers to annual spawning grounds, Red Grouper remain in their home ranges and spawn at a variety of depths in small, widely scattered groups less susceptible to concentrated fishing. Gag, reproductively active between February and April, gather over the sea floor in depths ranging from 150 to 375 feet and form into numerous subgroups numbering between 10 and 100 individuals. During the remainder of the year, females inhabit shallow inshore areas apart from the deeper-dwelling males. Scamp often share the same offshore reefs with Gag during breeding periods, but the two groups do not intermix. The reproductive behavior of both species is quite similar except for the longer spawning season of Scamp, extending from February until June.

Comparison of recent commercial landing data with catches made 15 to 30 years earlier revealed that Gag and Scamp, both aggregate spawners, have suffered a marked decline in numbers and mean size. But a substantial change in the sex ratios of both species is even more disturbing. Prior to 1980 the Gag population off South Carolina contained 17 percent males, but by the mid 1990s the number of males had dropped to less than four percent. The patterns for Scamp are similar but less dramatic. At the same time Red Grouper, just as heavily fished, show negligible decreases in size and almost no difference in sex ratio. Although the exact mechanisms creating the imbalance are unknown, it appears that unrestrained access to and targeted fishing of spawning aggregations is a key factor in the declining populations of both Gag and Scamp.

Theoretically, protogyny should compensate for the loss of males by increasing the transition of females into males. But evidently this isn't happening with Gag and other aggregate-spawning groupers. One possible explanation is that fishing pressures select for males, removing them faster than they can be replaced. Both males and transitional males are known to take a hook more aggressively than females. It is generally believed that social mechanisms rather than size or age influence the timing of sexual transition. If this is the case, the rapid removal of so many individuals from spawning aggregations could disrupt the cues necessary to initiate sexual transition. And because males and females coexist only during reproduction, the cues are most likely passed during the brief period when the sexes are together in aggregations.

With the sex ratio of Gag populations so out of balance, more and more females, even though well fed, are unable to reach their reproductive potential. Increasing numbers of females remain inshore during spawning periods. Others migrating to breeding grounds fail to have their eggs fertilized because of a shortage of males or they spawn incompletely or not at all. Atresia, the reabsorption of unspawned eggs, common in underfed populations, has been detected in healthy Gag and Scamp, possibly another damaging effect of fewer males.

Three large shallow-water species of snappers, Mutton Snapper, *Lutjanus analis,* Dog Snapper, *L. jocu,* and Cubera Snapper, *L. cyanopterus,* also gather in large exploitable spawning aggregations. In the mid-1990s we counted more than 60 small fishing boats anchored over a Mutton Snapper aggregation just west of Bimini, Bahamas. Fishers worked the site from morning until two hours after sunset, with a reported peak in harvesting activities just before dark. Unable to find sufficient ice for such a large catch, a significant portion of the snapper taken rotted on docks. In 1991, researchers visiting a Mutton Snapper aggregation site in the Dry Tortugas Islands, Florida, known as Riley's Hump, which had been heavily fished by commercial vessels for several years, found only a scattering of fish. Although the specimens collected were heavy with sperm or hydrated eggs, no spawning formations were observed. By the time the area around the 100-foot-tall seamount became part of the Tortugas Ecological Reserves within Dry Tortugas National Park in 2001 its Mutton Snapper aggregation had been fished to near extinction with less than a dozen fish sighted during the annual June aggregation period. In 2009, eight years after being designated as a no-take reserve, roughly 4,000 Mutton Snapper were counted. The prolific spawning area that also hosts several other targeted food-fishes including Dog, Cubera, Gray, *Lutjanus griseus,* and Yellowtail Snapper, *Ocyurus chrysurus,* and a number of grouper species, is favorably located for transporting larval recruits up the Florida Keys reef tract and possibly as far north as Florida's east coast. Even after the impressive rebound of Mutton Snapper, following the implementation of the protective status in the Tortugas, another well-known Keys' Mutton Snapper aggregation site, experiencing similar declines, remained open to fishing due to strong pressure from the local fishing lobby. In 2017 the Florida Fish and Wildlife Conservation Commission, did approve several Mutton Snapper management changes including an increase in the size limits from 16 to 18 inches (the minimum reproductive size of female Mutton Snapper) and lowered bag limits during the species' spawning season from May through June. In the same year NOAA (National Oceanic and Atmospheric Administration) closed all fishing at a Mutton Snapper spawning aggregation in the U.S. Virgin Islands. Mutton Snapper aggregations in Cuba and Belize have persisted since the turn of the century by limiting access during the spawning season. The Cubans allow only targeted fishing with hand lines by artisanal fishers from nearby villages. Similarly, in Belize permits are given out to a limited number of local hand line fishers for the Gladden Spit aggregation site during spawning months, while the four remaining spawning sites retain full no-take status throughout the year.

The Role of Artificial Reefs

In the past half-century more than 3,300 artificial reefs have been deployed as fish havens in state and federal waters of Florida. The original premise for the huge undertaking was to provide additional structural refuge for larval fishes to congregate with the hope of eventually increasing adult population numbers. In the early years almost any scrap material such as old tires ganged together with chains, automobile bodies, railroad cars, and airplane fuselages were dumped onto coastal sea floors to attract fish. Unfortunately, the corrosive effect of seawater quickly weakened high profile, lightweight materials causing them to collapse. Water movement also plays havoc breaking unsubstantial frameworks apart and scattering debris far from where it was initially placed. Even gigantic structures like the 510-foot U.S.S. *Spiegel Grove,* sunk in 160 feet of water six miles off Key Largo, Florida in 2002, are not impervious to storm surge. Three years after the huge 10,000-ton hull settled on her starboard side she was rolled upright from surge generated by Hurricane Dennis passing 200 miles away.

Today, fisheries managers, far more sophisticated in the use of artificial habitats, only grant deployment permits for heavy, stable, durable and non-polluting materials that more closely imitate natural reefs. If utilized properly, advocates of the program contend that artificial reefs work well for testing ecological theories and models, provide structure on low-relief hard-bottoms, increase the density of juveniles when designed for specific species, reduce predation and benefit the foraging efficiency of targeted species. But after years of study it appears to most researchers that the additional structures do little for conservation or increasing food-fish populations. This in turn has led to what is known as the "attraction-production debate."

In the last decades marine ecologists have come to the conclusion that populations are limited more by fluctuating larval recruitment than by a lack of habitat. At the same time artificial reefs have not proven to help fish grow faster or increase reproduction rates. A Japanese study published in 1989 concluded that artificial reefs increased populations only for octopuses. It also found that recruitment of larval fish was minimal. This supports the thinking of other researchers, who contend that artificial reefs act as nothing more than fish attractors by concentrating fish into one area, which gives the impression of increasing populations. This certainly appears to have been the case in the Gulf of Mexico Gag study—a grouper that actually uses reef habitat to reproduce.

Working with the idea that artificial reefs can be designed and advantageously placed to help young groupers survive to breeding age, federal fisheries researchers built a series of 22 experimental artificial reef sites in the Gulf of Mexico. Each reef consisted of 24 or 96 prefabricated 36-inch concrete cubes with open horizontal cores arranged in hexagonal formations. The structures were deployed between 1990 and 1993 in 40 feet of water 16 miles west from the mouth of Florida's Suwannee River. An extensive survey of the flat sand bottom prior to the project found few Gag in the area. After five years it was estimated that more than 400 Gag resided on each of the larger artificial reefs. In 1996 nine sites were opened experimentally to public fishing. As predicted, the Gag were hit hard. In less than eight months 40 percent of the resident populations, accounting for 77 percent of the biomass, were gone, leaving few reproductive-sized Gag on several reefs. Lessons learned from the Gag project include the importance of designing small, complex structures scaled to the target species, spacing structures far apart, and above all, not advertising their location to the public.

Commercial fishermen tend to be an independent lot with little patience for rules, regulations and paperwork. As fishing methods mechanize, competition multiplies, regulations encroach and fish stocks plummet, they become more defensive. Commercial and recreational fishers, more than any other groups, have witnessed the sharp decline of fish stocks in recent decades. In their heart of hearts, both groups must realize that no-take reserves are not only inevitable but also necessary to protect their long-term interests. If breeding stocks continue to be exploited at the present rate, even more fisheries will collapse. If this occurs it could be a long time, if ever, before the species make a comeback to sustainable numbers. However, with wise fisheries management, it is possible that limited-yield fishing could continue indefinitely.

But what about marine fishes that do not have commercial value? What about the damselfishes, the trunkfishes, the gobies, the blennies and the hundreds of other irreplaceable fishes? It is hard to believe that their fates, as well as those of all other creatures on the reef, are not intrinsically linked to game and market fishes. A system of protected and properly administered no-take marine reserves, extending from inshore habitats to offshore depths would not only be a boon for the "ol' mamas and papas" and spawning aggregations, but would also provide reef fishes with the diverse habitats required for survival of their larval recruits.

Many questions remain concerning the benefits to both fisheries and conservation from marine reserves. But if we wait for all the answers before taking action, it will be too late to save much of our remaining marine wildlife. Systems of no-take reserves are the last best hope for reef fishes around the world.

Angelfishes

Family Pomacanthidae

Rock Beauty

Large, graceful and bedecked with fins, angelfishes—the quintessential tropical reef residents—constantly attract the diver's eye. Even though western Atlantic waters host only 7 of the world's 91 Angelfishes, the seven are spectacular species from three genera. One of the most exquisite, the Queen Angelfish, *Holacanthus ciliaris,* is regularly sighted on coral reefs across the region, while the large and lovable Gray Angelfish, *Pomacanthus arcuatus,* the spirited Rock Beauty, *H. tricolor,* and the delightful little Cherubfish, *Centropyge argi,* add a touch of class to any reef scene. Interestingly, each of these Atlantic species, with the exception of the Rock Beauty, has a closely related and similar-appearing counterpart. Queen Angelfish are so closely allied with Blue Angelfish, *H. bermudensis,* that the two species often interbreed, producing fertile hybrids. French Angelfish, *P. paru,* and Grays are not only similar in size and appearance, but also share habitats and many of the same foods. Cherubfish are comparable to the Flameback Angelfish, *C. aurantonotus,* in size, behavior, diet and basic color scheme.

Until the mid-1970s, angelfishes were considered to be a subfamily of butterflyfishes, Chaetodontidae. The two groups certainly have many things in common: both are round, laterally compressed, have continuous dorsal fins, swim alike and even behave in much the same way.Today, a well-developed preopercular spine on the gill covers of angelfishes differentiates family members from butterflyfishes. Fish watchers can make a quicker identification

Queen Angelfish

Blue Angelfish

Gray Angelfish

French Angelfish

Cherubfish

Flameback Angelfish

by noting the extended rear dorsal and anal fins common to most angelfishes but absent on butterflyfishes.

At first glance, slow-moving angelfishes would seem to be easy targets for predators, but the family's typically large size and thin, round bodies present awkward mouthfuls for those interested in a quick meal. On the other hand, the darting, reclusive nature of the small juvenile Rock Beauty and Cherubfish attests to their vulnerability. French and Gray Angelfish juveniles have taken an alternate route to self-preservation by becoming cleaners that move freely in open water picking annoying ectoparasites from reef fishes. Their conspicuous black and golden markings—strikingly different from the patterns they will wear as adults—advertise their role as cleaners, a strategy that greatly increases their chances of escaping predation.

Only 11 Caribbean fishes are known to feed regularly on sponges; five are angelfishes.

Sponge-eating species, including angelfishes, evolved during recent geologic time, possibly as a response to increased competition for food on coral reefs. Sponges by their nature are a distasteful food. The primitive organisms are laced with tiny internal supports of indigestible and irritating calcareous and siliceous spicules. Repellent chemicals and tough, fibrous exteriors add to their repugnance. Nonetheless, the tissue from a wide variety of sponges makes up 95 percent of the food consumed by species in the genus *Holacanthus* (Rock Beauty, Queen and Blue Angelfish) and 70 percent in *Pomacanthus* (Gray and French Angelfish). Two of the most frequently eaten sponges contain the least amount of spicules, but many of the angelfishes' choices are packed with the mineralized structures. Several species toxic to humans are also regularly eaten. Angelfishes seldom devour an entire sponge. Instead, they follow timeworn paths leading from sponge to sponge, taking a few bites from each as they go. The sponge-eaters have developed strong, protracting jaws lined with overlapping rows of tightly packed teeth to shear through the sponges' tough fiber. A thick coating of mucus secreted around lumps of food helps to protect their stomachs during digestion. Algae is the primary food of juveniles and adult Cherubfish and Flameback Angelfish. Juveniles acting as cleaners obtain only 25 percent of their food by volume from picking parasites off fishes; the remaining 75 percent is made up of filamentous algae.

The five larger species establish territories to defend mates rather than to protect food supplies. Trespassing juveniles and adults of the same species are quickly chased, while other fishes, even with similar diets, are generally ignored. Even though they brim with sponges and other foods, uninhabited areas just outside established territories often go ungrazed. The two little *Centropyge* species constantly guard their turf from other small algae-eating species but, because their small territories are widely spaced, they seldom have reason to defend their harems from neighboring males.

Not much is known about the reproductive behavior of Atlantic angelfishes, and the little information available is often based on contradictory observations. Family members share a few similarities: all spawn at sunset during a limited period; spawning always occurs between a single male and female; pelagic eggs are shed in the water column after a spawning rise; spawning periods are not based on the lunar calendar and possibly take place during all seasons. Gray and French Angelfish are believed to be monogamous, perhaps forming lifelong bonds, but there have been a few reports of promiscuous spawning events for both species. It is possible that each species follows two reproductive agendas, or perhaps populations in one region spawn differently from those in another. This may be the case with Cherubfish, which maintain multifemale harems dominated by a single male in the southern and eastern Caribbean; while in Florida and the Bahamas many individuals appear to live as monogamous pairs.

The three species of *Holacanthus* and two species of *Centropyge* are protogynous hermaphrodites that typically live in harems. There, the largest female changes sex after the disappearance of a dominant male. Large sex-changed males mate exclusively with from one to four females living within defended territories. To keep tabs on their mates, males visit each in turn throughout the day and briefly forage at their sides before continuing on their rounds. Boundaries are regularly patrolled to maintain territorial sovereignty.

Species that form harems participate in similar courtship and spawning behavior. Every evening, dominant males spawn with females living in their harems. Courtship begins with males repeatedly presenting themselves laterally while displaying erect fins. Subsequent visits intensify and lengthen. Before actually mating, females make several false spawning rises, but each time the males attempt to position next to their sides, the females dart back to the bottom. When finally ready to shed their eggs, females ascend higher than before. The males approach from below and place their snouts on or near the females' genital openings and gently nudge them up into the water column. A sudden push on the females' abdomens forces the eggs to discharge. In the same instant, the males roll on their sides, placing their genital openings close to the females and shed sperm. The pairs quickly separate and dive for the bottom, where the females briefly nip or chase the males. The transparent clouds of gametes mix and disperse with the currents. Fertilized eggs hatch into slender larvae within 15 to 20 hours. Three to four weeks later, they settle in shallow water habitats and quickly transform into juveniles.

The sex of Gray and French Angelfish remains unchanged throughout life (gonochorism). Similar-sized pairs form extended, possibly lifelong, monogamous relationships. The lovely courtship of Gray Angelfish seems affectionate—during their slow spawning rises, the females gently flutter their tails against the faces of their mates.

Gray Angelfish
Pomacanthus arcuatus
&
French Angelfish
Pomacanthus paru

There is not a more amiable character on the reef than the stately Gray Angelfish. The slow-swimming, dinner-plate-sized fish seem to enjoy the companionship of divers, often following them closely for minutes at a time, snapping at bubbles or peering at their image in a face plate or camera port as if admiring their regal reflections. The frugally colored, grayish fish, accentuated

A juvenile French Angelfish cleans White Mullet.

A mated pair of French Angelfish.

by white lips, are reported to reach a length of two feet, but few grow to over 16 inches. Mated pairs live within large territories averaging more than 2,400 square yards that often slope seaward from 10 to 80 feet. Territories are protected from members of their own species, which are challenged and chased. The species is known to congregate occasionally in groups of up to 20 individuals that move about together in loose formations within a customary area. The function of these gatherings remains unexplained.

Gray Angelfish seldom take shelter as they roam their territories in the company of their similar-sized mates, which typically remain by their sides for nearly half the day. When reunited after extended separations the pairs circle enthusiastically, as if pleased to be back together. They often swim well off the bottom, dropping down occasionally to forage at the leisurely rate of three or four bites per minute. Four terraced rows of slender, pointed, tightly packed teeth, shear chunks of fibrous tissue from a variety of sponges—their favorite food. Gorgonian polyps also make up a significant portion of their diet, and other invertebrates, including zoanthids, bryozoans and hydroids, are occasionally eaten. Although algae forms less than 10 percent of their stomach content, the food source requires over a third of their foraging time to harvest.

Reproductive seasons and behavior have not been well documented. The similarity in size between mated pairs and the exclusion of juveniles from territories lead to the conclusion that the species is usually monogamous and does not change sex. On two consecutive September evenings, we observed a pair of Gray Angelfish participating in a lengthy courtship and spawning ritual above a low-profile reef bank in 30 feet of water. The pair arrived on site about 30 minutes before sunset and immediately began slowly circling and cavorting together near the bottom. Every few minutes, the presumed female would

During their slow spawning ascent, the female Gray Angelfish flutters her tail constantly against her mate's face.

gradually ascend three to five feet, closely followed by the male, which attempted to place his snout close to the female's rapidly fluttering tail fin. After rising together for only a few feet, the female prematurely broke for the bottom, followed by the male, and they again began to circle and chase. This routine continued uninterrupted for over 20 minutes, with subsequent false spawning rises taking the pair further above the flat sea floor. Finally, just after sunset and with little light remaining, the pair continued their rise 15 feet into the water column. During their slow ascent, the female's constantly fluttering tail gently caressed the male's face. At the apex, the couple abruptly rolled to their sides in an apparent spawn, although a gamete cloud was not visible. Soon after making a quick dash to the sea floor, the fish left the area together.

A promiscuous spawning behavior in waters off the San Blas Islands, Panama has also been documented. A late afternoon gathering of 8 to 14 Gray Angelfish was monitored for a two-week period. Beginning an hour before sunset several individuals, believed to be males because of their slightly larger size, peacefully circled over a group of unsettled females that milled about the bottom, chasing each other and making grunting sounds. The males developed a dark vertical bar behind their gill covers. Eventually one or two females, displaying darkened bellies and pale backs, moved off, followed by a single male. These subgroups were at times aggressively pursued and disrupted by other females. When a pair finally was able to isolate itself, they hovered calmly, with the male positioned slightly above the female for as long as 10 minutes. Eventually, the male began a slow circling ascent followed by the female. Charging males often broke up the spawning rises. If uninterrupted, the pairs would rise 10 feet before separating and returning to the substrate. No gamete clouds were observed.

French Angelfish, occasionally confused with the similar-appearing and closely related Gray Angelfish, can be readily differentiated by their bright, golden-rimmed scales. The juveniles are much more difficult to distinguish and can be identified to species only by subtle differences in markings on their snouts and tails. Young Grays have a clear, flat tail margin, while those of the French are rounded and lined in gold. The two angelfishes are also closely allied in their social, feeding and reproductive behavior. Mated pairs of French Angelfish also live in large territories spanning a broad range of depths, where they feed on many of the same sponge species eaten by Grays. French Angelfish, however rely less on gorgonian polyps and algae in their diet.

Juveniles of both species act as cleaners until about three inches in length. The smallest of the tiny black and gold-barred youngsters move about with an extravagant fluttering of oversized fins. Filamentous algae are their primary food, with parasitic copepods picked from client fishes making up only 25 percent of their diet. The little fishes also eat a few free-living copepods. As adults they switch roles, spending nearly 10 percent of their time attending cleaning stations.

Although French Angelfish do not change sex and are believed to be monogamous, the details of their reproductive behavior have not been established. Two observations of apparent spawning have been published. The first, occurring late one afternoon off Grand Bahama Island, Bahamas, involved five to eight similar-sized pairs and three or four smaller individuals. Each pair aggressively chased other pairs or individuals that came within 20 feet of where they swam side-by-side just off the bottom. This hostile behavior, often with the aggressors rolling laterally during an attack, kept the pairs evenly spaced across the sea floor. One closely observed couple displayed no obvious courtship behavior or distinctive color pattern, but 10 minutes after sunset they rose together at a low angle that carried them 20 to 30 feet forward and 7 to 10 feet above the bottom. Throughout the rise their abdominal regions were angled slightly toward one another. After separating at the end of their rise, no cloud of spawn was visible. The other pairs behaved in a similar fashion, without interference. What appeared to be spawning behavior between a single pair of French Angelfish also was observed in Puerto Rico. The two fish rose in the water column for several feet, with one fish positioned just below the other. At the top of their rise, the pair parted and returned to the bottom; again, no gamete cloud was observed.

Two conflicting reports examine the relationship between Gray and French Angelfish. One documents pairs of both species peacefully sharing overlapping home ranges in Florida and the Bahamas. However, aggressive behavior by Grays toward a French Angelfish was observed in Panama. In this instance, from one to three Grays were seen chasing a solitary French Angelfish whenever it approached within 100 feet. No sightings of mixed groups or pairs have been documented, indicating that the two species are incompatible. And, unlike Queen Angelfish and Blue Angelfish that regularly crossbreed, no wild hybrids of French and Grays have been documented. Hybrids of the two species have been artificially produced in aquariums, however, demonstrating the possibility of such an occurrence.

Queen Angelfish
Holacanthus ciliaris
&
Blue Angelfish
Holacanthus bermudensis

Queen Angelfish, considered by many to be the most beautiful reef fish in the Caribbean, wear a striking array of blues, greens, yellows and oranges accentuated by streaming dorsal and anal fins. Unfortunately, the species frequently seeks shelter during the day and is somewhat skittish when closely approached. Lengths of 18 inches have been reported, but individuals over 14 inches are considered rare. Although primarily a shallow water fish, Queens have been recorded below 200 feet, but are commonly found below 50 feet. Their sexual development has not been studied extensively, but both Queen and Blue Angelfish are assumed to be protogynous hermaphrodites. Females forage in separate areas of the males' large home ranges. During the day the dominant males regularly visit the solitary females in turn, approaching rapidly and circling before settling down and feeding by their sides for 5 to 10 minutes. Ninety-seven percent of their diet comes from 33 species of sponges. The remaining tidbits include algae, tunicates, hydroids and bryozoans. Solitary juveniles primarily eat algae until they reach sexual maturity and join harems.

The species' reproductive behavior has not been well documented. The few reported spawnings involve a single male courting members of his harem just before sunset within territorial boundaries. The observed male rolled laterally and repeatedly flicked his pectoral fins, then held them extended as he circled above two females. Much like Rock Beauties, *Holacanthus tricolor*, and Cherubfish, *Centropyge argi*, a female, when ready to release eggs, moves into the water column where she is joined by a male, which nudges her from below with his snout.

The Blue Angelfish, similar-appearing and closely related to the Queen Angelfish, is primarily a continental species inhabiting the tropical and

The Townsend Angelfish, a regularly occurring hybrid of the Queen Angelfish and the Blue Angelfish.

temperate waters of Florida and up the eastern seaboard to the Carolinas. The species also occurs in Bermuda and along the western fringes of the Bahama Bank. In south Florida, where the range of Blues overlaps that of Queens, the two species occasionally interbreed, producing a hybrid that was once described as a separate species, the Townsend Angelfish, *H. townsendi.* The hybrids are fertile and often cross-spawn with Queens and Blues and other hybrids, producing an infinite variety of color variations.

Courtship between a Queen Angelfish (right) and a Blue Angelfish.

In Bimini, Bahamas, seven days after the full moon of May and one hour before sunset, we followed a Blue Angelfish obviously distended with hydrated eggs. After foraging for 20 minutes, the fish moved across the bottom toward deeper water, where a pair of Queen Angelfish joined her. The Blue rose six feet off the bottom and jerked her body several times. One of the Queens left its partner and swam off with the swollen Blue female. Each time the pair paused the Blue lifted off the bottom, tilted her head up, extended her fins and shuddered but appeared to be ignored by the feeding Queen. After following a circuitous route for 10 minutes, the pair briefly joined a single Blue before moving off together once again. Five minutes after sunset, the female Blue rose 10 feet off the sea floor, where she was approached from below by the Queen, which nudged her abdomen with his snout before the fishes parted, swimming hurriedly away in different directions. There was no visible sign of a gamete cloud.

Rock Beauty

Holacanthus tricolor

Rock Beauties, with their distinctive yellow and black markings, add a bold splash of color to nearly every high-profile reef of moderate depth in the Caribbean. Known to be protogynous hermaphrodites, Rock Beauties live in harems dominated by a single large male that mates exclusively with from two to four smaller females. Juveniles less than two inches in length are brilliant yellow with a single black spot ringed in blue on their sides. The solitary youngsters are extremely shy, spending much of their day in hiding. As they mature, the black enlarges until it covers much of their bodies. At approximately three and a half inches, females become sexually mature and join harems. The transformation to male occurs at about five inches. Older males, reported to reach a foot in length, develop streamers from their dorsal, anal and tail fins.

The territories of males, averaging just over 1,000 square yards, encompass the smaller territories of females. The large territories are randomly distributed

A male Rock Beauty nuzzles a female's vent during their slow spawning rise.

across the sea floor, often leaving substantial, uninhabited gaps. Such large, untapped areas filled with plentiful food and shelter indicate that these resources are not a factor in limiting population size. However, males are quick to protect their turf from encroaching males that often display darkened faces.

Interestingly, when two females differ in size by more than 15 percent, they often share all or part of their territorial area without conflict. Their peaceful coexistence appears to be related to the social order of haremic life, where only the largest female can transform into a male. The larger fish seem willing to accept the smaller females, which may one day become part of their harems. On the other hand, similar-sized females, which might at a later date challenge the right to dominate the harem, do not tolerate one another and never share any portion of their territories.

Unlike most protogynous species, whose sex reversal is prompted by the disappearance of a dominant male, the largest females in harems of Rock Beauties regularly change sex before there is an opportunity to dominate females. In other species, bachelor males would be promptly driven from the territory and, as competitive outcasts, would be unable to join another harem. However, the fish hide within their territories, well out of view of the dominant males while undergoing their premature sex change. It is believed that, even though unable to reproduce, the new males, unhampered by the daily burden of egg production, undergo a rapid growth spurt. This puts them in an advantageous position to take control of harems when opportunities arise.

While moving about their territories, individuals feed periodically as they follow specific paths. For an unknown reason, Rock Beauties spend as much as half their time hiding in one of a dozen or more shelter sites. Two females sharing an overlapping home range typically use the same hiding place. Rock Beauties primarily feed on sponges and algae. They forage at a slow rate, taking fewer than three bites each minute. Even though nearly half of their feeding time is spent scraping algae, stomach analyses reveal that more than 95 percent of the diet is sponge and only four percent is plant material. The large amount of time and effort required to harvest such a small amount of algae indicates that plants, even in small quantities, must play a significant role in their nutritional needs. The remaining one percent of their food is made up of invertebrates and eggs.

Males regularly visit their females during the day and casually forage at their sides without making courtship overtures. About thirty minutes before sunset during most evenings of the year, females move to an established spawning site inside the males' home ranges. These sites are typically associated with

Males regularly develop black faces during courtship.

A male assesses a female's readiness to spawn.

prominent structures such as coral mounds or gorgonian bushes. While the females meander near the bottom, the males, displaying darkened faces, court with increased enthusiasm as daylight fades. Each female is approached with an exaggerated swimming motion, then circled rapidly and repeatedly presented with a lateral fin display. Before spawning, a female often rises several feet off the bottom as if ready to mate; but when approached from below by a courting male, she bolts back to the bottom, followed closely by the eager male, which again begins to circle and display. After one or two false rises a female finally holds still while the male positions himself underneath and pushes his snout firmly against the female's genital region. The pair slowly rises a few feet before suddenly dashing upward, quivering and shedding a visible cloud of gametes. The pair immediately separates and dives to the bottom, where the female chases the male briefly. Males typically spawn with several or all their mates each evening. Spawning activities are completed by 10 minutes after sunset, when females return to their territories and bed down for the night inside coral cover.

Cherubfish

Centropyge argi

Cherubfish

Cherubfish, delightful miniature angelfish seldom exceeding two and one-half inches, are distributed throughout much of the tropical western Atlantic. In the southern Caribbean where most common, the species is found in shallow depths from 15 to 50 feet, but in the Bahamas and southern Florida, they typically live below 80 feet. In the southeastern Caribbean, from Curaçao to Barbados, their range overlaps with a similar-appearing genus member, *Centropyge aurantonotus,* the Flameback Angelfish, that replaces the Cherubfish south to Brazil. In the northern region, Cherubfish typically live in pairs or small groups composed of a large male and female and several smaller, possibly immature, individuals. These groups often make their homes in the rubble piles of Sand Tilefish, *Malacanthus plumieri,* on the sand plains extending seaward from coral-fringed drop-offs. In the U.S. Virgin Islands, harems reside within 5- by 12-foot territories established along coral-shrouded banks. Each social unit includes a single dominant male and from one to four smaller females.

The lovely little blue angelfish with golden trim flit and dart about the bottom, often dodging inside one of their many shelter holes for minutes at a time. They are typically shy and difficult to approach. The species' color pattern varies slightly among individuals but not between the sexes. Juveniles are identical to adults except for a dark spot just above the pectoral fin base. Within harems single dominant males are only slightly larger than the females, but may be smaller than the females in an adjacent harem, leading to speculation that sex change is socially induced in this species.

Females, which show little hostility toward other members of their sex, set up housekeeping near the centers of their small territories. Throughout the day males busily follow established routes, leading from one female to another. During rounds, they occasionally wander out to territorial boundaries to reassert authority. Possibly because harems are seldom established close to one another, males have little reason to display aggression toward neighbors. Most hostility is directed toward small damselfishes that share the Cherubfishes' algal diet. Nearly half of the little herbivores' stomach contents is made up of fine sand particles picked up while grazing during the day in territorial pastures.

Male Cherubfish begin life as females. As with most protogynists living in harems, when a dominant male is lost, the largest female changes sex and takes his place. Spawning occurs each evening throughout the year. As darkness settles, males decrease foraging and pick up their pace of female visitations, following the same paths every two to six minutes. Females wait in or near favorite shelter holes along the males' routes. On initial approaches males do not linger, but if females are hidden, they take the time to search them out. Females are seldom coy, often swimming enthusiastically toward oncoming males, which proudly present them with lateral displays of erect fins and audible chirps, believed to be answered by the females. After brief interludes lasting but seconds, males proceed to their next mates. During subsequent visits, courtship efforts become more intense and prolonged. With fins spread and bodies curved laterally, frenzied males swim in tight circles around females. They frequently stop abruptly in mid-circle with their heads angled upward and flutter their outstretched pectoral and ventral fins. At this stage of courtship, anxious females occasionally pursue the males to their next mates. On rare occasions, two, or even three, females converge at a single site. This so exhilarates the males that they attempt to spawn prematurely but are aggressively interrupted by competing females.

Spawning generally takes place within 10 minutes of sunset. After several courtship encounters and false spawning rises, females, finally ready to release eggs, spread their fins, become blanched and, with paired fins fluttering, rise several inches. Males ascend from below at a 45-degree angle and place their snouts near the genital openings of the horizontal females which extend their pectoral fins downward. The nuzzling males slowly push the females' quivering bodies upward 4 to 12 inches with rapidly beating pectoral and ventral fins. A mass of eggs occasionally can be seen bulging from the females' genital openings. At the top of the rise, males push their snouts into the females, forcing the release of their cargo. Faster than the eye can follow, males flip laterally, position their ventral regions next to the females' and bathe the eggs in sperm. With males leading, the pairs dive back to the bottom. Spent females immediately take cover, while males move on to their next partners. An entire spawning episode, beginning with the females' rise and ending with their rush back to the bottom, takes between four and six seconds; the explosive discharge of gametes requires less than a quarter-second. After completing their reproductive duties, males calmly patrol their territories for a few minutes before retiring for the night.

Females discharge just over 100 eggs with each spawn. The thrust of the tail fins from the pairs' dive for the bottom swirls the gamete mass, aiding fertilization and dispersal. The tiny eggs, only 0.6 millimeter in diameter, are among the smallest planktonic eggs yet recorded for marine fishes. Yolk sac larvae hatch in 30 to 32 hours. Little is known about settlement or the juvenile stage of Cherubfish. From aquarium observations it is believed that adults can live at least eight years.

Basslets

Family Grammatidae

Like most basslets, Fairy Basslets swim upside-down beneath ledges and right-side up when away from shelter.

The Fairy Basslet, *Gramma loreto*, also commonly known as Royal Gramma, is one of the most exquisite fishes inhabiting Caribbean coral reefs. The common, two- to three-inch, violet-and-yellow fish reach their population zenith at depths of 40 to 50 feet, but occasionally inhabit clearwater reefs as shallow as 10 feet, and have been reported at depths of 200 feet. The small planktivores, living in colonies of a dozen or more individuals, hover near high vertical corals, beneath overhangs or just inside coral caves and fissures. Like many cave-dwelling fishes, Fairy Basslets orient their bellies to the ceiling when below ledges. The species ranges from Bermuda south through the Bahamas, to Venezuela and west to Central America but, for some reason, is absent in Florida waters.

Colony members seldom venture far from small well-defended territories encompassing their personal hiding-holes. Small species, including Masked Goby, *Coryphopterus personatus*, and Threespot Damselfish, *Stegastes planifrons*, believed to be potential competitors for their sleeping holes, are frequently chased. Colonies typically consist of one or two large breeding males that each mate with three to nine females. Young males, conserving their energies for rapid growth, do not spawn. Large males assert their dominance over colony members by chasing smaller individuals and occasionally engaging in mouth-to-mouth power struggles with similar-sized males. Although its social system is similar to that of many sex-changing protogynous species, the Fairy Basslet has been determined to be gonochoristic, with individuals remaining the same sex throughout life.

The species is reproductively active from January or February until June and possibly for a brief period in mid-October. To prepare their nests for eggs, reproductive males pluck tangles of algae, which are blown in and out of the mouth, possibly to remove microinvertebrates, then added to cushionlike mats built in the deepest section of small holes or narrow crevices used for nesting. Harvested algae is also used to conceal the entrance ways to nesting holes. Mouthfuls of silt are periodically removed from the nests and spit into open water.

Members of a Fairy Basslet colony feeding on plankton.

On consecutive days beginning just before dawn and continuing for an hour, three to seven females lay eggs in a male's algal nest. The negatively buoyant one-millimeter eggs, clinging together by tendrils and intertwined with algae, form flat, one- to four-square-inch, multi-layered masses. Guarding males hover in front of their nests or rest inside the entranceway with only their heads protruding. If approached, they generally retreat inside and gaze out the opening. On rare occasions approaching fishes are nipped. Based on the behavior of other egg-laying species, in-nest care would include fanning and the removal of infertile, infected or dead eggs. Field-collected eggs raised in well-aerated tanks suffered a high mortality rate, suggesting the importance of parental care. Three- to four-millimeter larvae with silvery eyes and well-developed jaws hatch in 10 or 11 days.

The Grammatidae family, which includes 17 species in two genera (*Gramma* and *Lipogramma*), inhabit the waters of the tropical western Atlantic. Four of the *Gramma*: the Fairy Basslet, the Blackcap Basslet, *G. melacara*, the Yellowcheek Basslet, *G. linki*, and the recently discovered Golden Fairy Basslet, *G. dejongi,* are partitioned by depth, often with as little as a 10-foot overlap between species. Depending on location, the Blackcap Basslet typically displaces the Fairy Basslet at 40 to 100 feet. It, in turn, is supplanted by the Yellowcheek Basslet below 250 feet. In Puerto Rico and Haiti, where the Blackcap Basslet is absent, Yellowcheek Basslet have been collected as shallow as 80 feet.

The Threeline Basslet, *L. trilineatum*, which has been collected between 25 and 200 feet but typically lives in the 60- to 70-foot range is the only species of

In a rarely observed behavior male Fairy Basslets compete for colony dominance.

genus *Lipogramma* expected to be encountered in safe diving depths. The reclusive three-quarter- to one-inch fish customarily swim upside-down near the ceilings of caves. Below 200 feet, the species is replaced by the beautiful, but elusive Heliotrope Basslet, *L. klayi*, one of several genus members that live between 200 and 400 feet. Two new species of *Lipogramma* were recently discovered during submersible dives at 1,000 feet off Curaçao.

Blennies

Blenniidae – Combtooth Blennies, Labrisomidae – Scaly Blennies, Tripterygiidae – Triplefins, Chaenopsidae – Tube Blennies

Tessellated Blenny, family Blenniidae.

Blennies are small, elongate, bottom-dwelling egg layers that live in a variety of habitats and depths ranging from tide pools to deep reefs. Because of their appealing little faces and animated behavior, blennies have become favorites with fish watchers. The large, diverse group has undergone two major revisions and still has taxonomists scratching their heads. Today, blennies are generally separated into four families by anatomical features that can't be easily differentiated underwater. The 19 combtooth blennies, family Blenniidae, living in the Caribbean don't all share the same diet. Many are shallow-water herbivores using a single row of close-set teeth to scrape fast-growing algae and detritus. Others procure algae with long, flexible teeth. The carnivores among the group dine on both benthic and pelagic invertebrates, primarily crustaceans. All combtooth blennies have blunt heads capped with large eyes capable of rotating 180-degrees and functioning independently. The 56 local species of scaly blennies, family Labrisomidae, tend to have pointed snouts indicative of invertebrate eaters, in this case crabs, shrimp and copepods. Triplefin blennies, family Tripterygiidae, true to their common name, have three-part dorsal fins. Although they flit about in the open on dock pilings and coral mounds in search of crustaceans, primarily amphipods and copepods, the nine small Caribbean species, all in genus *Enneanectes*, are easily overlooked because of their size. The 50 members of the exotic, hole-dwelling

Fleshy, single or multi-branching pairs of filaments, known as cirri, crown the heads of blennies. Their size and structure varies between species, and occasionally within a species. Longhorn and Medusa Blenny (above). Seaweed Blenny (below).

Blennies

Chaenopsidae family are famous for dorsal fins designed for seduction and mouth-to-mouth combat. These long-jawed carnivores consume a variety of crustaceans from open water and the seafloor. Males of several species in the family have been found to have a surprising number of eggs in their guts. It is possible they came from the guardians' own nests, either by the routine removal of non-viable eggs, or to supplement their diets during incubation.

Fleshy, single or multi-branching pairs of filaments, known as cirri, crown the heads of blennies and to a lesser degree their snouts. These trademark appendages, believed to disrupt head profiles from the keen eyes of predators, come in many sizes and shapes. Blennies can be easily distinguished from similar-sized, bottom-dwelling members of family Gobiidae. Gobies can be differentiated by their straight bodies, two-part dorsal fins and lack of cirri, which contrast with the blennies' flexible postures, single dorsal fins (except for Tripterygiidae) and the presence of cirri.

It typically takes an effort to find blennies, but they're around.

Glass Blenny male—normal color

Glass Blenny male—courtship color

Hairy Blenny male—normal color

Hairy Blenny male—courtship color

A male triplefin displaying a yellowish head indicative of courtship.

Comprehensive field surveys in Belize and Honduras identified 60 species hiding out in a patchwork of habitats. Each species was found to live almost exclusively within one of 11 habitats—drop-offs, spur and groove reefs, shallow forereefs, patch reefs, rubble/algae, tidepools, mangroves, shallow seagrass, deep seagrass, mainland rocks and offshore rocks. The largest number was associated with live coral in shallow surge-swept forereefs.

Many blennies are quite adept at changing colors and patterns for camouflage, and conveying intra-species communication during reproduction. One of the most radical examples is the courtship wardrobe worn by male members of the Hairy Blenny complex, *Labrisomus* spp. These large, shallow-water, lie-in-wait carnivores that inhabit rocky coastlines can reach an impressive nine inches in length. Like most blennies, the males establish nest sites inside sheltered areas that they vigorously protect from rival males and egg predators. During reproductive periods, the ordinarily dark males exhibit bright orangish-yellow heads and bright white bodies marked with black bars. The heads of males in genus *Emblemariopsis*, which includes glass blennies and banner blennies, turn jet black during courtship and spawning. The heads and forebodies of reproductively active triplefins acquire a yellow cast.

Blennies have large eyes capable of rotating 180-degrees and functioning independently.

Left: *Glass Blenny peering out from crowns of Christmas Tree Worms.* **Above:** *Male Seaweed Blenny guarding eggs.* **Below:** *A pair of spawning Molly Millers.*

Spinyhead Blenny pops out of its shelter tube to nab a passing speck of zooplankton.

Chaenopsidae – Tube Blennies

Tube blennies live exclusively in the Western Hemisphere. The family is made up of many delightful characters, including nine tiny, animated hole-dwelling planktivores in genus *Acanthemblemaria* as well as the free-ranging, fish-eating Arrow Blenny, *Lucayablennius zingaro,* and the stealthy Wrasse Blenny, *Hemiemblemaria simula,* that mimics the behavior and coloration of the initial phase Bluehead, *Thalassoma bifasciatum.* Last but not least, the high-flying Sailfin Blenny, *Emblemaria pandionis,* and flamboyant, spaghetti-thin pikeblennies round out one of the most fascinating groups of fish to be found.

Acanthemblemaria, that seldom reach more than an inch in length, make their homes inside abandoned worm tubes in dead and living corals. The Spinyhead Blenny, *A. spinosa,* and the Roughhead Blenny, *A. aspera,* are two of the most common and widely distributed genus members. Each species prefers to live in colonies to ensure an adequate supply of mates. Often mounds of dead reef rock and concrete mooring blocks are so peppered with protruding blenny heads that the sites are fondly referred to as blenny condos. Every few minutes, and even more frequently when currents run, the tiny fishes pop partially out of their holes to nab passing specks of floating zooplankton, primarily copepods. On rare occasions they dart across the rock to grab tiny bottom-dwelling crustaceans passing nearby.

Individuals regularly attempt to displace neighbors occupying larger, more favorable holes higher on the coral rock where planktonic food is more plentiful and the threat of predation is less. With dorsal fins extended, challengers grab the heads of hole-owners in their jaws and spin wildly. These ferocious takeover attempts, often lasting several minutes, are seldom successful but appear to be fundamental to the social structure of densely clustered communities. Females, that are smaller and less numerous but otherwise similar in appearance to males, lay eggs inside the males' shelter tubes. After hatching larvae spend an extended stint in offshore waters. Recently settled juveniles take cover in small crevices until they can find a vacant hole associated with a colony.

To study the nature of *Acanthemblemaria* communities, researchers removed residents, transplanted individuals and drilled additional holes in existing colonies. They found that the larger individuals dominate the more spacious worm tubes near or at the top of the colony, while underlings competed for positions at lower levels. It is believed that the larger serpulid and worm snail tubes provide more space for the males' egg nests than holes created by other organisms. The studies confirmed that both sexes inhabit dead Elkhorn Coral, *Acropora palmata,* in nearly equal numbers, while far more males occupy live coral colonies. When superior holes became available, residents participated in their own version of musical holes by each resituating in slightly larger homes. Vacated holes near the lower boundaries of the communities were quickly filled by immigrants leading to the conclusion that a surplus of "floaters" lived nearby in less desirable locations.

Within two weeks following the removal of 11 of the 22 resident Spinyheads inhabiting a coral head, three moved to larger holes and six new blennies immigrated to the site. The new blennies averaged three-quarters of an inch in length, compared to the one-inch average size of removed residents. When

Colony of Spinyhead Blennies find refuge inside abandoned worm holes.

all 15 blennies were removed from a second coral head, only five new fish moved into the vacated colony during the same time period, indicating the species' preference for saturated colonies.

Possession of a shelter hole appears to be necessary for survival. When researchers added 30 new Spinyheads to three colonies lacking vacant holes, only three were able to acquire a shelter tube. Within an hour, Redspotted Hawkfish, *Amblycirrhitus pinos,* and lizardfishes consumed 10 of the homeless arrivals. To test the Spinyheads' homing instincts, 11 individuals were moved 16 feet from their homes. Seven reoccupied their prior holes within two days. However, when a number of individuals were moved 65 feet from their homes, none made it back to their previous holes.

Most chaenopsid blennies live in specific habitats, typically within a restricted depth range. In the U.S. Virgin Islands, Spinyheads inhabit depths between 12 and 65 feet, whereas the majority of Secretary Blennies, *A. maria,* live in shallow surge-prone rocks, often only inches below the tide. Occasionally Roughhead Blennies co-occur with Spinyheads at the same depths and, at times, even share the same coral rocks. In these joint habitations, Spinyheads occupy the upper strata with Roughheads relegated to lower positions. After researchers removed the Spinyheads from shared rocks, the Roughheads moved into the preferred upper holes. In Belize, similar microhabitat partitioning occurs between Spinyheads and Stalk Blennies, *A. greenfieldi,* with the larger Stalks occupying the horizontal surfaces and Spinyheads living below.

The Arrow Blenny breaks somewhat from its family's hole-dwelling tradition, spending most of the day hovering inside shallow undercuts at the

Spinyhead Blennies compete for preferred shelter tube.

An Arrow Blenny with a captured shrimp.

base of reefs in depths from 40 to 100 feet. Either as solitary individuals or in small, undefined groups, the splendid, one- to two-inch fish are generally found hovering beneath ledges with their favorite food, Masked Gobies, *Coryphopterus personatus*. With their tails perpetually cocked for immediate thrust, the blennies move through the water slowly in a start-and-stop fashion, powered by pectoral fins. When within four to five inches of prey their tails spring straight, shooting them toward their targets. The slender blennies have been seen consuming gobies half their length. One goby was struck with such force that the temporarily stunned prey drifted toward the bottom, where it was seized mid-body, repositioned and swallowed headfirst. More in tune with other genus members, Arrow Blennies occasionally rest in holes or small sponge openings with only their heads exposed. The Wrasse Blenny, another free-swimming chaenopsid, is discussed in the mimicry section of the Symbiosis chapter (page 144).

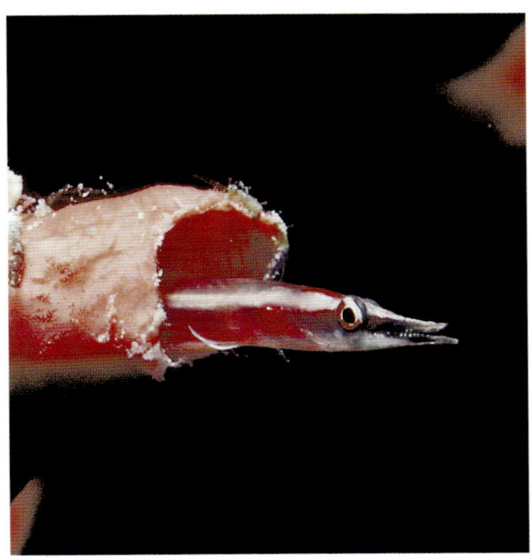

An Arrow Blenny resting in a sponge.

Male Yellowface Pikeblenny reacts aggressively to his reflection in a mirror.

Yellowface Pikeblenny

Chaenopsis limbaughi

We made our first sighting of soda-straw thin pikeblennies while lying on the sand searching for signs of life. An unfamiliar movement off in the distance near a stand of Manatee Grass, *Syringodium filiforme,* caught our attention. After inching forward several feet, we spied the well-camouflaged Daffy Duck profile of a Yellowface Pikeblenny staring back from the hollow end of an abandoned worm tube. Every now and then, the wispy two- to three-inch blenny vanished inside the tube, only to reappear seconds later. After that first encounter it was easy to understand why these rather common little fish are so difficult to locate. Soon the blenny ventured out of its tube and swam a few inches away where it pecked in the sand for buried invertebrates and snapped at clear specks of zooplankton spiraling past overhead. After a minute, the little predator scampered back to its tube, turned, located the opening with its finely finned puff of a tail and slipped inside backwards, once again leaving only its head exposed. The pike blennies' surprisingly large dorsal fins usually lay collapsed along their slender backs. To provoke a reaction we placed a hand mirror inches from the hole. The cartoonish fish instantly underwent a radical attitude adjustment ascending from its tube like a cobra from a basket. With its dorsal fin spread and gill covers flared, it arched forward and swayed from side to side. In a flash, the blenny shot forward and confronted its image in fierce mouth-to-mouth combat. The instant the mirror was removed the blenny calmed and snaked its way back to its shelter tube where it slipped out of sight.

Unfortunately, we made our pikeblenny find in June on the last day of our stay in Bimini, Bahamas. Adding to our dilemma, the blenny was in 55 feet of water, greatly limiting observation time. When we returned the following spring, pikeblennies topped our underwater agenda, but after two one-hour searches, none were found.

Back in Bimini in May, we hit the jackpot, locating 13 Yellowface Pikeblennies spaced randomly about the sand along the edge of the seagrass bed exactly where we had sighted the lone individual the previous year. Each of the five males and eight slightly smaller and less ornate females rested with their heads and bodies extending at various heights from parchment-like worm tubes. During frequent visits to the site during the summer, we discovered that the colony was far more extensive and individuals more mobile than expected.

Out of their holes the fish appear delicate, vulnerable and easy targets for fish-eating predators that prowl the seagrass and nearby sand flat. Fortunately, the keen-eyed blennies were well aware of their vulnerability. When Yellowtail Snapper, *Ocyurus chrysurus,* or Bar Jack, *Caranx ruber,* patrol nearby, the pikeblennies remain inside their tubes. However, when the coast was clear they were regularly out and about, often making long forays that would take them 10 to 20 feet from their original shelters. During outings they swam in eel-like spurts just above the bottom, expertly blending with their surroundings. When danger appeared, they always seemed to know where to find an unoccupied shelter tube nearby. Occasionally the slender carnivores perched on algal tops to survey their surroundings. On one extended excursion a female stalked and captured a juvenile razorfish feeding at the edge of the seagrass. The species consumes primarily planktonic copepods, but unidentified fish remains, shrimp larvae, shrimp and a single polychaete worm were found during a feeding study examining stomach contents.

We quickly discovered that the most rewarding pikeblenny watching took place from mid-to-late afternoon. During these hours, the females swim free in search of food and a bit of social life. And when the gals were out, the guys went nuts. Whenever the sand-colored females were on the move ever-watchful males, which make it their business to know such things, flush black, brighten their yellow beaks, pop their impressive dorsal fins taut and bob up and down while maintaining contact with their shelter tubes. If a wandering female comes anywhere near, anxious suitors soar from their tubes in graceful, high-flying 8- to 10-inch arches. The most enthusiastic remain aloft for seconds at

Male Yellowface displaying.

a time. Unhappily, their acrobatics are seldom rewarded. We never saw a female approach a displaying male but, according to aquarium observations, an obliging female simply follows a male into his tube and quickly deposits eggs for the male to fertilize and tend. The eggs hatch within days. Weeks later the larvae settle back to the sand and metamorphose into transparent half-inch juveniles.

Throughout the summer additional pikeblennies appeared until there were well over 50 individuals living in the colony. As the numbers increased confrontations between residents began to erupt at an increasing rate. For a few weeks, the blennies were in such a high state of agitation that even our outstretched fingers elicited attacks from males. When the females were on the move the dorsal fins of males unfurled like battle flags across the sand. Gray fins proclaimed aggression, black defense. It never became clear to us whether individuals were competing for dwindling food resources, favored mates, choice shelter tubes or all of the above, but for almost a week in August, afternoons were packed with antagonistic encounters.

The first altercation took us completely by surprise. While we were watching a large male frantically attempt to attract the attention of running females, a second, equally large male made a beeline for the boldly dancing blenny. The defender instantly arched from his hole to counter the onrush. For several seconds the pair clashed in a mouth-to-mouth slinging match before the attacker retreated to a shelter tube two feet away from a third, smaller male that also was busily attempting to lure a female his way. Like lighting the troublemaker charged, bowled the youngster backwards, and flung him out of his shelter tube. Somehow during the ensuing melee the smaller blenny regained his tube and hunkered down while the attacker, poised and ready for round two, glared from inches away. In a brazen show of misplaced ego, the small blenny popped from his hole, spread his black dorsal fin in defiance and danced in his rival's face. The assailant pounced, sending the upstart scrambling deep inside his tube. Nearly a minute passed before the rounded tip of the yellow beak of the blenny reappeared. Without waiting to see more, the attacker dived into the tube, forcing his way deep inside with a lashing tail. Unable to turn within the narrow chamber, the intruder looped his body, inserting his tail into the hole for leverage, which allowed his head to slip back into the open. Three times the hole-owner reappeared; three times the large blenny drove him back before finally retreating to a nearby shelter tube to catch his breath. Moments later, the embattled youngster poked his head out once again and peered about tentatively. Failing to see his assailant nearby, he doubled down unfurling his blackened dorsal into its full glory and began displaying with even more exuberance. Enraged once again, the attacker rose from his tube. Even this slight bluff was sufficient to send the defender inside his tube. But with no assault forthcoming, he reappeared and continued his taunt. This was more than the larger blenny could endure. In an attack faster than our eyes could follow, he arrived at the site in a cloud of sand just as the upstart once again disappeared from view. With time running out, we reluctantly lifted off the 55-foot bottom and drifted in the direction of the mooring line while keeping our eyes glued to the ongoing action. The following afternoon we found the little blenny perched high in his hole, intently scouting the terrain for running females as if nothing had happened. His brawny rival was nowhere to be found.

After nearly four months of weekly observations, we left Bimini at the end of August. When we returned two weeks later the water temperature was at its annual peak, and with the heat came patches of stringy red algae. We found the number of residents cut in half. The females were especially hard hit. Although the remaining males occasionally displayed, it was obvious that the colony's vim and vigor were gone. During the two years that we randomly surveyed the pikeblenny colony population numbers increased for a few months then when into decline. We also noticed a similar trend in a colony of closely related Sailfin Blenny, *Emblemaria pandionis,* (discussed in the next section).

A high-flying male Yellowface Pikeblenny attempts to attract a mate.

Male Yellowface Pikeblennies compete for a shelter hole close to females.

The only published study describing the ecology or behavior of pikeblennies involved a 13-month program monitoring a population of Orangethroat Pikeblennies, *Chaenopsis alepidota,* in the Gulf of California. They found that Orangethroats are quite mobile and that low-density, noncombative populations, rather than high-density, territorially aggressive colonies, are the norm. There is no mention made of sudden increases of females and, contrary to our observations, the Orangethroat Pikeblennies spent more time foraging in the morning than in the afternoon.

A Yellowface Pikeblenny reacts to encroaching finger of a diver.

Sailfin Blenny

Emblemaria pandionis

Our attention was so intently focused on the tiny, dark head protruding from sea floor that we hardly took notice of the Spotted Eagle Ray, *Aetobatus narinari,* patrolling the reef line overhead. We were waiting for an equally impressive wildlife show from a fidgety two-inch male Sailfin Blenny balancing on elongate ventral fins at the edge of its hole in the seafloor less than three feet from our face masks. Our presence and thundering bubbles seemed to have no effect on the typically wary fish; at that moment, in the late afternoon, he had nothing but sex on his mind.

Without warning, the cryptic, crustacean-eating bottom-dweller shot skyward in a flurry of fins instantly transforming from sheepish little blenny into a high-flying Casanova, vying for the affections of a female perched a few feet away. As gallant as his efforts were, the little Sailfin needed all the moxie he could muster. Before he could settle into his hole a neighboring male soared up from the bottom with his great dorsal fin spread. Farther away, across the limestone plain, other rivals began popping up, seemingly out of nowhere.

That first encounter with the small, charismatic blennies provided us with a search image that made it much easier to locate Sailfins in other shallow inshore habitats we later surveyed in Florida and the Caribbean. Most individuals spend the better part of their day inside small holes in the limestone bottom formed by worms, bivalves and snapping shrimp. At other locations we found colonies established in sand and rubble patches where individuals occupy the hollow ends of dead Staghorn Coral, *Acropora cervicornis,* rubble. Even though we spent hours observing the Sailfins we never tired of their entertaining antics. However, the more we watched their comings and goings the more perplexed we became by their behavior. Our initial survey, in Bimini, Bahamas, located 28 resident blennies living inside a 50- by 75-foot plot of algae-tufted hardpan terrain. The majority appeared to be males, which could be identified by their larger dorsal fins and dark nuptial appearance. We noted that although sporadic display jumping began at dawn and continued throughout the day, most displays occurred between 4:30 p.m. and dusk. If females were holed-up nearby or on the move, males

Male Sailfin resting inside shelter hole.

launched a series of jumps that carried them from three inches to two feet above the sea floor where they hovered for seconds before suddenly collapsing their fins and diving back into their shelter holes.

It soon became apparent that even though most individuals tended to occupy a favorite hole, some movement was going on. An explanation for a few changes of address became apparent late one afternoon when we saw a Sailfin making a mad dash across the seafloor. The little fellow looked vulnerable and out of place in the open. Once out of his hole his color changed from black to light brown to better match the bottom, and the magnificent dorsal lay awkwardly collapsed across his back. A round, rapidly fluttering tail fin propelled the slender body just above the bottom. This was our first observation of what we later named runners. After several minutes of crazy zigzags the runner arrived at a hole occupied by a slightly smaller male.

With dorsal fins flared and mouths agape, both the intruder and defender paled gray with rage. No quarter was given. Sand and algae flew as the adversaries locked jaws and slung each other about. After a brief pause, not even long enough for sand to settle, another attack was underway. But this time the defender made a quick move to the side, slipped past the interloper and backed into his hole. Undeterred, the aggressor plunged in after him driving both Sailfins out of sight except for the tip of a thrashing tail. With a puff of sand, they reappeared. Another charge, then both disappeared inside the hole once again. By the time they reappeared it was obvious that the tide of battle had turned.

The smaller blenny made one last half-hearted lunge before retreating to the top of a nearby algae clump where he perched, panting for some time before slipping away. The victor backed into the repossessed hole, turned jet black and raised his dorsal in triumph. The loser tested several nearby depressions before settling into a hole in the bottom a few feet away and began housekeeping chores. After his home improvements were finished, his dark color gradually returned and soon he was making display jumps as if nothing had happened. Of the several clashes we witnessed between runners and residents, this was the only successful eviction observed.

A female Sailfin approaches the shelter tube of a male. **Right:** *A male Sailfin display jumping.*

Female Sailfins clash in a territorial dispute.

The Sailfins did not seem to spend a lot of time feeding. Occasionally an individual dashed across the bottom and appeared to grab something off a nearby algal bush. One male shot straight up over four feet and nabbed a speck from open water. When we arrived at the scene, he held a shrimp crosswise in his mouth. Although at first we had a bit of difficulty distinguishing the females from the males, we soon learned to make distinctions. Male Sailfins when perched in their holes are nearly always black. The females, are mottled brown while both inside and outside their shelter holes, and have smaller, but still prominent, dorsal fins marked with a series of thin, broken diagonal lines. Even with these guidelines we were frequently uncertain of an individual's sex when distinguishing small males and male runners that instantly switch on their camouflage coats when out of their holes.

The true Sailfin gender test came when we placed a hand mirror in front of a blenny. Surprisingly, males that fought so valiantly among themselves seldom showed interest in their reflections; however, females went ballistic. Their color paled and gill covers flared as they lunged from their holes and darted along the mirror with their gaping mouths pressed against their image. Seconds later, the befuddled females briefly settled back inside their holes for a breather. Then out they came for another round until once again running out of steam. Occasionally females ran off the end of the mirror and stopped dead in their tracks at the sudden disappearance of their imaginary rivals. Females occasionally left their shelter tubes and moved around through the colony. As they prowled the bottom, nearby males danced like demons, but to no avail. The females had eyes only for trouble, and it came each time another female was encountered. After furious but brief clashes, they moved on only to engage in similar scuffles each time they approached a female-occupied shelter hole.

Even though we made several dawn dives when mating has been reported to occur, we never observed anything even remotely resembling foreplay. To the contrary, the brief encounters between males and females we witnessed were more along the lines of professional wrestling than romance. Finally, after weeks of anticipation, when we had begun to doubt there would ever be a new generation of Sailfins, it happened.

Sailfin with a captured shrimp.

During the last light of dusk, we spotted a rather small male sitting high in his hole, excitedly fluttering his great fin. As suspected, we found a female perched nearby. After several minutes, the female approached the dancing male. We expected a scuffle, but to our delight the male moved aside, allowing the female to disappear inside his shelter tube. The male immediately followed. After several seconds his dark head reappeared. He sat briefly in the opening, as if in a daze, before moving clear of the entrance. The female immediately popped out, and without even a backward glance, scampered off into the approaching night.

Unfortunately, after our two summers monitoring the Sailfin sanctuary, we had more questions than answers about the blennies' nature. For instance, why on earth do males tend to ignore their reflections in the mirror, but won't tolerate the approach of another male? And exactly what were the prowling females up to? But our biggest puzzler concerned display jumping. What little we gleaned from the few mentions of *Emblemaria pandionis* in scientific literature their energetic signal jumps are only employed to entice breeding females inside their shelter holes. But after our brief time among the Sailfins, it appeared to us that the male's most spirited displays were directed at rival males more often than females. But for whatever reason, the soaring jumps, that can last for seconds and occasionally carry the blennies more than two feet above their holes, are one of the most delightful behaviors observed by divers.

A runner attempts to evict a combative hole-owner.

Redlip Blenny—pale phase

Redlip Blenny—reproductive colors

Redlip Blenny—juvenile

Redlip Blenny—settling size one inch

Redlip Blenny

Ophioblennius macclurei

Redlip Blennies, common residents of shallow fringing rock reefs and coral crests, spend their days grazing on filamentous algae and detritus within their small territories. The charismatic, two- to four-inch, chocolate-brown fish with blunt heads and large red lips also come in pale morphs with shell-white bodies and reddish brown heads. Two decades of research in Barbados have extensively documented the social structure and behavior of the island's abundant Redlip population.

Males and females, represented in equal numbers in a given area, are indistinguishable except during reproductive periods when nuptial males display a conspicuous white and brown barred mask. Both sexes patrol and defend territories averaging five square feet during February and March, when the island's population numbers are low. Following heavy recruitment in May territories shrank to half the previous size. Individuals seldom venture far from their personal patches of reef rock for fear of predation or the possibility of losing territory to competitors while away. A persistent series of nips usually oust intruders.

Males prepare nesting sites within their territories by removing sand, bits of rock and debris from protective crevices. Reproduction peaks a few days before and after the full moons of spring, but spawning probably occurs year-round. The three-hour courtship and spawning periods begin just after sunrise when females leave their territories to lay eggs in the nests of preferred males. These treks can take ripe females up to 16 feet from the safety of their territories —a substantial distance in blenny terms. While females are on the move, eager males hover briefly in front of their nests with their fins spread and nuptial pattern on display. Females arriving at the males' boundaries elevate their bodies at a 45-degree angle to exhibit their ripe, swollen bellies. As many as three females may display before a male at the same time. After giving a quiver, the males dodge inside the cover where their nests are located.

When males find females to their liking, they approach with their fins erect and peck them on their heads before escorting the chosen females to their nest sites. The females disappear inside and begin moving back and forth across the walls while laying a batch of adhesive eggs. From time to time the males dart inside and fertilize the eggs in a similar manner. During the 15- to 20-minute egg-laying episodes the males periodically dash outside, and display an exaggerated S-posture. If another female approaches while spawning is underway, males chase the latecomers away. After finishing, the females may attempt to spawn with other males but usually swim directly back to their territories, provoking chases by territory-holders as they pass.

During each multi-day reproductive period, males receive from zero to 21 clutches, most frequently from multiple females. On some mornings, as many as three females laid eggs in the same nest. Brooding males repeatedly fin-fan and pick at the eggs, but their biggest concern is guarding their nests from roving packs of egg-predators, such as young Bluehead, *Thalassoma bifasciatum.*

Both female and male Redlip Blennies prefer large spawning partners. It was found that females approach the nests of larger males more frequently, spawn with them more often and deposit more eggs in their nests. By chasing away smaller females and accepting only the larger more prolific egg-layers, males were calculated to increase their reproductive success. In turn, females select for larger suitors that spend more time guarding their nests than smaller rivals.

The spherical, one-millimeter eggs hatch during the night of the fifth day after fertilization—one of the shortest incubation periods known for blennies—and rise toward the surface where they are carried to the open sea on outgoing tides. Approximately 45 days later the surprisingly large, three-quarter-inch, still-translucent settlers arrive at night. Within days they transform into cryptic, rarely seen, juveniles displaying mauve bodies overlaid with thin white bars. In Barbados, 90 percent of recruits arrive in May. Interestingly, the researchers were unable to find Redlip juveniles outside areas already colonized by adults demonstrating the recruits' ability to locate and settle where they had an increased chance of survival. The secretive juveniles attempt to establish territories within established colonies, even though their efforts are met with attacks from residents. Those unable to secure territories quickly fall victim to predation. Even so, the species' 66 percent rate of survival at the end of the first month following settlement is quite high when compared with recruits of other species. Surviving Redlips become sexually active about two months later, at a length of two inches.

Butterflyfishes

Family Chaetodontidae

Banded Butterflyfish regularly feed together in, what are believed to be, mated pairs.

Butterflyfishes generally flit about the bottom in pairs, using keen eyesight and protruding terminal mouths lined with fine teeth to pluck dinner from the substrate. The sexually indistinguishable mated pairs, which live together inside large home ranges, are believed to be companions for extended periods, possibly for a lifetime. Territorial defense is seldom evident except late in the day, close to the courtship period, when the customarily mild-mannered fishes occasionally chase neighbors.

The Caribbean is home to only seven butterflyfishes from a worldwide count of 129 species. The remaining species reside in the tropical Pacific and Indian Oceans. Although the Caribbean butterflyfishes primarily eat the crowns of polychaete worms, and polyps from gorgonians and zoanthids, each species tends to have its own favorite foods. The Foureye Butterflyfish, *Chaetodon capistratus,* ranges widely over the reef platform, compulsively nipping at nearly every bottom structure it passes. This common species is particularly fond of gorgonian polyps. Banded Butterflyfish, *C. striatus,* spend their day harvesting exposed polychaete appendages on shallow reef crests and patch reefs. Reef Butterflyfish, *C. sedentarius,* and Spotfin Butterflyfish, *C. ocellatus,* live slightly deeper, generally near high-profile reefs, where they dine on tubeworms, hydroids and small crustaceans, including shrimp. The protracted snouts of Longsnout Butterflyfish, *Prognathodes aculeatus,* are used to nip the exposed radioles of Christmas Tree Worms, *Spirobranchus giganteus,* the tube feet of echinoderms and various small crustaceans. Though usually found on deeper drop-offs and complex coral structures from 40 to 200 feet, in the southern Caribbean, Longsnouts often live as shallow as 20 feet. The rarely sighted Bank Butterflyfish, *P. aya,* seldom found in less than 100 feet of water, ranges from eastern Florida to North Carolina and the Gulf of Mexico. The Threeband Butterflyfish, *P. guyanensis,* inhabits Caribbean reefs below 200 feet.

Butterflyfishes also consume bits of tunicates and eggs. During annual coral spawns Foureye Butterflyfish and Banded Butterflyfish flock around spawning coral heads, gobbling down fat-filled gamete bundles as they emerge. We have observed Foureyes greedily pulling gametes from day-spawning brain corals as they were being released.

Butterflyfishes depend on their laterally compressed bodies and intimidating arsenal of dorsal spines to deter predators. Dark, eye-concealing bands and false eyespots, common on juveniles and the Foureye, help to misdirect attacks. Although they are slow swimmers that seldom seek shelter, their defenses must work well—few butterflyfishes have been found in the stomachs of piscivores. As night falls, most adults bed down deep within protective structures. Only the Foureye, a restless sleeper that develops a vague pattern of bars at night, is regularly sighted after dark.

Little is known about the spawning behavior or reproductive seasons of the Caribbean butterflyfishes. The majority of reported spawnings occurred at sunset during the cool-water months from February until May, but a few ripe females and sporadic courtship behavior have been noted in the fall. (Noting the swollen bellies of egg-laden females is the only way to visually distinguish between the sexes.) Only isolated pair spawning events between what appear to be monogamous pairs have been documented to date. During reproductive peaks Longsnouts are believed to spawn every other day at alternating sites within their home ranges. Ten minutes before sunset, a pair, or occasionally a male with two mates, rendezvous near a prominent feature rising from the sea floor. Courtship is brief and unspectacular. While slowly rising, the male laterally circles the female with his snout caressing her tail fin. Just before gamete release, the female pauses in a head-up stance while the male positions his snout next to her vent from below. After simultaneous release, the male moves forward,

Courting Longsnout Butterflyfish.

Foureye Butterflyfish dine on the gamete bundles of spawning brain coral.

leaving the female and a gamete cloud behind. When two females are involved, the male spawns with his second partner less than a minute after spawning with the first.

On three consecutive evenings, a researcher observed a savage wound-inflicting rivalry between Longsnouts when a solitary male challenged a pair attempting to court. The paired-male, at times assisted by his swollen partner, repeatedly attempted to drive the stranger away before the males finally squared-off inches apart. With their sharp foredorsal spikes directed at one another, the fish attacked repeatedly driving needle-sharp spines into their opponent's fins and body until both fish, showing obvious injuries, retreated. The next evening the trio arrived at the same site, but the female, minus her bulging abdomen apparently was not worth fighting over. On the third evening, with the female once again ripe, it was war. After another bloodletting the challenger abruptly departed, allowing the pair to spawn unmolested.

Pairs of Foureye Butterflyfish have been observed spawning on alternate evenings in much the same manner as Longsnouts. However, their spawning rise carries them somewhat higher above their spawning landmark and, instead of explosively releasing clouds of gametes, the eggs discharge in a thick stream. In regions with concentrated populations Foureye Butterflyfish may also participate in group spawns. On numerous occasions, we have observed up to a dozen individuals moving about the reef together in Bimini, Bahamas. Similar gatherings also have been reported in Puerto Rico, Jamaica and the Cayman Islands. The lively groups swim rapidly about in loose formations, nipping, pecking and chasing. From time to time pairs and foursomes break away, only to rejoin the constantly moving assemblies later. This gregarious behavior seems to be most prevalent in late afternoon shortly before the traditional pair-spawning period, but we have also seen groups congregate in the morning.

The less frequent reproductive activities of Banded Butterflyfish were observed by researchers once in February and again in March in Puerto Rico. Two solitary pairs, each with a visibly swollen female, swam steadily about the bottom of a large area, gradually moving closer to one another as darkness settled. The males occasionally brushed their mates' tail fins with their snouts. Neither pair associated with a raised bottom structure, and both made several false starts before finally ascending more than 20 feet above the bottom to spawn. The Spotfin Butterflyfish and Reef Butterflyfish have not been observed spawning. A partially distended Spotfin female was captured at dusk in May in the Bahamas. Her eggs were found to be only partially hydrated, leading to speculation that the species might spawn in the early morning.

Depending on the size of the female, several hundred to a few thousand eggs are released during each spawn. The buoyant, spherical eggs, less than one millimeter in diameter, hatch into tiny larvae with large yolk sacs between 24 and 48 hours later. Three days after hatching pigmented eyes are evident, and the yolk is gone. As the larvae mature they develop deep, laterally compressed bodies and a series of clear armorlike plates covering their heads and forebodies. The amount of time the different species remain pelagic has not been determined.

The button-sized larvae settle to the sea floor at night and make a rapid transformation into juveniles. Most juvenile butterflyfishes are creamy white with yellow fin highlights, distinctive ocellated eyespots and barred patterns. The skittish youngsters generally inhabit shallow habitats, taking up residence in sand pockets, around isolated coral and sponge formations, in seagrass beds and among mangrove roots. They lead solitary lives until they reach sexual maturity in approximately one year at a length of about four inches.

The purpose of Foureye Butterflyfish aggregations is not understood.

Damselfishes

Family Pomacentridae

The threat of nips from a Threespot Damselfish turns away a Queen Angelfish.

Any diver who has explored a Caribbean reef is well aware of the disagreeable disposition of damselfishes. Several species of the palm-sized fishes are driven by a compulsion to attack everything that comes near their personal pieces of the reef, including divers. Fortunately, their tiny mouths lined with fine teeth are designed for nipping algae, and even their worst bites fail to leave a mark. Of the 15 species from the tropical western Atlantic, only four are primarily responsible for giving the family its infamous reputation, though every damselfish is territorially assertive to some degree, especially males guarding egg nests. Three characteristics define family membership: a single nostril hole on each side of the snout rather than the usual two, scales that extend onto the fins, and an interrupted lateral line. Over 390 damselfishes have been classified worldwide. While the majority inhabit the tropical reefs of the Indo-Pacific, a few species have invaded temperate waters. Males and females of every species are difficult, if not impossible, to distinguish visually except during breeding periods when one or both sexes, but usually males, exhibit distinct spawning patterns and behavior.

Caribbean damselfishes are divided into four genera: *Stegastes*, the largest group, represented by seven species; *Abudefduf,* with two species, the Sergeant Major, *A. saxatilis,* and Night Sergeant, *A. taurus, Microspathodon,* which includes only the Yellowtail Damselfish, *M. chrysurus,* and the semi-pelagic, plankton-picking *Chromis,* with five species, including the slender

Brown Chromis, *C. multilineata,* and Blue Chromis, *C. cyanea.* Four species, the Dusky Damselfish, *S. adustus,* the Longfin Damselfish, *S. diencaeus,* the Threespot Damselfish, *S. planifrons,* and the Yellowtail Damselfish, farm and defend permanent territories of filamentous algae that surround their centrally-located hiding holes. These are the notorious nippers, and there is good reason for their aggressive nature: the reef is rife with algae-eating parrotfishes and surgeonfishes that covet the damselfishes' well-kept, fast-growing lawns. Even wrasses and other bottom-pickers will, if given the chance, gladly savage a damselfish garden for the wealth of tiny invertebrates living within the tangled algal strands. The bulk of the damselfishes' diet does not come from eating the lawns, but from the garden of tiny epiphytic cyanobacteria, known as blue-green algae, and associated detritus that clings to the filaments. Incidental animal matter, taken while grazing, is believed to supply the remainder of their nutritional needs. The hyper little fishes often go to such lengths protecting their turf that a four-inch defender will chase a three-foot parrotfish six feet into the water column. However, Blue Tang, *Acanthurus coeruleus,* one of the reef's most

Threespot Damselfish—genus Stegastes

Sergeant Major—genus Abudefduf

*Yellowtail Damselfish—*Microspathodon

Brown Chromis—genus Chromis

The damselfishes' luxuriant turf of filamentous algae supports a secondary garden of appetizing diatoms and other epiphytes.

abundant algae-grazers, have learned to penetrate their defenses by feeding in large, swarming groups, often including a variety of other bottom-feeding species. Algae-farming damselfishes hover in horror as the mob of snapping jaws makes it way down the reef line. When under siege, the overmatched tenant dashes about frantically, biting first one invader and then another until it gives up, moves to the side and watches its prized crop being ravaged. In less than a minute all is over as the interlopers move on to their next victim.

Never-ending border wars played out daily between homesteading damselfishes and free-ranging herbivores create a mosaic of heavily matted oases surrounded by rock scraped bare of filamentous algae. Defended territories are stable through time, with one generation after another tending the same sites. The damselfishes' thick algal mats prevent the growth of coralline algae and corals—organisms indispensable for bonding the reef's framework. Over the years, the reef rock underneath loosens and crumbles from the constant forces of erosion. At the same time, the lush growth is responsible for much of the reef's productivity through nitrogen fixation, a process necessary to form proteins in marine animals and plants, including zooxanthellae, the symbiotic algae essential for coral growth.

The ubiquitous Sergeant Majors and the rarely observed Night Sergeants also consume plant material, but they, like most family members, do not make their living by defending feeding territories. Night Sergeants live exclusively in rocky inshore surge zones with plenty of rapidly growing algae and few competitors, so there is no need to expend energy protecting crops. Sergeant Majors have a diverse diet with only 10 percent from plants and the remainder coming from benthic invertebrates, small fishes, eggs, and zooplankton picked from the currents. Plankton is also the primary food of Bicolor Damselfish, *S. partitus*, the lone plankton-eating *Stegastes*, and of all five species in the genus *Chromis*.

Damselfishes inhabit nearly every type of reef structure, from the shallows to depths well beyond safe diving limits. Although most species are found in a single habitat, others readily adapt to a variety of surroundings, and a few coexist with one or two other damselfish species. For example, Threespots prefer to establish their territories deep within the three-dimensional maze of Staghorn Coral, *Acropora cervicornis,* branches but regularly reside in Lettuce Coral, *Agaricia agaricites,* and large boulder corals. Cocoa Damselfish, *S. variabilis,* which live in extended home ranges too large to defend aggressively, favor massive boulder corals in deep water but also thrive in shallow Staghorn Coral groves. Large female Yellowtail Damselfish are typically associated with Elkhorn Coral, *Acropora palmata,* on reef crests, while the males occupy deeper rubble zones.

Brown Chromis often pick zooplankton from open water in thick aggregations.

An interesting example of cohabitation involves Yellowtails, which always reach their greatest abundance when sharing feeding resources with smaller Duskies or Threespots. Large Yellowtails, which can grow to more than eight inches, superimpose their broad feeding areas over the territories of several Duskies. Both species protect the same resources from the same herbivores, the smaller, submissive tenants are expected to perform the lion's share of the defensive duties. A similar, size-related dominance occurs with smaller Threespots; however, large adults are able to expel the less-aggressive Yellowtails from their turf.

Unlike their reef-oriented relatives, the Brown and the Blue Chromis have adapted to a semi-pelagic way of life to gain access to the rich plankton streams that flow above the reef. Both species feed much of the day in small to large groups, maintaining ties with the reef only to elude predators, sleep, spawn and frequent cleaning stations. Interestingly, these two species and Sergeant Majors, the region's only schooling damselfishes, seldom visit cleaning stations. The less numerous Blue Chromis tend to collect in small groups in shallower water and remain closer to the reef, while Brown Chromis assemble in massive aggregations that often fill the water column high above outer reef banks. Occasionally the species integrate and at times are joined by Creole Wrasse, *Clepticus parrae,* the region's other abundant plankton-picker. The three remaining members of the genus generally inhabit

Aggregation of juvenile Sunshinefish.

deeper water. The two- to three-inch adult Purple Reeffish, *Chromis scotti,* and Sunshinefish, *C. insolata,* and their brightly colored juveniles typically congregate in small groups in shallow pockets on walls and reef slopes. Yellowtail Reeffish, *C. enchrysura,* also live deep but prefer flat, heavily contoured bottoms and artificial reef sites. All three rely on plankton diets but never feed far from the reef's protection.

Half of the damselfish species, the seven *Stegastes* and the lone *Microspathodon,* patrol and defend permanent territories throughout the year. However, the breeding males of the remaining seven species, the five *Chromis* and two *Abudefduf,* only temporarily defend spawning sites during reproductive periods. Though the size of permanent territories varies considerably among species, each is the exclusive dominion of a single adult and functions to protect food resources, shelter holes and, in the case of males, nest sites and eggs. But not every adult in a population oversees a personal domain. These so-called wanderers, including both females and non-reproductive males, suffer a higher predation rate. Their presence, however, provides a pool of individuals ready to repopulate vacated territories, helping to maintain a constant population density.

Permanent territories have three basic components: a core, peripheral areas and pathways. The core is usually a small, intensely defended feeding area around the most frequently used shelter hole. The same hole also serves as a nighttime retreat and occasional nest site. Encompassing the core is a much larger irregularly shaped peripheral sector, defined by the limits of a territory-holder's defensive efforts. Territories adjacent to those of members of the same species are patrolled to maintain sovereignty. Paths, closely following protective ledges and outcroppings, are habitually used as routes of travel throughout the periphery.

Not all fishes are equal in the eyes of territorial damselfishes; during non-breeding periods the most forceful attacks are directed at their own species and algae-eating surgeonfishes and parrotfishes. When males guard eggs the intensity of attacks elevates dramatically, and known egg-bandits are encountered further from the core than at other times. While brooding, even normally mellow Sergeant Major males will sail up to 10 feet above their nests to confront passing divers.

All damselfishes are benthic egg-brooders that follow a standard reproductive regimen. It begins with males selecting and preparing nest sites, followed by courtship, egg-laying, and males tending and protecting the nests until hatching. Spawning occurs throughout the year, but the pace picks up in the spring, reaching its pinnacle during summer months, then tapers off as fall progresses. Monthly peaks, lasting several days, are linked to lunar or semilunar cycles. Nearly all courtship and spawning by species holding permanent territories takes place during the first hour of daylight, while those maintaining temporary nesting sites become reproductively active in the afternoon.

Intra-species communications are important to the social organization of damselfishes. Each species employs color changes and body language to signal aggression, submission, territorial dominance and sexual readiness; and all, except members of the genus *Chromis,* communicate with sound. The majority of messages come into play during reproduction. While courting, for instance, male Bicolors develop a striking black mask and tail pattern and perform a series of swooping dips, producing a chirping sound with each plunge.

Before courtship begins, nests are prepared to receive eggs. Most territorial damselfishes select a hard surface inside their shelter holes. Algae, invertebrates and debris are removed by mouth. Long-spined Urchins, *Diadema antillarum,* are chased away by biting off the tips of their spines. *Chromis* never clear surfaces but repeatedly fan open areas on the reef, in algae bushes and even on the sand in preparation for spawning. When all is ready, males assume their courtship colors and set about wooing mates during the early hours of the morning.

Spawning patterns vary among species, but typically the upper region of the back from behind the head to the tail lightens noticeably. If males are lucky, females approach their nests; but in most instances, the suitors must travel outside their territories and employ a mixed bag of tactics to coax potential partners back to their nest sites. The standard device is the signal jump, a common courtship behavior employed by members of several fish families: the males, in full spawning colors, dash almost straight up, turn and quickly swim back to the starting point. These eye-catching signals are usually performed in

Sergeant Majors in their reproductive colors laying and fertilizing eggs.

A male Longfin Damselfish displaying reproductive colors while tending nest.

a series over the nest site but also, on occasion, outside their territories. In an attempt to encourage females to follow them to their nests, courting males circle females, perform figure-eights, lead, shove, bump, nudge and, from time to time, nip flanks to get them moving in the proper direction. Each solicitation is executed with an exaggerated swimming style, characterized by increased body flexion. Once at the nests, quite a few females leave without spawning, forcing the undaunted males to repeat their high-spirited nuptial routine.

Compact clusters of eggs, called clutches, are laid in a single layer by females repeatedly dragging their extended cone-shaped genital papillae over the nests. This maneuver, also used by males to fertilize eggs, is known as skimming. When not in use the white, fleshy papilla, which can be seen during spawning, retracts into a recessed abdominal groove between the anus and the first spine of the anal fin. Occasionally pairs spawn in the same nest at the same time, side-by-side, but in most circumstances, the males fertilize the eggs while their mates are away from the nests. When, as often happens, more than one female spawns in a nest during the same day the fresh eggs are added to the same clutch. Eggs laid the following day are nearly always placed contiguous to those

A male Sergeant Major keeping his nest tidy.

of the previous day. However, if clutches have been in the nest for more than one day, new eggs customarily are deposited on the opposite side of the nests. Males can fertilize every morning throughout a reproduction cycle, but females must take a few days off before spawning again.

Once eggs are in the nests, the responsibility for their well-being is left entirely to the males. Eggs are attentively picked at to remove debris, bacteria and infertile capsules. Occasionally males place their mouths around small groups of eggs and gently agitate them. Clutches are regularly fanned to move water over their surfaces. Some damselfishes turn laterally to the eggs and scull with a single pectoral fin, while others face the eggs and direct water forward by alternating pectoral fins. A few males sweep across the eggs with their anal

Sergeant Major laying a row of eggs.

Territorial Longfin Damselfish

Juvenile Longfin Damselfish

Plankton-picking Brown Chromis without need to protect feeding territories display the same brownish gray color pattern from settlement through the juvenile stage and into adulthood.

Juvenile Brown Chromis

Settling Brown Chromis

and tail fins. The *Chromis,* whose eggs are often scattered in algae bushes, wedge themselves between the blades and sway back and forth. For an unknown reason, swarming clouds of tiny, transparent Mysid Shrimp, *Mysidium* spp., commonly hover in the entranceways of nests containing eggs.

The number of clutches received by males varies considerably. Why one male is more successful than another is not completely understood. Unlike other fish species which provide parental care, female damselfishes do not favor mates because of their larger physical size, quality of nest preparation or the number of eggs already in place. In fact, the only attribute correlated directly with mating success for any species of damselfish is an increased rate and number of dips performed by courting Bicolors.

Researchers recently documented a surprising new dynamic at work in the little Bicolors' mate selection process that probably applies to other family members as well. Often, egg-laden females pass several courting males en route to a preferred partner. It has also been noted that once reproductive cycles begin, female Bicolors, as well as females of other damselfish species, routinely inspect nest sites during nonreproductive hours of the day. Further observations established that approximately 20 percent of unhatched eggs, in some cases entire clutches, disappear from the Bicolors' nests. A quarter of these losses could be attributed to cannibalism by guardians. Such behavior, typical of many damselfishes, only occurs after new eggs are laid in nests already containing two or more clutches. However, the majority disappear at night, when males no longer guard their nests.

The culprit was discovered to be the nocturnally active Blunt-spined Brittle Star, *Ophiocoma echinata.* These five-armed egg-pirates hide in the reef and wiggle out to dine only after dark. It now appears that the female Bicolors daytime excursions are used to scout neighboring nest's susceptibility to egg predation. Females generally show a marked fidelity to a single partner, spawning with him 70 percent of the time. However, those females that discover a clutch missing or detect the scent of a brittle star almost always find a new nest to deposit their clutches.

A microscopic strand secures each elliptical egg in its nest. Throughout spawning, both parents periodically fan the clutch to coax the eggs into an upright position, which allows them to move slightly on their individual holdfast. Egg sizes vary among species, with those of the Brown Chromis and Blue Chromis the smallest, measuring just under 0.6 mm in length, while Beaugregory, *S. leucostictus,* eggs the largest, measure 1.28 mm long. Eggs generally begin to hatch at dusk three to seven days after fertilization. The tiny, transparent larvae immediately enter the pelagic stage of their life cycle. The time damselfishes spend as plankton is shorter and fluctuates less when compared to other families of reef fishes. The duration between hatching and settlement back to a shallow-water environment ranges from 17 to 34 days.

With the exception of the Sunshinefish, the juveniles of plankton-feeding damselfishes closely resemble and commonly live in association with adults. By contrast, the juveniles of the territorial bottom-feeders exhibit bright patterns distinctly different from their typically drab elders. This dichromatism between the age groups is believed to suppress the aggressive drive of adults. The colorful youngsters spend their early months in isolated, shallow-water areas well away from established feeding territories. Only with sexual maturity will they venture into adult habitats. It is believed that *Stegastes,* other than Bicolors, become sexually active by their second or third year and probably live for six to eight years.

Bicolor Damselfish
Stegastes partitus

Bicolor Damselfish are well represented in nearly every part of the tropical western Atlantic. Where prolific, the two- to three-inch, black-and-white fish establish territories on virtually every available coral structure, sponge, conch shell and rocky outcrop from shallow back reefs to the edge of plunging depths. Bicolors form distinct colonies comprising every size individual and both sexes. The species' ability to communicate with a broad repertoire of movements, color patterns and sounds makes it an interesting subject for behavioral biologists and fish watchers.

Throughout much of the day adult Bicolors feed in the water column, plucking tiny copepods, pelagic tunicates and larvae from the currents. The feeding height is limited by an individual's size, with even the largest males rising no further than five feet above the protection of their personal shelter holes. When predators or divers approach, the fish drop to the bottom and remain near cover until the threat passes.

Colonies are strict, size-related hierarchies governed by a dominant high-spirited male. His quasi-submissive subjects typically include two or three lesser males holding much smaller domains, a few females with limited, less-defined territories and a scattering of half-inch juveniles that spend their time cloistered near protective shelters. The social order is maintained by a seemingly endless succession of chases and challenges. Chases are nearly always directed at the next smaller individual, while challenges consist of rather weak-willed stands of defiance by subordinates after being driven back to their hiding holes. The more passive females have their own pecking order with

A Bicolor Damselfish—typical color phase.

Black mask pattern of courting males.

White body bar of aggression.

fewer and less-intense chases aimed at other females and juveniles. It is believed that the smallest female acts as a lightning rod for the frustrations of the entire colony. During reproductive periods, when the males' movements are limited by egg-tending, and competition for mates is running high, everyone vents anger at the lowly female, especially higher-ranking males. Only her ability to maneuver rapidly within restricted areas keeps her alive.

Even though Bicolors do not have to protect their planktonic food source, both males and females routinely defend their territories from prowling fishes. When eggs are in nests, attacks from guardian males increase in both number and intensity. The most frequent recipients of their wrath are the notorious egg-eating Bluehead, *Thalassoma bifasciatum,* and Slippery Dick, *Halichoeres bivittatus.* Occasionally, a large Bicolor male from a neighboring area takes

Bicolor Damselfish picking zooplankton from the currents.

A male Bicolor Damselfish protects his egg nest, inside the conch shell, from a hungry swarm of Blueheads.

temporary refuge in another colony. Such a serious breach of sovereignty is handled exclusively by the colony's dominant male. In a show of bravado he immediately switches on a vertical white bar of aggression and confronts the rival with a ritualistic suite of intimidating behaviors, including close parallel swimming, frontal and lateral displays and tail-backs, where he turns and backs toward the intruder with spread fins and a wildly beating tail.

The Bicolors' impressive inventory of communication skills is divided into three basic categories: body postures, color patterns and sounds. At crucial points in every significant interaction at least one, but in most instances a medley, of communicative exchanges comes into play. Depending on circumstances, Bicolors produce chirps, grunts or pops. For instance, the face-off described above between two dominant males would possibly include as many as 6 to 10 pops. Unfortunately, nearly all of the sounds are undetectable by underwater observers.

Both male and female Bicolors reach sexual maturity within a year following settlement and, if they survive, breed for two to three additional years. The prolific species spawns in all but the coolest months, with lunar peaks occurring during seven-day periods following the full moons of May through October. Even though females spawn only in the early morning and on specific days, males exhibit courtship behavior in every season and throughout much of the day. Their persistent displays attest to the females' control over the mating schedule. During spawning periods courtship begins at sunrise, when males acquire distinctive "black mask" patterns that turn their faces and tails jet black and their midbodies a bright contrasting white. Decked out in their eye-catching, two-tone wardrobe, males swim up to 20 feet from their territories to lure females back to their nests. As they approach a prospective mate, males execute a series of tilts and dips, a behavior similar to the signal jumps of other damselfishes. When a male starts to dip, no matter what time of day, others in the vicinity quickly follow suit, and with good reason—those males performing the most frequent dips receive the most clutches. Dip sequences are typically repeated three to five times in a row, with each moving the males closer to prospective mates. Chirps accompany each series. Experiments reveal that females are not only able to identify the sounds of individual males but can also distinguish the size of the singing males.

When females show interest by moving toward dipping males, the suitors fall in behind and encourage them to move forward with a nudge or two. They then zip around in front and eagerly lead intended mates in the right direction. Once at a nest, most females leave without spawning, forcing the males to repeat the entire tilting-dipping-nudging-and-leading cycle. It ordinarily takes repeated attempts before partners are persuaded to spawn, but in some cases females travel to nest sites and spawn without prompting.

Having decided to spawn, the females darken noticeably, and the males position themselves nose-down next to their nests and quiver. The pairs spread fins and circle side-by-side for 5 to 15 seconds while the males make occasional grunting noises. The males dart to their nests and briskly skim the surface. Like many damselfishes, the nests of Bicolors are hidden from view inside small holes, narrow crevices or conch shells. Once spawning begins, it generally takes the female from 20 to 40 minutes to lay between 500 and 5,000 eggs. The number of eggs corresponds to a female's size. Now and again females move briefly from the nests, allowing males to fertilize. While waiting, the males busy themselves by circling the sites, chasing away all comers and occasionally courting nearby females. Generally only one female spawns in a nest, but nearly a third of nests contain the eggs of two or even three females. During a weeklong reproductive cycle, females spawn once every two days and, in 70 percent of cases, at the same nest site.

Parental males do not spend as much time grooming and fanning eggs as other damselfishes, but they are accomplished nest-defenders. When first laid, the eggs appear white, changing to dull yellow by the end of day one, pink on day two, and finally purple on the third day. At dusk, three and one-half days after fertilization, the eggs begin to hatch, and the larvae enter the pelagic realm for 27 to 31 days before ready to settle to a shallow sea floor. Clusters of recently settled juveniles regularly inhabit the rubble mounds of Sand Tilefish, *Malacanthus plumieri.*

Threespot Damselfish

Stegastes planifrons

Threespot Damselfish lay claim to virtually every substantial thicket of Staghorn Coral in the Caribbean. However, if a reef lacks their favorite habitat, these highly adaptable, three- to five-inch fish willingly establish their permanent feeding territories on patch reefs, massive coral mounds, Elkhorn

Adult Threespot Damselfish

Reproductive male

Yellow phase of Threespot Damselfish

Juvenile Threespot Damselfish

Coral, dock pilings and fields of large-sized rubble. Males dominate the Staghorn Coral groves, with the larger individuals controlling territories in the center while smaller males and a few females reside in the less desirable outer branches. Excluded females and subordinate males spill over into surrounding habitats. Many Staghorn colonies are filled to capacity with contiguous Threespot territories, each with its own protected algae garden growing on dead interior branches. Although space is at a premium, conflicts between neighbors are limited.

Algae-eating intruders, however, are confronted throughout the day. Because of their more vulnerable situation, defenders on the outskirts challenge prowlers more frequently. Continued advances by a surgeonfish will provoke a squadron of high-strung Threespots to rise above their territories in a united front. Smaller residents from the periphery regularly scout the colony's interior for vacant or poorly defended territories, but the newcomers are not allowed to remain unless they are equal in size to neighboring territory holders. Through this slow process, long-lived individuals eventually gain access to the favored real estate in the colony's interior.

Like most damselfishes, spawning occurs in every season, with increased activity during warmer months. Peak periods begin four to seven days after the full moon. Prior to courtship males prepare nest sites on living Staghorn by persistently biting the underside of branches until an elongate channel surrounded by a close-cropped mat of red algae is formed. In other habitats nests are located on the ceilings of overhangs and rock caves. At the first light of dawn, either females or males initiate spawning. Females with territories in coral mounds often cross open terrain to mate in Staghorn colonies. Courting males acquire a dark diffuse blotch across their sides, their cheeks flush pale, and the spot at the base of each pectoral fin darkens. In large colonies, courtship is contagious; the first signal jump sets off a frenzy of jumping males across the maze of branches. Once aroused, males leave their territories with only one thought in mind. Females are approached and circled with an embellished swimming style, accompanied by clicking sounds. Males energetically lead, herd and nudge their choices toward their nests. Once near, they dart back and forth from nests to mates, perform signal jumps and mock spawning runs, indicating where eggs are to be placed. When females initiate spawning, their backs pale to brownish yellow and their bellies lighten as they approach the males. Once positioned next to their partners the females are led to the nest sites, where the pairs simultaneously make slow circular spawning passes. In this manner, between 2,000 and 10,000 eggs are laid in a single nest.

The males fan and tend the eggs every few minutes, but their best parental attribute is nest-guarding. Nothing that comes near the Threespots' egg-laden nests leaves unchallenged. Smaller fishes are immediately chased, while larger intruders are forewarned with lateral displays, exaggerated swimming, headstands and clicking sounds before a barrage of well-placed nips is delivered to the eyes and fins.

The fresh yellow eggs turn green as they mature. Hatched larvae remain in their planktonic stage for 17 to 25 days. The bright yellow juveniles stay close to cover, feeding on a diet evenly divided between algae and bottom-dwelling copepods. With growth, they become more dependent on plant material. Unlike most damselfish juveniles, they seldom inhabit shallow inshore areas, choosing instead small patch reefs and the shelter of fire coral ridges in deeper water.

The Staghorn Corals' dense network of branches provides Threespots ideal protection from roving predators. The branches also serve as substrate for algae gardens, which regularly cover 20 percent of a thicket's surface. Threespots build their gardens by relentlessly nipping at the same bit of tissue until it can no longer exclude invasive algae, which, once it gains a foothold, begins sending up shoots within days. From start to finish, gardens take about a month to complete, but the job doesn't end there. Throughout their lives dominant Threespots are continually at war with the corals expanding their lawns size 5 to 10 percent a year. The Threespots' unending battles with living coral along their boundaries are marked by white-capped projections indicating overgrown lesions. As callous rings grow upward, encroaching algae appears as small tufts at the ends of cylindrical protrusions, known as chimneys. Algal lawns spread over the tops and sides of branches leaving their pointed tips undamaged and able to grow. With time and the help of boring crabs and sponges and parasitic

Threespot Damselfish regularly nip Staghorn Coral to expand their algae gardens, which form white-capped projections indicating overgrown lesions.

snails, the branches degrade and collapse, but continue to grow where they settle. This seemingly destructive process actually helps the thicket prosper, which in turn increases habitat for Threespots.

Comparisons of the chimney-like protrusions on fossilized fragments of Staghorn with contemporary specimens traces the relationship between Threespots and Staghorn thickets back 125,000 years to the Pleistocene. The recent die-off of Staghorn across the tropical western Atlantic has had a marked effect on the historic alliance forcing Threespots to adapt to dwindling habitat. Some have shifted to Lobed Star Coral, *Orbicella annularis,* a mound-building coral with far fewer places to hide. Others simply overpopulate the few remaining Staghorn thickets. This has proven to be a serious problem for the well being of small, marginal colonies significantly slowing their growth, and often smothering coral stands leading to total mortality.

Dusky Damselfish

Stegastes adustus

Dark, olive-brown Dusky Damselfish inhabit a variety of bottom structures from shallow to moderate depths, but prefer shallow, heavy surge habitats with complex coral structures including Staghorn Coral and wave-resilient Branching Fire Coral, *Millepora alcicornis.* The four- to six-inch adults also prefer shallows with large rubble, and breakwaters. Smaller juveniles with bright orange backs frequent shallow shoreline rubble slopes from inches to a foot in depth and bare rock in back-reef areas. Both males and females aggressively defend permanent feeding territories less than a half-yard square. The species reaches its highest population density inside Elkhorn Coral gardens, where it shares feeding turf with large Yellowtail Damselfish, which dominate the smaller Duskies.

Duskies spawn year-round between the new and full moon, with synchronized peaks occurring sporadically every two to four days. Females spawn several times within each lunar cycle, at intervals of from one to nine days. It is not yet understood what prompts spawning peaks. At daybreak females travel to prepared nest sites in secluded areas of the males' territories. The backs of courting males pale as they perform a series of signal jumps that occasionally attract multiple mates. In rare cases as many as three females lay eggs together, but most prefer waiting their turn. Eggs are deposited in several installments, each lasting from 1 to 20 minutes. Between bouts, females often return to their territories for brief periods while the males fertilize. The final product is a compact, single-layer clutch. Males frequently defend and tend as many as four clutches in a nest, each laid on a separate day. Eggs begin hatching at dusk four days after fertilization, and the larvae settle back to the reef between 20 and 23 days later.

Beaugregory

Stegastes leucostictus

Beaugregories typically inhabit sheltered inshore bottoms of sand and rubble. They are quite common in boat channels, bays, jetties and around dock pilings where they often find shelter in empty conch shells and cans. Those inhabiting reefs generally take sanctuary in rubble-strewn back-reef surge zones or occasionally reef fronts, most frequently in depths of less than 20 feet. Juveniles

A young adult Beaugregory makes its home inside an abandoned Queen Conch shell.

often take refuge inside small, isolated structures on open sand, while adolescents live in areas of sand and rubble; adults prefer rubble bottom. Because of the large amount of eggs, worms, fishes, crabs and shrimp in their diet, and the limited volume of algae (just over 20 percent), the species is considered to be carnivorous, with no need to protect algae patches.

The bright, horizontal, two-tone blue and yellow color pattern of the juveniles remains through sexual maturity. Individuals take on olive-brown overtones only after moving into rubble with older adults. Half of the males maintain large, weakly defended permanent territories with several widely spaced shelter holes. The remaining adult males move freely from place to place. A shortage of good locations to build nests in rubble possibly accounts for the surplus of unstable males. Females are even more mobile and far less aggressive than territorial males.

Spawning occurs to some degree year-round, with peaks around the full moons of January through October. Breeding males prepare and defend nest sites in small rubble caves, conch shells, concrete blocks and cans. Hard interior surfaces are picked clean of algae and sponge. Thrashing tails remove sand from entranceways and often create trenches encircling nests. Occasionally unsolicited females come to spawn, but in most cases males leave their territories and travel up to 30 feet to woo females. Courting males lighten noticeably across their backs as they signal jump, circle prospective mates and herd them back to their territories. The closer they get to their nests the more hyperactive males become—darting, dipping and attempting to herd the females inside by swimming next to their sides. Spawning is brief, lasting only 10 to 20 seconds. Typically nests receive multiple clutches over several days. Guarding males remain close to the nests' entrances, now and again darting inside to fan the eggs. Defense is spirited; wrasses and damselfishes are chased at distances of three to four feet from the nests. Male Saddled Blenny, *Malacoctenus triangulatus,* often take advantage of the Beaugregories' fish-chasing skills by locating their egg nests next to the clutches of brooding males.

The eggs, the largest laid by the damselfishes in the region, are bright yellow when fresh, later changing to green. Depending on the number of clutches present, a single nest may accommodate between 4,000 and 25,000 eggs at one time. A clutch begins to hatch four and one-half to five days after fertilization. The planktonic phase of the life cycle lasts from 19 to 21 days.

Cocoa Damselfish

Stegastes variabilis

Cocoa Damselfish weakly defend rather large home ranges in a variety of habitats, from inshore patch reefs to deep, coral-strewn sandy slopes, where they often wander about the bottom for hundreds of yards. The species' temperate disposition allows it to share habitat with Threespot Damselfish and Bicolor Damselfish and tend to remain segregated from their juveniles. Cocoas consume a mixture of soft bottom materials, including algae, crowns of feather duster worms, tunicates, anemones and isopods.

Breeding increases during summer months, with multiday peaks beginning two weeks after the full moon. During spawning periods, breeding males defend nests on the ceilings and vertical walls of small coral recesses. When females come near, courting males approach with a head-up display. If the females continue to advance, males perform a quick series of dips before rushing back and executing an impressive set of high signal jumps directly over their nests. Next, males tightly circle the females before herding them to their nests with an exaggerated swimming motion. After pairs enter nests, the males skim briefly before the females begin to lay eggs. Brooding males frequently chase small egg-eating wrasses, but don't seem disturbed by fishes too large to enter the nests. Eggs are regularly inspected, fanned and mouthed. Usually, only a single clutch is present at any one time. Hatched larvae settle 22 to 25 days later.

Yellowtail Damselfish

Microspathodon chrysurus

An Elkhorn Coral garden wouldn't be complete without a flourishing community of Yellowtail Damselfish. They also show preference for shallow areas of exposed reefs with a substantial coverage of algae species and Branching Fire Coral. Fortunately for divers, these large damselfish, which often reach a length of seven inches, are among the family's least-aggressive members. Their lovely juveniles, still referred to in the aquarium trade as "Jewelfish," display a scattering of iridescent blue dots over rich cobalt blue bodies. The tiniest have translucent tail fins that become yellow as they mature. Adults retain the basic juvenile patterns but in less vibrant tones. Males and females are visually indistinguishable except during courtship and spawning, when their colors change temporarily.

Large females reign over widespread territories of varying sizes on reef crests, while males typically occupy deeper zones of Elkhorn rubble. Deeper territories

Threespot Damselfish in yellow phase.

Juvenile Threespot Damselfish

Male Threespots display light bodies, bright yellow tails and dark faces during courtship and spawning activities.

tend to be larger, but the circumference of a Yellowtail's territory does not appear to relate to the individual's size. Boundaries regularly encompass several territories of either Dusky Damselfish or Threespot Damselfish. The smaller species share the harvest of epiphytic diatoms that grow in profusion on the protected algal lawns. The large, dominant Yellowtails rely almost exclusively on their more aggressive cohabitants to keep algae-eating intruders away from their mutual feeding grounds.

In areas with dense population, where there never seems to be sufficient habitat to go around, larger Yellowtails have a distinct advantage in staking territorial claims. Smaller individuals of both sexes are forced to become wanderers. These nonterritorial individuals graze within the boundaries of established residents that generally offer only token resistance as long as the visitors exhibit a submissive posture. The weak-willed defense of territory-holders is primarily directed toward neighboring Yellowtails and wanderers late in the day.

A peak four- to five-week spawning period occurs in February–March and again in July–August, with decreasing activity during the three following months. Males over four and one-half inches long with permanent territories begin readying nests at the onset of each breeding period. Sites are generally located in 20 to 35 feet of water, near the center of territories on the inclined surfaces of exposed Elkhorn rubble or around the base of gorgonian stems. Males use their teeth to rasp algae and invertebrates free from one or, in rare cases, as many as four nests. About 15 minutes before sunrise, ripe females migrate as far as to 300 feet to spawn. When they arrive, their light gray heads darken to a deep blue and their yellow tails brighten—colors that remain throughout courtship and spawning. By contrast, courting males wear dull gray coats that lighten as they become increasingly motivated.

Excited males attempt to lead potential partners back to their nests by swimming in a series of increasingly shorter figure-eight patterns. The maneuvers may take them up to 30 feet from their territories. If their first attempts fail, they begin butting and herding the females in the proper direction. Once close, the enamored males perform several signal jumps near their nests. With pectoral fins fluttering, cooperative females rub their abdomens in a circular path over the cleared surfaces, laying eggs as they go. If an area is too small for all the eggs, females continue at a second, nearby nest. While spawning, which can take from 3 to 90 minutes, the males chase intruders and frequently circle their partners. Every few minutes females swim to the side allowing males to fertilize the deposited eggs. After each pause the males once again signal jump and, if necessary, drive the females back to their duties. When females are finished, the males chase them from the area. As females flee back toward shallow water, smaller courting males enthusiastically vie for their attention. During peak breeding season, females typically spawn once every three days; however, their fidelity to a particular partner has yet to be established.

As many as four or, in rare cases, five separate clutches incubate at one time inside a male's nest or nests. During the day guardians continually fan, mouth and pick at the brood to remove algae, tiny invertebrates and infertile or hatched egg cases. With their nests full of eggs, the traditionally passive males become quite temperamental, often confronting egg-eating fish several feet from the nest. Grunts and occasional nips regularly accompany chases. The advances of neighboring Yellowtails pose little threat and are easily thwarted by brief displays of spread fins. Eggs failing to hatch with their clutch during the night are eaten by males the following day as they prepare their nests for additional spawns. Just over 10 percent of eggs disappear in this manner.

Three days after fertilization the reddish brown eggs turn gray, and the encased embryos begin wiggling in a circular motion. The following evening the tiny larvae break through tears in their egg cases and begin the 21- to 27-day pelagic phase of their life cycle. They are believed to settle in shallow patch reefs, where the brightly colored juveniles take up temporary residence near the flat branches of Blade Fire Coral, *Millepora complanata,* which makes up the bulk of their early diet. Although we have never observed the behavior, it has been reported that Yellowtail juveniles occasionally act as cleaners.

Sergeant Majors frequently gather in openwater aggregations.

Sergeant Major

Abudefduf saxatilis

A diverse diet allows Sergeant Majors to thrive in a variety of inshore habitats, from coral gardens to jetties and mangroves. The opportunistic feeders are equally at home picking plankton from currents, nibbling tender algae, plucking invertebrates and eggs off rocks, nabbing floating insects or, on occasion, taking tiny fishes. When hand-fed from a boat or dock the scrappy fish boil the surface like piranhas, and, during the annual coral spawn, they are first on the scene to gobble up the windfall. In fact, any shallow bottom with shelter and a source of food seems to suit their fancy.

Sexually mature adults tend to cluster in loose groups within home ranges covering a few hundred square feet of sea floor. Males reach sexual maturity at approximately four inches, and females about three inches. The boldly barred black-and-yellow males and females are visually indistinguishable except during courtship and spawning, which occur sporadically throughout the year with a slight peak in early summer. During multiday spawning bouts, triggered by unknown cues, groups move to nearby nesting areas. Mature males prepare nests on barren walls and beneath overhangs or, as a last resort, on the hard sea floor. Premium locations provide better protection from silting and predation. As soon as sites are selected, the once mild-mannered males turn steel-blue and aggressively territorial—a color and attitude that stays with them

During courtship and spawning males (above) develop dark blue bodies. As courtship intensifies, white patches appear on the heads and bodies of males.

throughout much of their six- to eight-day nest preparation, courtship, spawning and egg-brooding cycle. Each male first sets about the long, meticulous task of clearing a two- to three-foot circular surface for a nest. Living materials, such as hydroids, tubeworms, mollusks and algae, are removed by mouth. Large objects are dragged or head-butted to the side. Repeated fin-fanning sweeps away sand and debris. When nests are as slick as dinner plates, males begin advertising for egg-laden females.

Courtship occurs randomly throughout the day when blue males suddenly begin to dart erratically back and forth across their nesting territories. As courtship becomes intense, white patches appear on the suitors' heads and cheeks. During spawning additional white blotches often appear on the males' bodies, and similar markings occasionally occur on females. To attract distant mates, males perform repeated loops that carry them two to four feet off the

Male Sergeant Majors display dark courtship colors while guarding and tending egg patches. Recently laid eggs are reddish (left); mature eggs are silvery gray (right).

Sergeant Major egg nest. Recently laid, yolk-filled eggs are reddish. Mature eggs appear silvery gray from the embryos' developing eye pigment.

bottom. Persistence pays—frequently dozens of jumps are required to entice egg-laden females.

Usually single, but in some cases, multiple partners show up to spawn, but only one remains. Mating pairs skim in circular patterns over the nests, with the males positioned next to and slightly behind their partners' outside flanks. When males leave to chase an approaching predator, females frequently flee. Males pursue them up to 15 feet and once again perform a series of loops to rekindle interest. Occasionally, well-placed nips help prompt a quick return. It is common for females to be lured back repeatedly before spawning. Males generally remain near the nests, keeping watchful eyes on their partners, during the hour-long spawning periods. Now and again they dart in to shed streams of milt. When females slow, an encouraging peck on the fins usually quickens their pace. If neighbors approach, outraged males place themselves

After the male nest guardian beds down for the night a crab steals out of hiding to dine on the maturing embryos.

Ready for the launch.

One second old.

between mates and rivals with their bodies angled down and heads touching the bottom.

A single female can lay up to 20,000 eggs, and as many as four females may contribute to a single nest. Each fresh clutch is placed adjacent to the most recent batch. The elliptical pink-to-red globules are extruded in single, continuous side-by-side rows from an extended genital papilla that brushes along the bottom. Throughout spawning partners periodically fan the eggs to position them upright so each can sway on its individual holdfast. When finished, the females leave their broods to the exclusive care of the males.

Protecting a nest is a high-energy endeavor. Males not only fan and regularly remove damaged and infertile eggs, but must also remain on constant alert for egg bandits. Bottom scavengers like nothing better than a dinner of rich, plump fish eggs. The most frequent thieves are Slippery Dick, Bluehead, *Thalassoma bifasciatum,* Bluestriped Grunt, *Haemulon sciurus,* Redlip Blenny, *Ophioblennius macclurei,* and other Sergeant Majors. Even crabs, brittle stars and sea urchins are not to be trusted especially following sunset when nest guardians sleep.

On several occasions we have watched swarming aggregations of young wrasses overwhelm the males' defenses and storm their nests. Once underway, nearby fish joined in the routs. In each case it took the frenzied guardian minutes to restore order. It has been reported that males, unable to chase off attackers, occasionally join the feast. A marine biologist reported watching a Schoolmaster, *Lutjanus apodus,* lurk near a besieged nest picking off small wrasses on their way to the free-for-all.

Males tend their nests for six to eight days, depending on the number of clutches involved. As the yolk is absorbed and eye pigments enlarge, the reddish eggs turn green then silvery gray before hatching. As soon as males abandon their nests, the eggs' tattered remains are quickly picked clean by nearby fishes. Within days the sites are once again cloaked in a cover of algae and silt.

Three days after hatching, the large-headed, big-jawed, tenth-of-an-inch larvae actively swim and eagerly consume tiny copepods. Between 17 to 20 days after hatching, the planktonic larvae have grown to nearly a half inch and are ready to seek shelter in floating rafts of *Sargassum,* tide pools and mangrove banks. The tiny juveniles, miniature versions of their elders, group together for protection as they grow. In a matter of months they will be ready to join adult populations in shallow coastal waters.

Night Sergeant

Abudefduf taurus

If you want to see the largest of all Caribbean damselfishes, you will have to leave the reef and move inshore to rocky, surf-swept shallows or steep shore embankments where Night Sergeants live. These rather reclusive, six- to ten-inch fish subsist almost exclusively on a diet of rapidly growing filamentous algae nipped from rocks. The fish regularly associate with up to eight individuals. When divers arrive, the cagey fish dart under ledges or swim out of the area but usually return after a few minutes. They resemble their genus-mate, the Sergeant Major, except for their larger size and less colorful countershadings of browns and brownish greens.

From April through September, males prepare nests on the hidden undersides of rocks or inside small holes the day before spawning. They darken for courtship and mating and occasionally while guarding clutches. Females turn dark only while on the nest. When females approach, males perform several signal jumps. With great ceremony, interested females spread their fins, beat their tails briskly, and approach the males laterally with their bodies angled downward. While the females spawn, the males stand guard, chasing away other damselfishes and blennies. When the females leave periodically, the males fertilize until their mates return to lay additional eggs. Spawning, begins at first light and often continues for two hours. Occasionally, two females skim the same nest together. The two-and-a-half- to three-inch clutches, each containing from 2,000 to 10,000 eggs, are guarded continually by the males until hatching six or seven days later. After 20 to 23 days in open sea, the larvae settle inshore and promptly metamorphose into tiny, miniature versions of adults.

Night Sergeant

Genus *Chromis*

The five species of damselfishes in the genus *Chromis* that make their home in the region's waters can be separated into two distinct groups. The first includes a pair of the Caribbean's most abundant reef fish, the Brown Chromis, and Blue Chromis. Both of the four-inch fish have developed slender profiles and elongated scissorlike tails enabling them to feed on zooplankton high above the reef, where speed equals survival. The Yellowtail Reeffish, *C. enchrysura,* Purple Reeffish, *C. scotti,* and Sunshinefish, *C. insolata,* more closely exhibit the classic, laterally-compressed damselfish shape. All three favor deepwater habitats. The Purple Reeffish and Sunshinefish congregate in small mixed groups including juveniles on reef slopes below 80 feet, where they feed on zooplankton passing close to the reef. Yellowtail Reeffish are more commonly found on deeper patch reefs and around artificial wreck sites.

Where brisk currents run, Brown Chromis gather high in the water column to pick plankton. Small groups work the shallow reefs while vast aggregations, numbering in the hundreds, throng to the rich currents above deep reef slopes. Microscopic pelagic animals, mainly copepods, comprise the bulk of their diet. Synchronized beats of translucent pectoral fins shuttle them about as, now and again, their entire mouth structure shoots forward to nab floating specks of food. When openwater piscivores such as jacks or mackerels approach, the widespread clouds of fish close ranks and slowly ease away; if the predators advance to within 20 feet, the school dives as one for the reef. As evening approaches, the fish feed closer to the reef. Most find shelter for the night, while others bed down on the open sand and acquire blotched night colors.

The Brown Chromis has two reproductive strategies: pair spawning in temporarily defended territories and mass spawning aggregations. Reproduction occurs in all seasons with distinct peaks for 10-day periods following each full moon. Solitary males begin three-day spawning bouts by staking claims to

Brown Chromis feed high off the reef.

Blue Chromis feed just above the reef.

A pair of combative Brown Chromis lock jaws in an extended encounter.

sections of the reef and preparing crude nest sites on exposed ledges, crevice walls, gorgonian stems, sponges or *Sargassum* blades. The males repeatedly fan their future nests with rapidly-beating pectoral fins. When ready to spawn, they perform a few brief, three-foot vertical signal jumps to attract mates. Often as many as 10 fish of both sexes gather at the same site. At first every fish is chased, but soon the females reveal their gender by turning their tails toward the males and exposing their ventral regions. Partners are led to the nest by a series of quick dashes. When near, the females are herded, butted and nipped into place. Sometimes clusters of females hover above the nests, but only one at a time enters to lay eggs. The males frequently join their mates, but wait until the females finish before actually fertilizing. One busy male was reported to have

Mass spawning site of Brown Chromis.

A roving pack of Scrawled Filefish opportunistically feed on Brown Chromis eggs.

spawned 30 times with different partners within 45 minutes.

Males guard, fan and mouth the eggs for three days before they hatch. Rival males approaching the nest are challenged with lateral and head-on displays of spread fins and flared gill covers. Every now and then males face off only inches apart and joust back and forth, occasionally locking mouths. Wrasses, small grunts and other damselfishes are attacked as far as four or five feet from the nests. If a threatening aggregation of wrasses approaches, males assume defiant head-down positions at the entrances to their nests.

Group spawnings, sometimes involving a thousand individuals, occur once each lunar cycle in areas with large populations. One or two days preceding spawning, small groups of large adults settle to the reef to fan the bottom, dispersing in less than an

A male tries to protect eggs from a crab.

hour. By the morning of day three, the bottom is covered with vigorously fanning adults. After an hour, a few random signal jumps herald the beginning of a five- to six-hour orgy of wildly skimming fish. It is believed that females spawn in several nests. The hatched larvae drift off with the currents for 24 to 30 days before they are ready to settle to the reef. Recently settled juveniles, which look much like adults, often cluster just inside the lips of rock ledges. Roving aggregations of various opportunistic species temporarily gather during the mass spawning of Brown Chromis. The aggregations we witnessed on three occasions in three locations along the coral-rich reef slopes in Bonaire consisted primarily of large Scrawled Filefish, *Aluterus scriptus,* accompanied by a scattering of surgeonfishes and parrotfishes. One group consisted of 13 filefish. It appeared that the filefish were employing the same tactic as Blue Tangs and other surgeonfishes, where large groups attack damselfish gardens overwhelming the defensive efforts of a few defenders.

Unlike their more numerous counterparts, Blue Chromis typically inhabit slightly shallower depths and prefer to feed alone or in small groups no more than six to eight feet above the reef. More than half of their diet consists of planktonic algae; the balance comes from pelagic animal matter, primarily copepods, eggs, tunicates and larval fishes. The juveniles' diet contains only 20 percent planktonic algae, which increases to nearly 80 percent for young adults, and then decreases to under 65 percent with maturity. Adults maintain permanent territories around a single small hiding hole, where they flee from danger and sleep each night. The availability of hiding places tends to have a direct bearing on the species' distribution. Blue Chromis are usually well represented in reef areas with substantial structure and occasionally inhabit tangles of Staghorn Coral. Their prudent habit of feeding close to the reef and seeking safe sleeping quarters serves the species well. Stomach content analysis of Graysbys, *Cephalopholis cruentata,* a common, small grouper, revealed 50 Brown Chromis for every Blue Chromis.

Reproduction occurs year-round with two-week peaks straddling the full moons of March through July. Males seem unconcerned about where nests

Blue Chromis spawning.

The bi-colored courtship pattern of a male Blue Chromis.

are located and often select a different spot within their territories for each spawning period, showing no preference for protected areas. Individuals living in low-density populations often spawn on open sand in the vicinity of their hiding holes, while those in heavily populated areas prefer algae-covered coral. Spawning females turn blue-black; courting males darken their backs and sides, which contrasts sharply with their bright white bellies. Only pair spawning has been observed in this species. Often males have to travel 50 or 60 feet to find females to court. Males perform a series of signal jumps and dips before leading interested females to nest sites. Several females customarily lay eggs in a single nest. Like those of the Brown Chromis, the eggs are tiny, clear, scattered and impossible to see. Nests are guarded by males for three days before hatching. The pelagic larval stage lasts for 27 to 34 days. Clusters of juveniles often hover near the protection of the reef.

The spawning behavior of the three deepwater *Chromis* are quite similar. Each species pair spawns on leaf algae or the bare walls of rock depressions. Nest sites are probably prepared within the males' preexisting feeding territories, generally spaced at least 15 feet apart. Group spawning has never been reported for the three species. Nest preparation consists of little more than the males fanning sand off small algae bushes. During courtship and spawning, males develop pale blotches across their backs that, in the case of Purple Reeffish, often cover the soft portion of their rear dorsal fins and tails. The blotches become more intense with increased reproductive activity. Males leave their nest sites to solicit females by actively dipping, chasing and herding. Once back with mates, males perform one- to two-foot signal jumps above their sites. Several females are usually attracted during a single spawning period, but only one at a time enters a nest. The males customarily enter and circle the nest first, followed by the females, which deposit eggs for several minutes, often with males positioned by their sides. Spawning involves quite a bit of shaking and quivering causing considerable disturbance in the algae. When the females leave, the males continue to fertilize and fan the eggs. It is believed that all the eggs in a nest are laid in a single day. Guarding males usually hover a few feet from their nests, as if disguising the eggs' exact location. Chasing is vigorous; wrasses and large parrotfishes are challenged four to five feet from the nest sites.

Dragonets

Family Callionymidae

Male dragonets signal sovereignty by displaying their tall dorsal fins.

Lancer Dragonet

Paradiplogrammus bairdi

Territorial disputes and courtship behavior of three-quarter- to two-inch Lancer Dragonets take place on shallow sand flats and algae-tufted bottoms. During the day these small algae-eating bottom-dwellers are so cleverly concealed by their brown-mottled camouflage that only intermittent sprints across the sand give them away. Eyes set high on a wide, flattened head and a pointed snout lined with thin yellow lips give the odd little fish its appeal. Males can be distinguished easily by their larger size and a tall, golden foredorsal fin that snaps erect from time to time to intimidate rivals and entice females. The smaller, similarly patterned females, signal back with a less impressive foredorsal fin often edged in black.

Lancer Dragonets can be quite common in the sand and algae flats around shallow patch reefs, but require a bit of an effort to find. From a low perspective, look across the terrain for a glimpse of quick movement as the little fish scoot over the sand, stopping periodically to peck at algae. If you find one, others will invariably be nearby. Once sighted, they can be trailed closely without disturbing their routine. And if you happen to follow a male, he might just lead you to a good fight.

When an ever-watchful territorial male spies a smaller male, he usually pauses just long enough to flash his tall fin in a nonchalant show of sovereignty. However, when rivals are evenly matched matters often lead to classic confrontations of ritualistic pageantry dating to the time of dinosaurs. One afternoon, a male we had been following charged a male of comparable size. The rival stubbornly held his ground. Both slowed, stiffened and arched their bodies until trembling from the strain. With fins on high and tiny mouths puckered into tight little "O's," the adversaries, striving to look as grand as possible, slowly circled one other comparing size, beauty and tenacity in an attempt to establish or maintain dominance. After three tight turns, without so much as bumping a fin, the challenger broke rank and darted to a nearby algae clump where he picked at the greenery. Buoyed by success, his antagonist flicked his fin and followed.

Soon the fish were circling again, but when their ceremonial posturing failed to produce a clear winner the pair resorted to combat, facing off, locking jaws and applying torque. Within seconds, the challenger, lost purchase and was flipped on his back. Breaking free, he fled into the bush.

The dragonets' splendid fins are also used to beguile mates. Every full-sized male we encountered appeared to maintain a harem of two to four smaller females. During the day each member goes their own way feeding on algae within a restricted range defended by the male. In the Bahamas from May until September, the dragonets predictably began courting every afternoon an hour before sunset. During the day when a territorial male

The rival males circle one another slowly.

After locking jaws, males remain unmoving for seconds.

Still locked mouth-to-mouth, the combatants float off the bottom.

One dragonet gains control and flips his opponent over.

dragonet happens upon a female, he briefly paused long enough to casually snap open his golden fin a time or two before moving on. As the light fades, and harem members forage closer together the male's interest in the opposite sex escalates.

Male dragonets court each member of their harems multiple times before spawning.

From the instant a male's golden dorsal signals his growing enthusiasm females play their role with subtle brilliance. While their admirer prances, dances and sashays up to a female's side she indifferently continues to feed. Undaunted, the male spreads his fins, arches his back and goes rigid with passion. The female occasionally offers a modest fin-flick in return before scurrying off with the male trailing behind. This cat-and-mouse demeanor typically continues for 15 to 20 minutes before the female finally indicates willingness by holding her black fin erect. The male quickly positions parallel and pushes against her side. Still coy, she turns away again and again, making him repeatedly spin to remain close. Finally, as the sun slips below the horizon, the reluctant female allows the male to remain pressed against her body. With their small ventral fins churning, the pair, pasted side-to-side, slowly rise about a foot above the bottom where they pop apart, leaving a shimmering strand of eggs in their wake. When darkness falls, the little fish shimmy into the sand, often leaving their elevated eyes exposed.

Dragonet spawning rise.

Flounders

Family Bothidae

Territorial Peacock Flounder on the move as dusk-time spawning approaches.

Living on the open sand without ready access to hiding holes is risky business. To avoid being eaten, sand-dwelling fishes have adopted a variety of strategies. Jawfishes build burrows lined with rocks where they retreat at the first sign of danger. Sand Tilefish, *Malacanthus plumieri,* cover their shallow sand hideaways with mounds of stones and shells to keep out invaders. Garden eels and snake eels use their sharp tails to corkscrew backwards into the sand, and hatchet-headed razorfishes power-dive beneath the surface. But flounders, one of the oddest of the oceans' many oddities, are, far and away, the most masterful of all the sand-flat survivalists.

Flounders belong in the Order Pleuronectiformes, the flatfishes, a group that also includes halibuts, soles, whiffs, sanddabs, plaice, turbot and tonguefishes. Flatfishes have been placed in one of three families principally by the location of their eyes. Bothidae, or lefteye flounders, are mainly shallow-water tropical species with both eyes on their left sides; righteye flounders, Pleuronectidae, primarily live in temperate or deep waters, and Soleidae, the sole and tonguefishes, have eyes on their left sides but slightly out of alignment. A few flounders from both families have a scattering of "reversed" species whose eyes are located on the contrasting side.

In preparation for a hazardous life on the sand, flatfishes pull off one of the animal kingdom's most remarkable feats of developmental biology. The amazing transformation ultimately remodels a typical vertically swimming body into a horizontal posture so adults will be less noticeable on flat sea floors. The wizardry takes place during larval metamorphosis in the open ocean. During their first weeks after hatching the tiny, translucent offspring have standard upright bilaterally aligned bodies, conventionally aligned fins, and like all larval fishes, a single eye, on each side of the head. While in the larval phase, one eye migrates across the skull and ends up on the same side and next to their opposite eye. Whether the right or left eye migrates depends upon the species, however on rare occasions the alternate eye of an individual migrates. During the process, backs darken, bellies lighten, swim bladders disappear, and the dorsal and anal fins shift until they encircle the flat oval bodies. Metamorphosis, controlled by hormones from the thyroid and regulated by the pituitary gland, softens the skull as a rapid production of connective tissue inexorably forces one eye through the pliant bone of the skull. Following the two- to three-week transformation a lone pectoral fin remains near the center of the now horizontal flatfish back, exactly where one would think the dorsal fin should be. The pectoral fin on the males of many species grows quite long, and lies across the back unless raised to intimidate rival males or impress females. The exact timing of these energy-demanding alterations varies significantly among species as well as among members of the same species, lending credence to the idea that this momentous phase of metamorphosis begins only after larvae stockpile a sufficient store of food. From our limited observation of night-settling flatfishes, they all swim upright and their quarter-sized bodies remain translucent until settling on their sides and instantly vanishing from sight.

Partial eye migration of larval Eyed Flounder.

Eyes of Dusky Flounder almost in position.

The flatfishes' thin horizontal profiles not only hide flounders and other flatfishes from predators and prey but also allow them to bury quickly in the sand, leaving only their stalked, independently functioning, 180-degree-rotating, periscopic eyes exposed. But without question, the tropical flounders' best defensive strategy is rapid adaptive camouflage. When moving from one bottom location to another, they can change body patterns within seconds to match their new surroundings. Once settled, visual cues from their immediate environment are transmitted via nerves to thousands of irregular-shaped cells

Male Peacock Flounder with full complement of colors (above). Paling to almost white (below).

located in the skin, known as chromatophores. The flat color cells pale when pigments converge in their cores, and darken as the pigments expand throughout radiating cell branches. Red, yellow, orange, black and brown can be produced by activating overlapping chromatophores containing different pigments. Fishes capable of displaying the most complex patterns such as the Peacock Flounder, *Bothus lunatus,* a common shallow-water flatfish, have higher concentrations of chromatophores.

Although it has long been known that flatfishes change patterns to match backgrounds, little research supports the concept. The few studies of their camouflaging abilities have used species from temperate rather than tropical waters for their experiments. An early reference, published in 1911, reported that the backs of flatfishes placed on checkered patterns for several days appeared to mimic the checks. Two subsequent researchers, also experimenting with temperate species, refuted the earlier findings, concluding, "flatfishes simply have a universal texture allowing them to blend into any environment."

In the early 1990s, researchers from the University of California, San Diego, using the Peacock Flounder as their subject species, were the first to extensively investigate the camouflaging abilities of a tropical flatfish. They first tested the idea that the rapid color and pattern changes seen in the wild were simply illusions. Early photographic experiments put a quick end to this hypothesis. Five specimens rotated through a series of five bottom patterns—yellow sand, coarse gravel, one-centimeter and two-millimeter checkerboard squares and a solid gray sheet—closely matched the backgrounds in 91 percent of the trials. Even more impressive, most changes were complete in just two to eight seconds.

In the most dramatic experiment, flounders placed on the larger checkerboard instantly developed splotches closely mimicking the squares. However, the markings did not bear an exact spatial relationship with the pattern. In other words, the fishes' splotches were not aligned with the checks they rested on. Fish gently moved to a different position on the checkerboard retained their previous markings. When a piece of gray cardboard was slipped beneath the fish, the subjects' adapted wardrobes remained unchanged. It was also noted that each time a specimen was reintroduced to a particular design, it developed a corresponding pattern more rapidly than before.

Each time a Peacock Flounder settles in a new location, it quickly changes colors and pattern to match the bottom.

During another set of studies, flounders placed on a white background with widely spaced black polka dots paled significantly, leaving only solitary, dark, centrally located eyespots and a dusky blotch encircling their eyes. White dots of the same size on a black background caused the fish to darken, leaving only a series of white blotches along the edges of their bodies. These sparse patterns required minutes rather than seconds to complete. The researchers also observed that the fishes' entire bodies instantly blanched when the fish were frightened.

Microscopic analysis revealed that six sets of independently controlled markings are responsible for the myriad designs. Individual markings include a series of H-shaped splotches edging the body; small dark rings; light-colored rings encircling dark dots; thin scrawled lines; a central eyespot, and a spectacle frame around the eyes. By mixing and matching and varying the contrast of different sets, the fish are able to blend expertly with any natural background.

Peacock Flounder primarily eat fishes, usually small minnowlike species. On occasion they snap up a mantis shrimp or even octopuses. The stomach of one 12-inch individual contained two small four-inch jacks. The author of a feeding study found six herrings and two anchovies inside one specimen which was collected on the bottom 25 feet below a school of the small fishes swimming on the surface. He speculated that the flounder caught the minnows when they fed near the bottom, rather than making a feeding rush to the surface. However, we once observed a three-inch Eyed Flounder, *B. ocellatus,* powered by an undulating skirt of fins, swim eight feet off the bottom to nab an insect floating on the surface at night under a dock light.

Flounder watching is at its best at dusk. During the hour before sunset the Peacock Flounder, the Eyed Flounder and the Maculated Flounder, *B. maculiferus,* a common sand-flat resident of the southern and eastern Caribbean, court and spawn in harems controlled by large territorial males.

Flounders

The three species, all pelagic spawners, shed their gametes above the bottom in male-female pairs. Dominant Peacock and Maculated Flounder males are larger than females and also have much longer pectoral fins, prominently displayed during courtship and male-to-male rivalries. Both male and female Eyed Flounder have short pectoral fins, but the sexes of the small two- to three-inch species can be differentiated by the females' eyestalks that are set closer together than those of the males. Fortunately, during courtship flounders take little notice of inquisitive divers; flashlight beams, however instantly interrupt their behavior.

Two researchers from Texas A&M University compiled extensive observational data on the social and reproductive behavior of Eyed Flounder during two winter seasons in Bonaire. By coincidence, their shallow-water study site inshore from the old town pier was the same location where we observed and photographed the species a few years later. The researchers found that Eyed Flounder harems consist of from one to six females occupying small sub-territories within the larger 50- to 350-square-yard territories of dominant males. Typically the territories of several males lie side-by-side, often covering large expanses of shallow sandy terrain.

Females are jealously guarded from neighboring males that are always looking for an opportunity to increase the size of their harems and territories, even at the expense of a tussle. Male-to-male confrontations occur most commonly just before and during courtship. Trespassing males typically flee when chased, but on rare occasions they stand their ground facing each other across the sand with raised forebodies arching toward their opponents like striking snakes. Attacks are swift and vicious, with combatants locking jaws and slinging each other about. Mature rivals often battle for five minutes or more before matters are settled.

A Peacock Flounder slowly stretches its mouth, a common trait of lie-and-wait predators.

Flounder frequently bury in the sand, often with only their stalked eyes exposed.

Male Eyed Flounder spawn once or, in a few cases, twice every evening with each female in their harems. Courtship begins about 40 minutes prior to sunset, with males following established routes as they make repeated visits to each of their mates in turn. Initial meetings are brief, with the males pausing just long enough to inspect swollen bellies and possibly sniff pheromones to evaluate the females' readiness to release eggs. From time to time the males perch atop sand mounds to scout for predators. On a few evenings, lingering piscivores such as jacks or lizardfishes prompt harem members to dive under the sand for extended periods, preempting the evening's spawning activities. As an evasive tactic, the escaping flounder has the ability to flip completely around beneath the sand, ending up facing in the opposite direction with only their eyestalks exposed.

As courtship continues male Eyed Flounder become more impassioned, often touching their mates' sides with their snouts, laying forebodies across the females' backs and vigorously pursuing fleeing mates. Eventually the females comply, allowing the males to wiggle beneath their bodies. Stacked like pancakes and powered by fluttering fins, the pairs rise slowly off the sand. At some point, between 6 to 30 inches above the bottom, the males suddenly contract, releasing clouds of sperm, and immediately dive back to the protection of the sea floor. A second later the arching females discharge transparent eggs into the males' dissipating sperm and streak back to the bottom. As soon as the females are settled, the males inspect their partners' abdomens to be certain all of the eggs were released. If females are still swollen, a second spawning rise is initiated; if not the males zip off to spawn with other mates.

Males typically make repeated visits to each female in their harems before being allowed to slip beneath their partners' bodies, and even then, the females often break away prematurely during the spawning rise and return to the bottom. Occasionally a lucky male simply sidles up and the female obliges by lifting her flat body in acceptance or even by swimming onto the suitor's back. After shedding eggs, females leave their daytime feeding territories to sleep in deeper water. Males head off in the opposite direction, leisurely picking at bits on the bottom as they meander toward the shallows where they bed down in the same patches of sand each evening.

Flounders

A courting male raises his pectoral fin.

The larger Maculated and Peacock Flounders (males reach 9 and 12 inches in length respectively) court and spawn in a similar fashion. However, the males' tall pectoral fins add grand spectacle to the occasions. The males of both species appear to oversee harems of no more than two females. Late one August afternoon in Bonaire we documented a courtship and spawning event involving a large male Peacock Flounder and his harem of two females, each only half his size. The male, free-swimming with colors blazing and his pectoral fin set high, caught our eye. We followed him to the first of the two females where he settled to the sand and began nuzzling her side. Showing little interest, the female repeatedly turned away from his efforts. The male swam to the second female and repeated his routine with similar results. He then swam away, disappearing in the distance. Five minutes later he reappeared, courted each female in turn, and once again sped away. He returned for a third visit just after sunset, approaching a female with his pectoral fin raised high. Without hesitation she raised the front of her body and slipped onto his back. The male lifted off rapidly toward the surface with the gravid female centered on his back. At their pinnacle, 25 feet above the bottom, a cloud of gametes exploded from the pair. Caught by the currents, the brood of thousands vanished into the approaching night.

The high spawning rise carries the pair toward the surface.

Above: The spawning rise ends with a release of gametes beneath the surface.

Frogfishes
Family Antennariidae

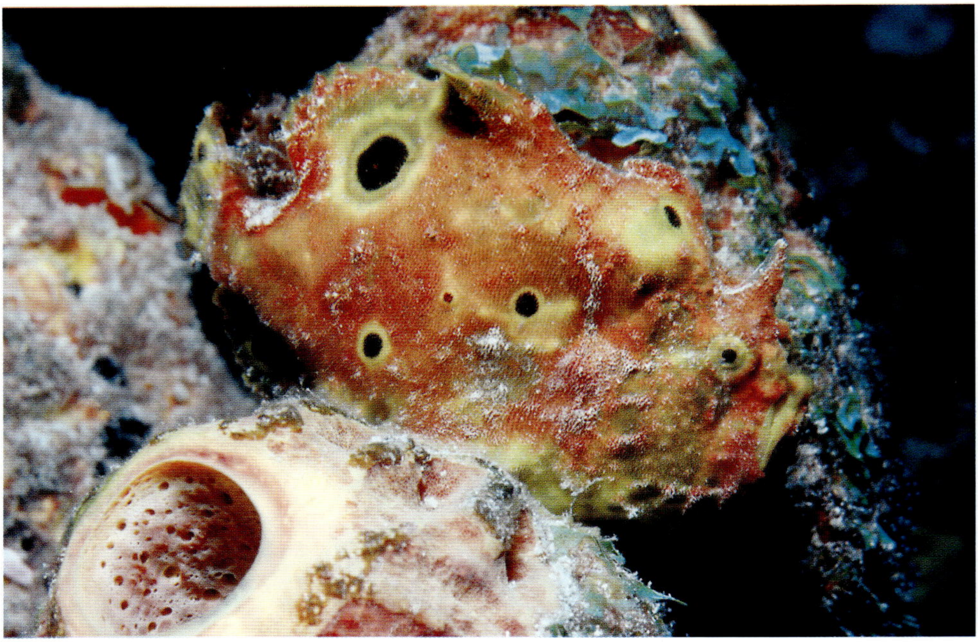

The occellated spots on Longlure Frogfish mimic the outcurrent siphons of the sponge colonies where they typically perch.

Although frogfishes are little more than fist-sized stomachs with large mouths, the oddball eating-machines are among the most sought after fishes on the reef. The trouble is finding them. The same lumpy texture and color-changing camouflage that conceal frogfishes from predators and prey also hide them from divers. Nestled within a stand of sponge, the Longlure Frogfish, *Antennarius multiocellatus*—the Caribbean's most commonly sighted family member—mimics its perch down to its ocellated markings resembling the sponge colony's outcurrent siphons. If the same fish moves to an algae-covered ledge, it will, over a period of hours to weeks, alter its color and pattern to mimic the turf-covered rocks just as cleverly.

The frogfishes' ability to change their appearance has led to the formal description of 165 species even though today taxonomists recognize only 47 species worldwide: six of which inhabit the waters of the tropical western Atlantic. Although found across the Caribbean, the largest frogfish populations occur in Barbados, Bonaire, Dominica, St. Croix, St. Lucia and St. Vincent. Interestingly, seahorses also reach their population peaks at many of these same islands. Frogfishes are typically globular, with rough skin formed by modified

Right: *Frogfishes periodically stretch their mouths open—a behavior common to most predatory fishes.*

An undamaged lure of a young Longlure Frogfish mimics a crustacean.

scales, tiny eyes and a small tubular gill opening behind each pectoral fin. The fin rays of the rounded dorsal, anal and tail fins are buried beneath thick flesh, and the two sets of paired fins (pectoral and pelvic) resemble stumpy legs complete with toelike appendages. Topping off the outlandish outfit, the first spine of the dorsal fin is modified into a thin, whiplike rod tipped with a lure. With the pole, known as an illicium, and the bait, called an esca, the lie-in-wait predators fish for dinner by dangling the baits enticingly in front of their jaws.

When potential meals are nearby, the rod swings forward from its resting place on the head and jiggles the bait. As curious prey enters its strike zone, the frogfish lowers slightly and, as if shot by a cannon, blows its mouth out to 12 times normal size. The victim, along with a considerable volume of water, is sucked through the mouth and into the stomach as the pectoral fins simultaneously clamp the tiny gill openings shut. Measured at six milliseconds, the strike is the fastest recorded for any vertebrate predator. Attacks are so swift that schooling fishes seldom react to the sudden loss of a neighbor.

If hunting is good, individuals remain for days or even weeks without moving more than a few feet. They frequently establish their ambush sites near beds of filamentous algae hoping to attract roving groups of young herbivores. As algae-eating fishes are consumed the surrounding growth flourishes, attracting even more prey. Frogfishes are not picky eaters. Although small fishes top their food list, they also consume crabs, mantis shrimp and other crustaceans. Not all

Right: *A Longlure Frogfish angles for dinner with a small, previously damaged lure.*

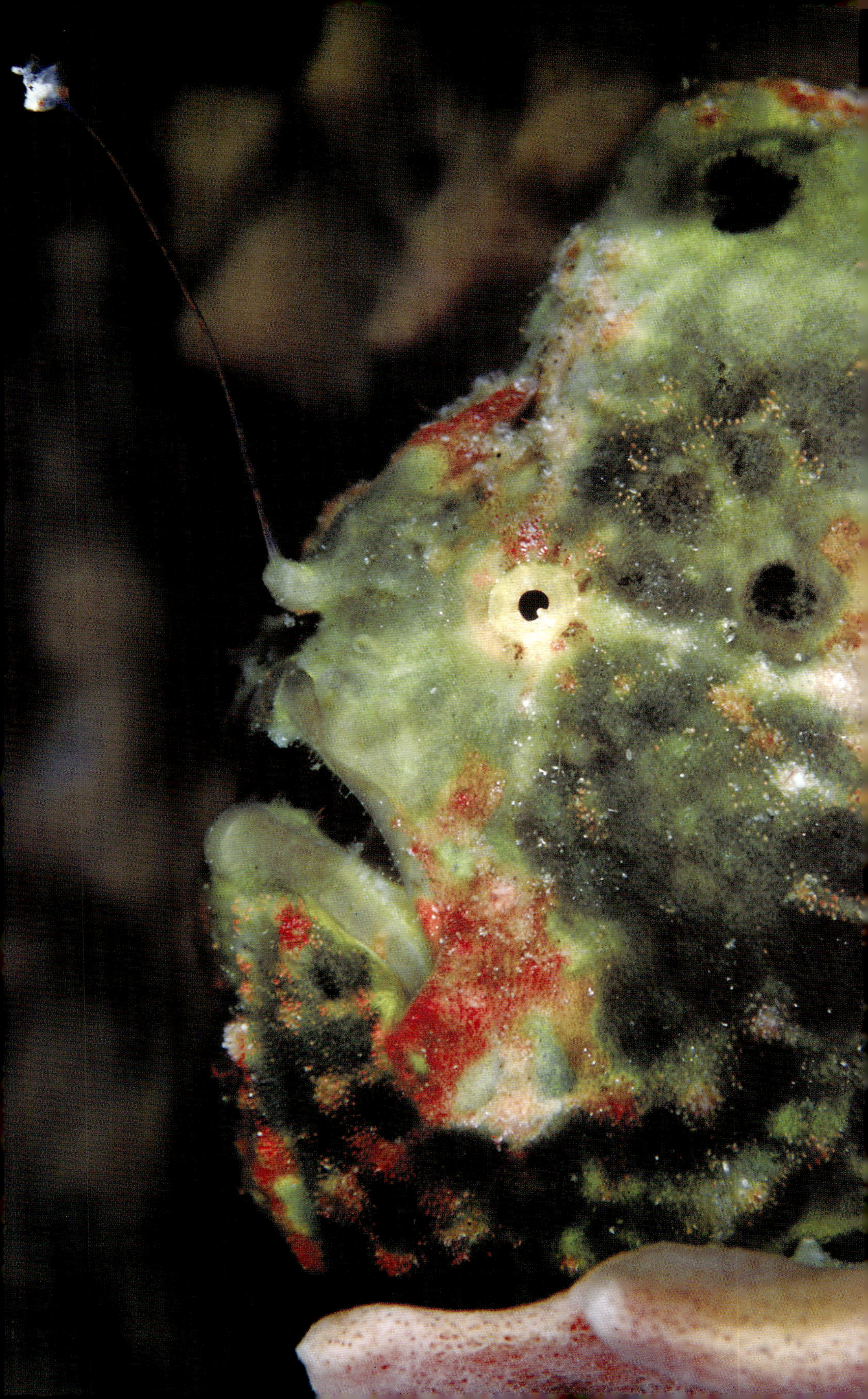

A frogfish perches on rope sponge, high above the reef.

victims are small. A Longlure Frogfish was reported to have consumed a squirrelfish longer than its own body. Captured prey too large to pass directly into the stomach is held by tiny, recurved conical teeth while being positioned for swallowing. Frogfishes are so voracious that they regularly grab fishes too difficult to handle. We watched an individual cough out a spine-laden butterflyfish just seconds after capture, and an aquarium specimen was found floating dead with an inflated porcupinefish inside its stomach. However, frogfishes appear to be immune to the poisonous spines of scorpionfishes, a regular item on their menu.

Fishing is used in about 80 percent of captures. In some cases, curiosity rather than hunger attracts victims to the dancing lure. At other times, small territorial neighbors associate the movement as a threat and attack. Some victims, believing that the frogfish is part of the substrate, unwittingly swim into its range. Surprisingly, frogfishes also fish at night. In the dark, the lateral-line sensors of cardinalfishes and other nocturnal feeders pick up the vibrations of the wiggling bait. Chemical attractants may also help to attract prey. By the nature of their use, the thin pole and fleshy lure take a beating. When not in use, the rod is laid across the head and the lure is tucked close to the dorsal fin. After an illicium and its attached esca were experimentally removed, the appendage showed signs of growth in five days; in six months it regained its original length, but the lure failed to regenerate to its prior shape.

Even though their success as hunters depends, in great part, on remaining motionless, frogfishes

A tiny, recently settled frogfish.

Frogfish occasionally swim quite rapidly.

The bodies of Sargassumfish match the color and shape of Sargassum *foliage.*

A courting male (right), the smaller of the pair, remains close to his mate for several days as she swells with her expanding egg raft.

occasionally stalk prey. With dorsal fins raised and walking on their paired fins, the hunters approach rapidly at first, but slow, crouch and lower their dorsals as they near the target. In rare instances frogfishes pursue prey while fishing with their lures. As many aquarists can attest, frogfishes are gluttons, consuming whatever food is provided and, if precautions are not taken, other fishes in the tank. One four-inch specimen readily gobbled down two to three goldfish each day for an extended period without showing ill effects. If given a chance, frogfishes will eat their own kind. Sargassumfish, *Histrio histrio,* are regularly found with a dozen or more smaller individuals inside their stomachs. In an aquarium, one of two equal-sized specimens ate the other.Except during

A large female (upper left) and two smaller suitors to her right each blend in well with their surroundings even though all three display different color patterns.

Right: *A female in a rare event, is accompanied by three males during a spawning rise (upper left). Females discharging egg rafts (bottom left & right).*

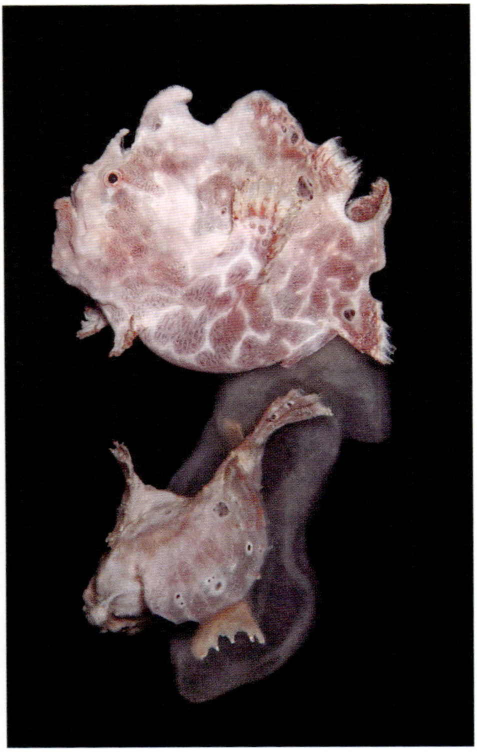

A Striated Frogfish moves slowly across a sand and algae bottom stopping occasionally to extend and twitch its large, three-lobed worm lure.

courtship and spawning, frogfishes tend to be solitary. If approached closely by another individual, the defender will take a position in front of its ambush site, raise on its pectorals, spread its medial fins, lighten, open its mouth, quiver and lean in the direction of the trespasser. In most confrontations the intruder will move away slowly, but occasionally will respond with a similar display. In rare instances, threatened frogfishes have been known to gulp water to increase their size. Likewise, if a frogfish is grabbed by a predatory fish, such as a lizardfish, it will draw in water to inflate its body making itself more difficult to swallow.

Given their awkward, sacklike shape, it is a wonder frogfishes can move about at all. In addition to hunting, they also move to find new ambush sites, search for mates and to flee when threatened. As one might imagine, their gaits are ungainly. The two leglike pectoral fins are used for force, and the much weaker paired pelvic fins, located close together under the throat, provide support. In "crutching"—the most frequently used walk—the pectoral fins push the body forward until the weight rests temporarily on the extended pelvic fins, allowing the pectoral fins to swing forward for the next stride. During a "modified walk" the pectoral fins alternate strides, much like terrestrial vertebrates, with the pelvic fins providing little more than support.

Frogfishes swim in three ways. The most frequently used style employs a rhythmic waving of the body and tail that moves the fish at a rate of a body length per second. Frogfishes can jet away at about the same speed by forcibly funneling large mouthfuls of water through their restricted gill openings. When panicked, frogfishes thrash their tails, force water out their gill openings and stroke furiously with their pectoral fins. The all-out effort causes the fishes to dart erratically at a rate of five body lengths per second.

Female frogfishes broadcast spawn releasing long, gelatinous rafts of eggs that unwind from the scroll-like shape of the ovaries into three- to four-foot ribbons. Within 48 hours the buoyant egg rafts, containing between 48,800 and 288,000 eggs, soften and disintegrate. Days later, the fertilized eggs hatch into one-millimeter pelagic yolk-sac larvae, which develop in the open ocean until viable to settle back to the reef.

Most species of frogfishes become reproductively active for a period of days

A gravid female (right) rests passively on the bottom panting while a frisky male repeatedly circles and caresses her with his extended fins.

After much effort, the male is finally able to wedge himself beneath her swollen body.

to weeks. During this period a single female may produce several rafts of eggs each released days apart. As a female's eggs mature she attracts suitors by releasing chemical attractants, known as pheromones, into the water. If multiple males arrive, one usually proves dominant chasing the others away, however at times two or three males remain nearby as the eggs ripen and may attempt to spawn simultaneously. As the time of spawning nears usually only a single male initiates courtship, nestling next to the female and rubbing against her sides while twitching and spreading fins. Courtship can last most of the day prior to spawning, which typically takes place shortly after sunset. At locations subject to a substantial tidal flow the release of the egg raft tends to be coordinated with outgoing tides. As the time of fertilization approaches, the male's courtship becomes more animated and the swollen female begins to rhythmically open and close her mouth as if panting.

The images (below and previous page) follow the courtship and spawning

The courting male, using the power of his tail fin, forces the larger female several feet off the bottom where she releases her bouyant egg raft.

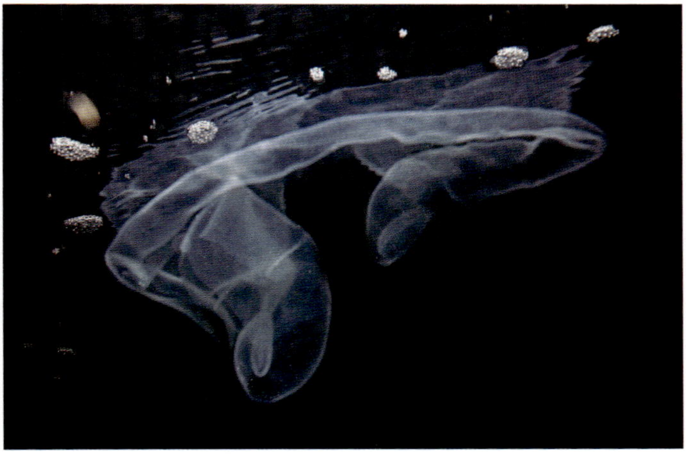

The released egg raft unfolds from the scroll-like shape of the female's ovaries as it slowly rises toward the surface.

behavior of a pair of Striated Frogfish, *Antennarius striatus,* found on a shallow sand bottom in Florida's Intracoastal Waterway. When we arrived just after sunset, an aroused male, with fins erect, circled a passive female, swollen to the size of a grapefruit. After struggling for some time to nuzzle beneath her tail the male was finally able to wedge his head under her body. In an instant the pair, propelled by the male's thrashing tail fin and stabilized by outstretched pectoral fins, lifted vertically a few feet above the bottom where, in a blur of dislodged sand, the female spun rapidly to the right simultaneously releasing a bouyant raft packed with eggs. As the gelationous ribbon exploded from her body the suitor spun in the opposite direction expelling an invisible cloud of sperm, which along with seawater is absorbed into the pores of the buoyant mass that slowly rises toward the surface. The spent pair immediately dived for the bottom where they unsteadily waddled off in opposite directions. From takeoff to touchdown the spawning rise required 13 seconds to complete.

As the female discharges her egg raft, the male spins in the opposite direction releasing sperm.

Groupers & Sea Basses

Families Epinephelidae & Serranidae

Black Grouper, a member of genus Mycteroperca, *primarily eat fish.*

In the tropical western Atlantic, members of the grouper and sea bass families range in size from Goliath Grouper, *Epinephelus itajara,* the largest reef fish in the Caribbean to the Bantam Bass, *Parasphyraenops incisus,* measuring less than two inches. In between are such curiosities as the small, cryptic reef basslets, genus *Liopropoma,* night-prowling soapfishes, genus *Rypticus*, and the enigmatic hamlets, genus *Hypoplectrus*—a group of anatomically similar fishes that, for yet inexplicable reasons, come in an array of color patterns. Although most sea basses are solitary carnivores feeding primarily on fishes and crustaceans, two species, the Atlantic Creolefish, *Cephalopholis furcifer,* and the small Chalk Bass, *Serranus tortugarum,* pick plankton from open water.

The word "grouper," previously describing medium and large bottom-dwelling serranids, comes from the Portuguese "garoupa," a derivation of a name native Brazilian fishermen used for large species of sea basses. Today, following a reorganization of grouper and sea bass classifications, the common name grouper primarily refers to species in genera *Epinephelus* and *Mycteroperca*. Most inhabit shallow tropical seas, but a few species survive in temperate waters and several, including Snowy Grouper, *Hyporthodus niveatus*, Misty Grouper, *H. mystacinus*, and Warsaw Grouper, *H. nigritus,* generally live well below safe diving limits.

The strapping, top-level predators are secretive but curious fishes that, in the days before spearfishermen, would swim confidently over to have a look at an approaching diver. A well-developed air bladder allows the bulky predators

Orangeback Bass—genus Serranus.

Shy Hamlet—genus Hypoplectrus.

Candy Basslet—genus Liopropoma.

Greater Soapfish—genus Rypticus.

Atlantic Creolefish, a planktivore.

Chalk Bass, a planktivore.

to calmly hover or maneuver easily through caves or overhangs within their home ranges, typically located around high-profile coral reefs and wrecks that attract food, provide shelter and have cleaning stations. Enjoying the services of cleaning shrimp and fishes is a favorite and often prolonged pastime for larger and mid-sized species. Mature groupers show a strong homing tendency; most tagged specimens, released far from their ranges, make their way back to the area where originally captured.

Nearly all groupers can switch colors and body patterns quickly and can also lighten or darken radically. These often striking transitions are triggered by changes in habitat, water depth, light levels or the fishes' dispositions. Changes in patterns frequently occur when groupers feed, attend cleaning stations,

A Tiger Grouper with mouth open and gill covers flared—a customary cleaning pose.

spawn, rest or signal members of their own species. Their many appearances can make identification tricky. For example, Yellowfin Grouper, *Mycteroperca venenosa,* most frequently display a series of tan body blotches and spots superimposed over a light background, but they can swiftly brighten to red or turn so dark the blotches disappear completely. The brown undercoat of Rock Hind, *Epinephelus adscensionis,* pales when the bottom-hugging species slips off the reef and moves over white sand. Red Grouper, *E. morio,* which normally wear tobacco-brown coats with a scattering of white dots, acquire irregular blotches of white across their heads and bodies when resting on sand. Coney, *Cephalopholis fulva,* a colorful moderate-sized grouper, can rapidly change

A juvenile Snowy Grouper hovers at the edge of a deep dropoff. Only the juveniles of this deep-water species make their homes in shallow water. Adults inhabit depths below 800 feet.

between a solid reddish brown and a distinct bicolor pattern. Their two-tone attire is generally displayed in shallow-water, during the early morning, again late in the afternoon, or while stalking prey. Several groupers modify their appearance during reproduction. The courtship patterns of Nassau Grouper, *E. striatus,* are the most dramatic. The species exhibits at least four distinct reproductive patterns, but the most impressive is the flamboyant black-and-white coat showcased late in the day when spawning is imminent.

Giant groupers begin life as pelagic eggs spawned annually. The eggs hatch into translucent larvae that quickly develop elongate spines and large mouths. After drifting with the currents for 25 to 45 days, the larvae, at approximately one inch, settle to the bottom in species-specific, shallow-water habitats where they hide while completing their metamorphoses into juveniles. Nassau Grouper begin life on the sea floor hiding inside algae-covered coral clumps in mangrove lagoons; Red Grouper in seagrass beds; Goliath Grouper in mangrove forests; while Gag and Black Grouper, *Mycteroperca bonaci,* swept inshore by tides, settle on oyster reefs in high-salinity estuaries.

Little is known about the juveniles' growth rates, feeding habits or habitat shifts as they mature. During early development, most species feed almost exclusively on crustaceans. One study of the early life stages of Nassau Grouper found that an increasing demand for food forces rapidly growing juveniles to leave their original settlement sites at approximately two inches. By six inches, the youngsters take up a second temporary residence in protective breaks in the sea floor. They next relocate to shallow patch reefs or conch shells in seagrass flats and later move to more complex reef structures. Most large, sex-changed males resettle in deeper water, leaving a virtually all-female contingent in the shallower end of the species' depth range.

Although difficult to establish, the maximum age of several grouper species has been extensively examined because of their importance to commercial fisheries. Goliath Grouper, the largest Atlantic grouper, have been documented to live for 37 years, Nassau Grouper 31, Red Grouper over 25, Scamp, *M. phenax,* 21, Red Hind 17, Gag 16, Yellowfin Grouper 15, Black Grouper 14, Graysby, *C. cruentata,* and Coney 9 years.

Mass spawning aggregations of Nassau Grouper numbered in the tens of thousands before overharvesting.

Groupers are ambush hunters that feed on fishes, crustaceans and cephalopods throughout much of the day, with spikes in activity during the low-light periods of dawn and dusk. Small grunts and parrotfishes make up the bulk of their fish diet, while crabs top their list of crustaceans. Members of genus *Mycteroperca* primarily eat fishes, whereas the species in genus *Epinephelus* prefer crustaceans. A quick forward lunge, accompanied by rapidly expanding jaws and flaring gill covers, vacuums prey into the mouth, where it is swallowed whole. The forceful suction of Goliath Grouper and other big groupers is sufficient to suck lobsters from crevices. Coneys and Graysbys regularly hunt with eels and octopuses. This behavior, known as nuclear hunting, is discussed in the Feeding chapter (page 72).

The majority of sea basses and groupers are protogynous hermaphrodites. This sexual development strategy allows individuals to reproduce first as females and later, at a larger size, to change into males when certain yet unknown social or environmental factors are met. This form of sex-switching is considered by many to be a superior reproductive tactic, enabling populations to adapt more readily to short-term environmental fluctuations. Groupers and sea basses have adapted two basic versions of hermaphroditism. Most larger species are sequential hermaphrodites, with the majority of individuals in a given population making a complete and non-reversible sex change. A few smaller family members are simultaneous hermaphrodites, able to function actively in both the male and female role. Within each of these methods are numerous variations on the theme. Hamlets, *Hypoplectrus* spp., and Harlequin Bass, *Serranus tigrinus,* two simultaneous hermaphrodites, spawn as both males and females each night in a monogamous relationship, known as egg trading. Lantern Bass, *S. baldwini,* reproduce in small harems, with one large dominant individual spawning as a male while several other hermaphrodites function only as females.

Several larger groupers believed to be sequential hermaphrodites—Nassau Grouper, Tiger Grouper, *Mycteroperca tigris,* Black Grouper, Yellowfin Grouper, Goliath Grouper and the intermediate-sized Red Hind, Gag and Scamp—travel many miles during one or two months each year to spawn in huge aggregations at traditional breeding grounds commonly used for centuries by their ancestors. A Nassau Grouper was documented to journey over 60 miles in the Bahamas to join a spawning aggregation. Another Nassau Grouper in Belize relocated nearly 150 miles, and a Gag traveled more than 350 miles along the coast of the southeastern United States in a possible spawning migration.

The migrations to the immense aggregations, often involving hundreds or thousands of fish, are synchronized by the onset of the full moon and by water temperatures, typically during the cooler winter months. The traditional rendezvous sites are generally located near promontories along outer reef shelves at depths over 100 feet. Many basic questions remain unanswered about the great "grouper runs" of Nassau Grouper, which were first reported in scientific literature in 1972. For instance, it is still unknown whether individuals migrate each time their species gathers to spawn, if additional spawning activities occur outside aggregations or even if adults reproduce each year.

For several days prior to the actual spawning event, incoming fish collect at aggregation sites and mill about nervously near the bottom. Females, obviously swollen with eggs, huddle close to the protection of bottom structures. Shortly before the full moon, courtship behavior and color patterns intensify in late afternoon. Spawning takes place at or just before sunset within subgroups whose sexual composition varies among species. Female Nassau Grouper and Red Hind attract multiple males just prior to spawning rises. These females possibly spawn with multiple groups. Scamp, a medium-sized grouper, assemble in subgroups with up to 20 females attended by a lone male. The same species occasionally forms into clusters of more than 50 females and several males. Male Tiger Grouper and Yellowfin Grouper, that often aggregate at the same sites as Nassau Grouper, defend small territories where they are approached by females ready to spawn. As the late afternoon spawning period approaches, males more aggressively defend their areas of bottom from other males. Not all common, shallow-water groupers aggregate to reproduce: Red Grouper, Coneys and Graysbys pair spawn on home reefs with each of several females in their personal harems.

A Goliath Grouper moves about the reef with an entourage of small Bar Jacks.

Goliath Grouper

Epinephelus itajara

Without question, Goliath Grouper are the largest reef fish in the region and one of the two largest species of groupers in the world. The American Fisheries Society's Committee on Names of Fishes changed the fish's previous common name, Jewfish, to Goliath Grouper in 2001. This marked the first time a fish's name has been revised for political correctness since the freshwater Northern Squawfish became the Northern Pikeminnow, *Ptychocheilus oregonensis,* in 1948. The behemoths, estimated to reach eight feet in length and weigh more than 1,000 lbs., were frequently sighted by divers until the late 1960s. During the last half of the 20th Century the great fish populations experienced a decline at the hands of commercial and recreational fishers using hook and line, traps, and trawls. Later, during the early 1970s the top-level predators began to fall victim to spearfishermen, armed with explosive powerheads, who could easily swim right up to the fearless fish. For nearly a decade, huge carcasses of the magnificent creatures, hoisted high on block and tackle, gathered sightseers and flies at marinas from Florida to Mexico. By the 1980s, the few Goliaths remaining on inshore reefs were so wary of divers that they fled when approached. During the same period, many of the species' annual spawning aggregations, unmolested for centuries, were overexploited to the point of collapse by line fishing. Aggregations targeted off the southwest coast of Florida were reduced from between 25 to 150 fish in 1982, when first discovered, to between zero and five by 1989. The heavy fishing pressure and loss of mangrove nursery habitat became so excessive that an emergency ruling

was enacted in 1990, making it unlawful to take or possess Goliath Grouper along Florida's southwest Gulf Coast, the historical population center of the species. During the same year the no-take moratorium was extended to all federal waters, and the U.S. Virgin Islands followed suit in 1993.

By 1995, the laws protecting the slow-growing, long-lived species began to show positive effects. Many inshore wreck sites in the eastern Gulf of Mexico were once again homes to small but growing populations. Along Florida's southeast coast and in the upper Keys, areas where the species had not been reported by recreational divers for nearly 20 years, sightings started trickling in. Once the population began to rebound, spawning aggregations also reappeared off the southwest coast, and on several shipwreck and artificial reef sites off Palm Beach County on Florida's Atlantic coast. Today, Florida's charter boat fleet is transporting a new generation of divers to see aggregations of the great fish numbering in the dozens. Fortunately, divers have traded their spearguns for cameras this time around.

Goliath Grouper surrounded by a ball of scad.

Historically, *Epinephelus itajara* ranges across the Gulf of Mexico, Caribbean and Bahamas, up the Atlantic coast possibly as far as the Carolinas and south to Brazil; however, the largest population occurs in Florida waters. The big fish can be identified quickly by their huge size and a scattering of dark spots over their bodies and fins. Except during late summer reproductive seasons, when the fish migrate to deep-water spawning sites, adults are not particularly gregarious, usually establishing claims to protected shallow-water locations near wrecks or caves where they remain for extended periods. Goliaths, the only grouper capable of surviving for extended periods in brackish water, frequently enter harbors, bays and estuaries. Fishermen have reported the large fish feeding in tidal mangrove swamps so shallow that their tails thrashed the surface.

No one knows exactly how the opportunistic predators stalk their prey. Although generally sedentary and sluggish, they are capable of short bursts of speed. Prey is drawn into their gaping mouths with suction created when they suddenly expand their jaws and flare gill covers. The force of water exiting the gills is so powerful that clouds of debris swirl behind the lunges of feeding fish. Slow-moving, bottom-dwelling invertebrates such as Calico Crabs, *Hepatus epheliticus,* lobsters and shrimp make up the bulk of their diet; however a stomach content examination of nine specimens from the U.S. Virgin Islands and Puerto Rico revealed an array of invertebrates and fishes, including Southern Stingray, *Dasyatis americana,* Honeycomb Cowfish, *Acanthostracion polygonia,* spiny puffers, crabs, octopuses and the carapace of a small sea turtle. Juveniles subsist mainly on shrimp, crabs and small fishes. Large Goliath Grouper have been known to grab a spearfisherman's catch. In the early 1970s, an underwater hunter related that he made the mistake of getting between a hungry Goliath Grouper and food. While trailing a mesh bag containing three lobsters around a wreck site off Tampa, Florida, a 200 to 300 lb. Goliath came up from behind and in a rush engulfed both the bag and the diver's lower right leg. In his struggle to free himself, the young man lost his crustacean dinner, a new flipper, and two-thirds of the skin from his knee to his toes.

The species is known to make loud "booming" sounds produced by its swim bladder and surrounding muscles. The booms, which have been recorded throughout the day and night, reach peaks from 1:00 to 3:00 a.m., during the reproductive season leading researchers to speculate that they are associated with the act of spawning, which has yet to be documented. The booms are believed to have several other uses including territorial warnings and keeping tabs on other members of their species. Resident Goliath Grouper are also known to emit the thunderous sound when startled by an approaching diver. Researchers have heard juveniles boom before taking bait, indicating a possible use to stun prey when feeding.

Research is beginning to expand our understanding of the species' early life. Large expanses of mangrove forest have been found to be vital for both settling pelagic larvae and juveniles until they reach 30 lbs. The scarcity of such habitat in the Eastern Caribbean is believed to be the reason for the species' limited population in the region. Ten Thousand Islands in the southwestern corner of the Florida Everglades, adjacent to the species' historic population center, is home to the largest concentration of young Goliath Grouper yet discovered. Juveniles, with large, pronounced spots and four or five dark bars, spend their first five to six years in the shallows where they seek protection beneath deep shelves created by the area's numerous mangrove islands. Juveniles have also been reported to inhabit brackish estuaries near oyster beds.

Subgroup of Goliath Grouper with courting males displaying pale forebodies.

Known to live for 37 years, but suspected to live longer, Goliath Grouper grow slowly at about four inches each year for their first six years; afterwards yearly growth declines to an inch and continues to decrease with age. Other than during spawning season, when the head region of some species visibly lightens, there is no visual difference between the sexes. There is some evidence of protogynous hermaphroditism in the species, with a few individuals possibly passing through an immature female stage before maturing as males. Males become sexually mature between four and six years at about four feet in length; females take between six to seven years to achieve sexual maturity at a slightly larger size.

Movement from home ranges to spawning aggregations is triggered each year by the full moon of July. Spawning is believed to peak around the new moons of August and September. Aggregation sites form in the vicinity of wrecks, artificial reefs and high profile rocky reefs in depths from 50 to 150 feet. During reproductive periods various sites attract between 30 to 100 fish. A few individuals reside on permanent aggregation sites; however most make annual migrations usually to the same spawning site. Several tagged individuals were found to migrate up to 240 miles from their resident territories. Larger Goliaths have a tendency to journey farther than smaller individuals.

While congregated, the fish either remain close to the bottom near protective structures or hover 10 to 20 feet up in the water column. An extended courtship period lasts from late morning to the middle of the afternoon. As males approach females their forebodies from pectoral fins forward turn pale, contrasting sharply with their dark bodies. Females exhibit no significant color change, remaining in a characteristic blotched pattern.

Since 2001, divers and researchers have been making regular visits to six of Florida's east coast spawning aggregations located off Palm Beach County from Jupiter south to Boynton Beach. Each aggregation annually hosts between 45 and 90 Goliaths that hang out in subgroups numbering a dozen or more individuals near high-profile artificial reefs. On rare occasions the aggregated fish coalesce into super groups. Unlike other large groupers that display distinctive patterns during reproduction, Goliaths, of both sexes, wear muted patterns composed of "a thousand shades of gray."

When not hovering above the bottom, or lazing like boulders on the bottom, Goliaths hunt, sometimes in resourceful ways. Using their tails and pectoral fins to fan sand and their heads to bulldoze debris, some of the fish excavate

Goliaths, of both sexes, wear muted patterns composed of "a thousand shades of gray."

Some Goliaths at aggregations fan shallow depressions in the sand where they settle for extended periods. The purpose of the behavior remains a mystery.

personal depressions along buried ledges around the aggregation sites where they settle for extended periods. Similar circular depressions, referred to as "fox holes," are dug in the open sand, normally away from the central sites. Researchers speculate that the sand pockets are used as blinds to help conceal the predators' bulky bodies from approaching prey; however their exact purpose remains a mystery. It is common for swarms of minnows and small jacks, known as scad, to swarm near Goliaths for prolonged periods. When the numbers become a nuisance the predators wander away from their subgroups with their pests in tow. Once in the open, marauding jacks dive bomb the baitfish forcing them into tight balls surrounding the Goliaths' heads—a predator's dream finished off in a few gulps.

Two apparent courtship behaviors are repeated from time to time throughout the period. The first, "vent nuzzling," occurs when a male moves slowly alongside a female from the rear and appears to inspect or nudge her vent area at the posterior of the abdomen. As the male approaches, the female rolls her body, angling away from the suitor. The male, usually persistent, swims underneath and tries from the opposite side. At this point, the female generally slips away. In another behavior the male approaches directly toward the side of a female that is either hovering or slowly rising. Once near, he positions parallel, and the couple slowly rotates before the female swims off. The male then moves to another female and repeats the process. Throughout the courtship, especially when approaching females, males emit low-frequency booming sounds clearly audible 100 feet away.

Unfortunately, short bottom times at depths of 100 feet or more severely restrict the study of the Goliath Groupers' reproductive behavior, and limits the chance of being on location when a spawning rise or release of gametes takes place. Recently, researchers determined that spawning occurs between midnight and 3:00 a.m. on the new moons of August and September. Evidence was based on nighttime collections of fertilized eggs, gonad biopsies to detect fully hydrated eggs ready for release, and recordings capturing an increase in booming calls during early morning hours. Spawning after dark is believed to limit egg predation by the schools of small planktivores so common on Goliath Grouper aggregation sites.

In the absence of visual cues, the calls during the nighttime are believed to signal an impending spawning rise, or possibly stimulate the hydration process. Because of the large quantity of milt produced, it's presumed that the grouper participate in group spawning rises with two to several males attending a single female—a common practice of aggregate spawners. Seasonal spawning is often synchronized with conditions to optimize larval transport to shallow mangrove nurseries where they settle after a 30- to 80-day pelagic stage. The spawning window insures settlement from September until early January, the beginning of the dry season, when heavy leaf litter is more quickly broken down from a rise in salinity, which promotes the production of prey for young juveniles.

Thick swarms of silversides or scad frequently cluster around Goliaths. When the numbers become a nuisance the predators occasionally feed on the fish.

Both male and female Nassau Grouper typically display a reddish brown barred pattern.

Nassau Grouper
Epinephelus striatus

Even though Nassau Grouper are the most widely studied grouper species in the tropical western Atlantic many important gaps in our knowledge of the species' ecology and behavior remain. The distinctive species, with five irregular, olive-brown body bars over a light background and a prominent saddle-spot on the upper tail base, have been documented to weigh as much as 55 lbs., reach a length of just over three feet, and live up to 31 years. The solitary and typically sedentary predators spend much of the year living within large home ranges in clear inshore waters with high-profile bottoms where larger males inhabit deeper depths than females. The fish have been observed regularly moving around over several hundred square meters of bottom encompassing overlapping ranges. Social hierarchy appears to be maintained by smaller individuals briefly flashing a submissive bicolor body pattern when encountering larger individuals.

The species is indigenous to the tropical western Atlantic and in recent history ranged from Bermuda (where once common, but now considered extinct) through the Bahamas, down Florida's east coast, and across the Caribbean. It is absent from much of the Gulf of Mexico and less prevalent in southern and eastern Caribbean waters. Historical records, established by examination of fossilized otoliths, indicate the fishes' presence in Florida waters as far back as 2000-1000 BC. Recently expanding research related to the behavior and ecology of Nassau Grouper has been driven by the species' unprecedented decline in population numbers and size.

Nassau Grouper have long been one of the most important market-fish harvested in the tropical western Atlantic. The first official survey of Puerto Rican fisheries, compiled in 1900, found the Nassau Grouper to be "a common and very important food fish, reaching a weight of 50 lbs. or more." As recently as the 1970s, it was the fourth most common shallow-water fish landed in Puerto Rico. A 1968 reef fish biodiversity and abundance study at Alligator Reef near Marathon in the Florida Keys recorded multiple Nassau Grouper sightings on every survey dive. Today, however, the species' numbers have plummeted so drastically that it is considered commercially extinct in Puerto Rico and the U. S. Virgin Islands, and so threatened in Florida that a no-take moratorium was imposed in 1993. In 1994 volunteer research divers from the Reef Environmental Education Foundation (REEF) logged only 19 sightings in 145 dives in Key Largo, where spearfishing has been banned since 1960, and five sightings from 158 dives in Key West. Only 18 sightings were made from 204 dives in the Dry Tortugas, a region considered to be one of the species' historic population centers.

There are no visual distinctions between the sexes outside courtship and spawning and even during reproductive periods color patterns aren't specifically diagnostic of sex. Nassau Grouper were once believed to be protogynous hermaphrodites (changing sex from females to males) like several other family members. However, because there is a significant size overlap between the sexes and at least some individuals are born as primary males, this view is now questioned. Recent studies have disclosed that juveniles go through a bisexual phase before reaching sexual maturity, then lose most of the gonadal tissue of one sex. Because individuals retain remnants of their bisexual phase, it is difficult to confirm a sexual transition in later life. Some individuals more than likely remain either male or female (gonochoristic) throughout life. A portion of females could possibly change into males at a large size, but this premise is difficult to substantiate with so few large females remaining due to fishing pressure. Sex ratios appear to be biased somewhat in favor of females. However, the majority of assessments were made from landings made during annual spawning aggregations perhaps also skewing numbers due to heavy fishing. It has been reported that larger sex-charged males more readily take bait. But a more likely scenario from data in the Cayman Islands shows males arriving earlier and remaining longer at aggregations (400 hours for males compared to 320 hours for females) exposing them to prolonged fishing.

Although all large grouper populations have decreased during recent decades, none has declined more quickly or dramatically than the Nassau Grouper. The downfall of this relatively large apex predator is directly related to its annual mass-spawning behavior, which makes it a sitting duck for overexploitation. An extensive gathering of large fish was first reported in Cuba during the 1880s, but the phenomena did not become common knowledge outside local fishing communities or appear in scientific literature until Dr. C. L. Smith observed a Bahamian spawning aggregation using scuba in 1971.

The following year he published an account of his observation, including dramatic photographs, in *Transactions of the American Fisheries Society.*

Clarence Lavett Smith, Jr. (Smitty), a prominent ichthyologist, worked at the time of the historic encounter as a curator at the American Museum of Natural History in New York, a position he held for 35 years. Over his distinguished career he wrote more than one hundred scientific books and papers including an early co-authorship of *The Grouper of Bermuda* followed with Synopsis of biological data on groupers, and *A Revision of the American Groupers:* Epinephelus *and Allied Genera.* By luck Dr. Smith just happened to be the right person in the right place when a colleague at the Lerner Marine Lab in Bimini, Bahamas, where Dr. Smith was engaged in field work, mentioned a report of a local fisherman indicating that Nassau Grouper were schooling at Cat Cay not far south of Bimini.

Along with three colleagues and a fishing guide, Dr. Smith loaded tanks in the lab's launch and set off for the site the following day. When the researchers arrived just past noon on January 22, 10 fishing boats bobbed at anchor over an area extending for 800 meters above a deep outer reef shelf. One boat crew reported that they began fishing the aggregation five days earlier. On the first day a single boat reported hauling in 600 lbs. of fish. Even though the winds increased and the sea deteriorated, vessels from as far away as Abaco and Miami continued to arrive at the lucrative slaughter for the next several days.

A "grouper run" of Nassau Grouper moving along a wall toward a traditional spawning site in the Bahamas. Photo from the early 1970s before fishing decimated the species.

Underwater, the grouper, ranging from one and a half to three feet in length, completely dominated the clear seascape extending beyond a sweeping, 125-foot-deep, reef-covered outer bank. While many of the grouper rested in scattered groups on the craggy sea floor, the majority hovered 10 to 15 feet up in the water column, while a few nosed around fish pots—the primary method employed for taking the fish. At the time of their afternoon dives nearly a third of the grouper displayed conspicuous black-and-white, two-tone coats and many were noticeably swollen with gametes, but no spawning, pairing or social interaction was observed.

Before heading back to the lab, the researchers estimated that the aggregation contained 30,000 fish, later amended to between 30,000 and 100,000. The mass of fish involved in such large aggregations is difficult to calculate even with multiple researchers conduct surveys during multiple dives on multiple days. Dr. Smith's underwater estimate was made during a single afternoon by multiplying a rough calculation of the aggregation's area (arrived at by noting the distribution of the fishing boats), by the density of fish (estimated to be one fish every five square meters). His increased assessment of possibly as many as 100,000 fish was arrived at later after taking into account that between 35,000 and 40,000 fish had been caught at the site during the 1970 aggregation season.

The day following Dr. Smith's dives, the sea of grouper disappeared as mysteriously as they arrived. From what we know today it is likely that Dr. Smith's team arrived at the site on the final day of spawning, but his team left before dusk when spawning takes place and the fish began dispersing back to their home territories. This was the first of numerous large fish species that have since been documented to spawn in such "spectacular abundance." Dr. Smith had the prescience to include a cautionary observation toward the end of his 1972 research paper describing his observations: "The existence of localized spawning sites where most of the reproduction takes place means that the grouper fishery is more precarious than we have heretofore suspected." So true, the Cat Cay aggregation was fished into oblivion in just a few short years after Dr. Smith's pioneering dives. Nearly 50 years later the largest Nassau Grouper spawning aggregation ever observed on scuba has failed to reform. Eighty-four Nassau Grouper spawning aggregations have since been documented across the species' range: 33 in Cuba, 18 in the Bahamas, 10 in Belize, 6 each in both the Cayman Islands and Mexico, with others scattered from Bermuda to the U.S. Virgin Islands and Honduras. Even more poignant, of the 84 spawning aggregations known to exist in the 1970s and 1980s fewer than a dozen remain in 2018. Of these only a single site off the west end of Little Cayman Island, protected and monitored for 16 years by REEF and the Cayman Islands Department of the Environment, has substantially increased in size.

The moon is believed to cue the annual grouper gatherings, but there are strong indications that surface water temperature of about 77 degrees also plays a critical role. Reproductive-sized Nassau Grouper (between 16 to 20 inches at approximately four years of age) migrate to traditional aggregation sites, where they routinely return to spawn each year. Aggregations are generally

Reproductive-sized Nassau Grouper migrate to traditional aggregation sites each winter where they spawn at sunset.

located near prominent topographic features, often off the ends of islands or over seaward projections from the reef shelf. Each winter, before fishing took such a heavy toll, it was estimated that as many as 500 grouper moved en masse along outer reef lines on their way to spawning sites, in what fishers dubbed "grouper runs." Some fish only have to travel a few miles from home while others, such as those in the Bahamas, used long shelf contours to guide journeys of up to 125 miles with some estimates of 205 miles. Bahamian fish have been documented remaining at aggregations for a day or two and staying away from their home territories for an average of 12 to 30 days. In the Cayman Islands much smaller runs, made up of 5 to 100 fish, have been sighted from the first until the fifth day after the full moon, traveling along the island's shelf edge at a depth of 65 to 108 ft.

Before their Nassau Grouper were fished to extinction in Bermuda, the northern extent of the species' range, aggregations formed from May to July, when the water reached a desired temperature. In the Bahamas, gatherings at "grouper holes" take place in December and January; but farther south in the warmer waters of the Cayman Islands, the fish do not typically gather until January and February.

A researcher who has monitored the Little Cayman West End aggregation, for several years predicts the local peak spawning month by counting the number

Minutes before spawning begins, most male and female Nassau Grouper exhibit the striking bicolor pattern.

of days between the winter solstice and the full moon in January. If the number is more than 30 days, February will be the primary spawning month, if less than 30 days, it will take place in January. When rediscovered by fishers in 2001 the Little Cayman aggregation hosted as many as 7,000 fish, before local fishermen removed 2,000 fish in a matter of a week. The following year (2002), the first year the aggregation was monitored by REEF in conjunction with the Cayman Islands Department of the Environment, the aggregation was estimated at 5,000 grouper. That same year another 2,000 fish were taken and suddenly in two years another spawning aggregation was heading for extinction. Then, at the end of 2003, a fishing ban was put into place and things began to change.

Those individuals displaying the "white-belly" pattern are presumed to be females because of their distended abdomens.

During the first five years of monitoring, the grouper were found to frequent the West End site for either two, or in some cases, three consecutive months. Months attracting the largest number of fish varied. February was the prime month three years and January twice. Each month's spike corresponded with the 30-day computation based on the winter solstice mentioned above. Then data started to come in and patterns began to form. Larger, older individuals tended to visit the site more regularly and remain longer than smaller individuals. And most grouper returned to their home reefs between spawning

Research diver visually assesses size and number of Nassau Grouper attending aggregation, while also recording previously tagged fish.

months. While away, some grouper investigated a fished-out site located at the opposite end of the island, approximately a six-mile journey; and one individual took seven turns around the island searching for action before ending up back at the west end. During the March full moons of two successive years between 200 to 300 grouper returned to the West End site for a third consecutive month, however spawning was never observed. The days with the largest concentration of fish varied from the first to the fifth day after the full moon. The yearly variations between prime spawning months and days after the full moon correlated well with favorable conditions for fertilization (slack tide) and offshore transport of eggs and larvae. Current drifters deployed at the Little Cayman West End site at the time of spawning revealed eddies forming just offshore, a condition conducive to retaining offspring in local waters. Extended models tracking the larvae's long 35- to 40-day pelagic voyage indicated a similar trend, following a long circuitous route before eventually heading back near where they were spawned.

Nassau Grouper arrive at spawning sites over several days, forming into a cloud above the bottom while others rest on deep sea floors often next to cleaning stations tended by cleaning gobies, *Elacatinus* spp. Fish gathered at aggregations and during runs exhibit one of four distinct color and marking patterns/phases that can be integrated or quickly switched from one to another. The first, the normal, year-round, five-bar pattern is commonly displayed in varying intensities during the morning hours. The second pattern, the "white belly," also most prevalent earlier in the day, is a noticeably blanched version of the five-bar pattern where the faded sidebars disappear as they spread onto the all-white belly. Because those exhibiting the white-belly phase frequently have swollen abdomens they are presumed to be females. The "dark phase" is believed to be worn by both sexes; however females in the dark phase consistently lead spawning rises. The "bicolor" is by far the most dramatic and symbolic pattern. Although exhibited by a few individuals throughout the day, it becomes more prevalent as the time to release gametes approaches. By evening, when spawning is imminent, nearly 80 percent of the fish in large aggregations display the unmistakable two-tone wardrobe that divides their bodies longitudinally. The upper half of the body, from the eye to the tail base, intensifies to black, while the lower portion, including the pectoral, ventral, anal and tail fins, lightens to snow-white. A white band running above each eye from the lips to the dorsal fin accentuates the striking presentation. The bicolor scheme, worn at times by both males and females, is believed to be a sign of submission, possibly reducing aggression between members of the normally territorial species.

The West End aggregation in 2018 included a large portion of young fish, probably offspring produced since Cayman Islands banned fishing on aggregation sites.

The gonads of male Nassau Grouper comprise 10 percent of their body weight—more than 20 times greater than other Caribbean family members.

Group spawning bursts/rises involving multiple males following a single female have been recorded on multiple evenings beginning as early as an hour before sunset and lasting until 20 minutes after sunset, but are usually of shorter durations. The days with the most recorded spawning bursts varied from the first to the eighth day after the full moon. In some cases limited spawning took place the day following peak spawning. It is speculated that these post-spawning episodes possibly involve larger more fecund fish or late arrivals.

On the first year of monitoring at Little Cayman (2002), preceding spawning, the majority of the fish gathered into an upward-tapering cone-shaped formation with its broad base near the sea floor. Spawning bursts originating throughout the cone began 100 to 36 feet below the surface. A similar cone formation was documented in the Bahamas and the U.S. Virgin Islands. However, during subsequent years the spawning pattern altered significantly at the West End site. Instead of massing into a cone the fish spread across the shelf lip and out to sea at a depth of about 100 feet. As late afternoon approached, the fish, which had been milling around randomly just above the bottom, rose 30 to 40 feet and moved toward deeper water. At this stage, nearly all the fish were in either the bicolor or dark phase. Those displaying the bicolor phase, believed to include both males and females, ascended in the water column, while a few dark-phase females hovered like blimps several feet above the sea floor. The males began to swirl in the dimming light as they tracked criss-crossing trails of pheromones toward swollen females. Within seconds of locating their targets, the males started circling, executing abrupt turns and aggressively nudging, causing the females to bolt horizontally, which attracted packs of frenzied males. Females easily shook off smaller subgroups, but eventually gave in to the pressure, blasting toward the surface with up to 20 suitors in pursuit. At the pinnacle of their rises the females erupted, releasing tens of thousands of tiny transparent eggs simultaneously engulfed within colossal clouds of sperm just before the fish sailed apart in a starburst maneuver and angled down toward the bottom. At times trailing fish arched over the gamete clouds, apparently shedding milt. On peak spawning nights as many as a hundred spawning bursts filled the sea with gametes, quickly lowering water visibility. During a single spawning rise, a one- to two-foot female is believed to release between 400,000 and 600,000 eggs. The males' prodigious amount of milt is an adaptation of their highly competitive reproductive strategy. Not surprisingly, the gonads of male Nassau Grouper comprise 10 percent of their weight—more than 20 times greater than other Caribbean family members. Pair-spawning groupers generally invest about one percent of their body weight in sperm production.

It has not been conclusively established if Nassau Grouper pair-spawn either at, or away from aggregations. However it appears that the species only breeds during once-a-year winter aggregations. It also seems that a yet unknown population threshold must be sustained before aggregations function properly. If cues necessary for congregating and spawning are learned behaviors they may be lost to a population after the removal of a substantial number of older individuals. Without the knowledge of site locations, timing and necessary behaviors young fish, just reaching sexual maturity, are left without guidance. If this is the case, the loss of old-timers from over-fishing coupled with the fact that the Nassau Grouper's larval offspring have a strong tendency to return to local reefs will make it difficult for remnant aggregations as well as population numbers to recover. This seems to be the case in the Bahamas where a heavily fished aggregation made up of fewer than 100 fish, was observed exhibiting a subdued courtship and spawning behavior, and less than half the assembled fish displayed the bicolor pattern. Most exhibited either the normal or white-belly phase and only a few changed into the dark phase. At sunset, bicolored fish followed a few barred individuals that, after a short run, retreated to the bottom and sought refuge without releasing gametes. This also seemed to be the case in the Cayman Islands and St. Thomas where a small number of fish arriving during spawning months at remnant aggregations also displayed fewer spawning patterns and exhibited unenthusiastic courtship behavior. It was also noted that a small aggregation of approximately 1,500 fish located off Cayman Brac, not far from Little Cayman, remained at their aggregation for longer periods of time than those at the larger West End site. Older individuals also tended to hang around aggregation sites longer than smaller fish, exposing these larger, more prolific spawners to extended fishing pressure.

Nassau Grouper are not the only species to use traditional aggregation sites for spawning. Researchers monitoring the Little Cayman West End site have documented courtship behavior and spawning for six additional species: Tiger Grouper, *Mycteroperca tigris,* Yellowfin Grouper, *M. venenosa,* Horse-eye Jack, *Caranx latus*, Bar Jack, *C. ruber,* Black Jack, *C. lugubris,* and Mackerel Scad, *Decapterus macarellus.*

After fertilization, the groupers' positively buoyant, one-millimeter eggs rise to the surface in three to five hours and hatch into yolk-sac larvae 23 to 40 hours later. Two days after hatching, the pigmented eyes of two-millimeter larvae are visible. With their yolk sac dwindling, larvae must begin feeding within 60 hours. In this early stage larvae filter-feed on microscopic unicellular plants sieved from the water by a sticky network of mucus filaments secreted from glands in their gill arches. Later they dine on shrimplike mysids and copepods, and with growth, fish larvae as large as nine millimeters. Their ability to capture such comparatively large food items is made possible by the extension of an upper jaw, which nearly doubles the width of their gape. Twenty five days into their pelagic voyage, the translucent, seven-millimeter grouper are trimmed with foredorsal and anal fin spines extending more than half the length of their bodies. During the last week of their 35- to 40-days at sea, the survivors transform into pre-juveniles nearly an inch in length and, along with cohorts, are carried by cross-shelf winds and currents through tidal cuts, where they settle in shallow inshore habitats.

A group of researchers in the Bahamas found recently settled Nassau Grouper hiding within algae-shrouded colonies of Finger Coral, *Porites porites,* scattered about a shallow, mangrove-lined bay. In different regions, similar habitats may also serve as nursery sites. At one and a half inches, the barred pattern appears and, except for size, the little fish resemble adults. During the next four or five months the secretive juveniles hide within algal masses while growing nearly a half inch each month. When the juveniles are three to four inches long, increasing demands for food force them to relocate to nearby breaks in the bottom. By late summer the fish, having grown to nearly six inches, find homes in abandoned conch shells and coral patches in shallow seagrass beds where they reside until ready for life on the reef. Young fish from 7 to 12 inches grow approximately 4.5 millimeters per month, while those measuring 13 to 17 inches slow to a growth rate of 1.9 millimeters each month. The fish continue to develop rapidly until they reach sexual maturity in four to seven years at between 16 and 20 inches.

As the fish mature they switch from eating tiny crustaceans and begin to hunt for larger prey, including lobsters, mollusks and fishes. Wrasses, parrotfishes, grunts and snappers are the most frequently eaten fishes, while crabs, mantis shrimp, octopuses, squids, shrimp and hermit crabs head a long list of favored invertebrates. One 23-inch grouper was found with an intact 25-inch moray eel inside its stomach. The soft bodies of Queen Conch, *Lobatus gigas,* are also consumed, but it is not known how the fish remove the animals from their thick shells. Hunting occurs throughout the day, with increased activity during the low-light periods of early morning and dusk.

A recently settled two-inch juvenile Nassau Grouper displays the same pattern as adults.

The species tends to be somewhat territorial, often living within a home range for a number of years. As an example of their territoriality, when startled by divers, especially near dusk, large individuals sometimes react with a loud, disconcerting boom produced by a violent contraction of muscles surrounding the air bladder. Tagging studies indicate a strong homing instinct based on visual clues. Although primarily a solitary species, small groups are occasionally found together inside caves or under ledge overhangs. Like most of the larger family members, Nassau Grouper spend a large amount of time at cleaning stations. The service is so ingrained that individuals raised in captivity without the services of cleaning organisms have been sighted at cleaning stations within an hour of their release.

After the peak of Nassau Grouper landings in the Bahamas during 1997, the islands experienced an 86 percent decline over the next 20 years. A similar scenario has played out throughout the species' historic range. Because the vast majority of Nassau Grouper landings occurred during the annual winter spawning season, restricting fishing on aggregations seems an obvious answer; however, fisheries management in most countries has lacked the will to impair a local industry even though the alternative is a collapsed fishery, which by 2018 has been the result in many areas. Creating more no-take marine reserves is another possible strategy for preserving the species. Unfortunately, closed fishing areas are few and far between. Many of the approximately 50 marine protected areas established in 21 tropical western Atlantic countries lack adequate funding, significantly limiting enforcement and compliance. Two of the largest and oldest marine reserves in the region have experienced success: Cuba's Jardines de la Reina, Gardens of the Queen, an 849 square mile national park stretching for 93 miles along the country's south-central coast, and the Bahamas' 176 square mile Exuma Cays Land and Sea Park established in 1958 and designated as a no-take reserve in 1986. The Exumas park has increased grouper biomass four-fold since fishing restrictions were implemented. Bermuda, the first Atlantic country to implement substantial policies to stabilize their dwindling reef fishing industry by banning fish pots in 1990 and fishing on all grouper aggregations in 1996, is a cautionary tale of too little too late. Out of 1594 REEF fish surveys conducted by volunteer divers from 2003 to

2013 only nine Nassau Grouper sightings were recorded. Not only have the island's four spawning aggregations collapsed, but the species is now virtually extinct from local waters. During the next five years of surveys not a single sighting of a Nassau Grouper has been substantiated. A sad story for a fish that less than a half century ago dominated Bermuda fisheries and glorified the country's seascape.

Even with the dramatic decline in Nassau Grouper landings in the Bahamas, only half-hearted measures to stabilize the fishery have been instituted. Closure of several of the species' 13 confirmed spawning sites was first implemented in 1998, and by the end of 2003 a two-month winter fishery closure went into effect that was extended to three months in 2005, but

Mass spawning aggregations of Nassau Grouper numbered in the tens of thousands before overharvesting.

enforcement has been limited and poaching documented.

As recently as the 1950s, Nassau Grouper were frequently sighted on South Florida reefs, but by 2018 the species is rarely encountered, and to date a spawning aggregation has not been substantiated in continental U.S. waters. A ban on fishing or possession of the species was put in place in 1993, but had no discernible effect on the declining population. Even with the ban in effect the species is still taken unintentionally as by-catch by fishers targeting the grouper/snapper complex. Although released, the fish too often fall victim to barotrauma from rapid decompression. No-take reserves are rare in South Florida waters. The only sizeable safe haven for fish is the Dry Tortugas Marine Reserve. But even today few Nassau Grouper are sighted within the protected area.

The only successful management story regarding the beleaguered species comes from the Cayman Islands, home to one of the last and the largest Nassau Grouper spawning aggregations known to exist. That the site survives today is due to a 15-year partnership between REEF and the Cayman Islands Department of Environment. Trying to save edible marine life is a tricky business: Cayman fishers, like others, protect their turf. For a decade and a half REEF and the DOE have banked on science, education and goodwill to win the day. Their efforts have paid handsome dividends, more than tripling the Little Cayman West End aggregation's size, to nearly 7,000 fish in 2018.

After witnessing declines in the numbers and size of Nassau Grouper landings over several years, in 1986 local fishers asked the DOE to implement a monitoring program. In 1995, following continued decline in population numbers, an "alternate year" fishing policy was proposed, but failed to be implemented due to lack of political will. By 1998 a series of policies began to be enacted. First, the three spawning aggregations off the end of Grand Cayman were designated as restricted areas requiring permitted access. This action was followed in 2002 with the implementation of the once-rejected alternate year fishing policy, as well as a 12-grouper catch limit. That same year an aggregation site off the west end of Little Cayman, believed to have been fished out in the late 1960s, was rediscovered and aggressively exploited. Two thousand, from what was estimated to be 5,000 fish visiting the site, were removed in the first year. Leslie Whaylen, a young scientist from REEF, who made the first assessment of the site, and witnessed the heavy-handed exploitation in 2003, initiated the collaboration between REEF and the DOE. At the end of the same year a moratorium went into effect banning fishing on all spawning sites for eight years, with the understanding that the aggregations would be scientifically monitored and documented during the period. At the time, the Marine Conservation Board stated that of the six known aggregations two were "fished out and three in serious decline." By the end of the first eight years of protection, the annual gathering at the west end increased to 4,000 fish. Again in 2011 the conservation measures were renewed for an additional eight years. Hopefully the success and knowledge gained in the Cayman Islands can foster similar initiatives in other countries to help save an priceless icon of Caribbean reefs.

Adult Tiger Grouper displaying typical color pattern.

Tiger Grouper

Mycteroperca tigris

Tiger Grouper, known for their thin white "tiger" bars, and ability to change colors and patterns in an instant, grow to be just over three feet, but are generally less than two feet in length. The rather slender, bottom-hugging, ambush predators inhabit rugged coral and rock terrain from 20 to 170 feet where they feed exclusively on fishes. The color pattern of bright yellow juveniles, with a dusky mid-body streak, look nothing like those of adults. Juveniles transitioning to adults display recently acquired bars over tan, yellow or black undercoats. Adults, typically tan with orangish spots on their lower head and gill covers, can lose their bars, fade, darken, or turn deep red when attending cleaning stations or resting on the bottom.

Tiger Grouper, strongly believed to be sex-changing protogynous hermaphrodites, form annual spawning aggregation 6 to 10 days after the full moon in January or February. A traditional aggregation site in Belize was monitored for three consecutive winter seasons from 2003 to 2005. The area, near Glovers Reef, covers 250 square meters of coral and sponge covered ridges ranging from 56 feet at their crest to 300 feet at a seaward dropoff. Grouper began gathering two to three days after the full moon and reached their population peaks four to seven days later. Over the three years, attendance

Juvenile color pattern.

Youngster displays "tiger" bars.

Youngster displays "tiger" bars.

Adult Tiger Grouper can pale or darken.

Often in red phase when near bottom.

Male courtship pattern.

Three male Tiger Grouper spawn with female (right) at peak of group spawning rise.

varied from 36 to 111 participants. The largest males showed up first and established territories, 15 feet across, on top of the tallest reef ridges. Slightly smaller males claimed areas around the outskirts while small sneaker males arriving last, hid in the bottom cover outside the primary spawning grounds. Females, easily differentiated by their small size and behavior, appeared on the same schedule as males, and settled beneath ledges near the bottom.

Once established, territorial males patrolled their invisible boundaries throughout the day displaying a striking pattern consisting of a greenish bronze head, and dark body with a large, white blotch on the lower rear quarter. The ubiquitous pattern, which varied in intensity, was switched on when confronting either males or females, but was invariably flashed whenever a rival male displaying the same white patch pattern approached a contested boundary. Intruders were dispatched with slow escorts out of the territory or aggressive chases. Males of equal size were challenged with prolonged, mouth-open face-offs, or engaged in ritualistic confrontations by rolling to their sides and cautiously swimming around each other. When subtle dominance tactics failed to settle disputes, vicious tissue-ripping attacks directed at a competitor's flank settled matters. By the end of the gathering, the majority of males displayed open wounds.

During the three years of the study in Belize, spawning events took place on consecutive evenings from the fourth through the eleventh day following the full moon in January. On the days of spawning, the grouper became restless about a half hour before sunset. The males rose off the bottom and hovered in loose groups 10 to 15 feet up in the water column above or just outside their territories causing boundaries to blur and aggression to increase. Large dominant males commanding the best real estate appeared far more concerned with smaller intruders than their similar-sized neighbors. Emboldened, sneaker males slipped in from the outskirts prompting the larger males to give chase. As the males gathered, females started rising and settling back to the bottom, but soon were swimming calmly through hovering groups, increasing the larger males' aggression toward rivals. Males often followed the females closely, showing off their courtship pattern. As minutes ticked by the activity increased until all the grouper were in the water column.

Eventually, a female would pause among a group of males, elevate her body at a 45-degree angle and quiver attracting a rush of males to her side. During the early stage of courtship, females would often ease away without initiating a spawning rise. Finally, when eggs were hydrated and ready for release, the female would suddenly rocket up a few feet with three to six males in hot pursuit and discharge her eggs, which were immediately engulfed in a cloud of sperm. Interestingly, with all the effort males expended defending territories, every spawning rise attracted multiple competitors including sneakers. After releasing their gametes, larger males immediately returned to their territories and waited for the next gravid female to show interest, as sneakers sped off in search of another party to crash. The majority of females spawned once, but a few have been observed spawning multiple times at other locations within the site during a single evening. Females seldom showed a preference for the dominant males even after their persistent efforts to protect territories. The entire event, from the first spawn until the last, took approximately 13 minutes.

Smaller female and larger male dive toward bottom after pair-spawning.

Equal-sized territorial male Tiger Grouper in prolonged standoff.

At Little Cayman's West End aggregation site, Tiger Grouper aggregate during the same winter moon phase as Nassau Grouper, *Epinephelus striatus,* and Yellowfin Grouper, *Mycteroperca venenosa.* Although primarily monitoring Nassau Grouper, the researchers also recorded the Tiger Grouper's behavior. Their observations of spawning differ somewhat from those made in Belize. In Little Cayman, where the spawning activities cover a larger area including multiple ridges, just before sunset groups of about 20 females rather than solitary females, gathered in the midst of male territories (spaced approximately 100 feet apart), and initially pair spawned. As females began pair spawning in rapid succession, females and surrounding males contract into smaller areas,

Vicious attacks between larger males become common as spawning approaches.

allowing multiple males (in some cases up to nine individuals) to participate in each spawning event. With the increased action sneaker males were able to join in the spawning rises without being chased by the territorial males.

Tiger Grouper have also been observed sharing a spawning aggregation at a traditional site off the island of Guanaja in Honduras with Nassau, Yellowfin and Black Grouper, *Mycteroperca bonaci.* There are also reports of Tiger Grouper possibly spawning outside aggregations. Two days before the full moon of February, a researcher in the Bahamas observed a group of three Tiger Grouper displaying courtship patterns and showing agonistic behavior. In August, of two separate years, we watched several large male Tiger Grouper wearing courtship colors aggressively defend lek-like territories an hour before sunset on an 80-foot coral slope in Bonaire. On both occasions we had to leave the area before sunset when spawning might have taken place.

Unlike the majority of protogynous groupers, which tend to have a higher ratio of females than males; Tiger Grouper turn the table. There were estimated to be nearly twice the number of males as females at the Belize aggregations. Similar male-dominant numbers were reported from aggregation sites in the Caribbean, Mexico, and Gulf of Mexico. The anomaly could be related to a lack local fishing pressure that tends to select for males. However, in Mexico after fishing pressure increased significantly, there was still a greater number of males attending aggregations. Perhaps female Tiger Grouper transition to males at a faster rate than normal, or possibly the percentage of the population born as males is larger than other protogynous groupers.

Red Hind

Epinephelus guttatus

Red Hind are moderate-sized grouper reported to reach over two feet, but generally average less than a foot in length. The solitary hunters, with attractive coats of reddish brown spots spread over a whitish undercoat, live near the bottom close to the protection of the reef's structure. They have little fear of divers and customarily allow a cautious observer to approach within two or three feet before slowly easing away. Crabs are the most common item on their menu, while fishes, such as Bluehead, *Thalassoma bifasciatum,* Boga, *Haemulon vittatum,* goatfishes and small morays, follow closely behind. Mantis shrimp comprise over 15 percent of their diet, with shrimp and octopuses rounding out their preferred foods.

Females, which cannot be distinguished from males outside of reproductive periods, occupy broad home ranges that overlap with those of 1 to 18 other females. They live on reefs of shallow to moderate depths, well within safe diving limits, while larger males inhabit much deeper waters. Even though the fish live in close proximity to each other, there appears to be little social interaction, with most preferring to spend much of their time alone, partially hidden in a favorite location. Not much is known about the species' early life stages. On rare occasions shy one- to two-inch juveniles, with widely spaced red body spots on a bright-white background, are sighted sneaking about near cover on patch reefs in moderate depths. Although most adults live for 10 or 11 years, one 18-year-old specimen has been documented. Research from the St. Thomas region determined that Red Hinds reach approximately eight inches in length the year after settlement with many reaching sexual maturity by their third year.

Red Hinds are protogynous hermaphrodites, changing from females to males at some point in their life cycle. What triggers sex change is unknown, but individuals in transitional stages are found year-round. Spawning aggregations in Puerto Rico and St. Croix and St. Thomas in the U.S. Virgin Islands have been studied for nearly three decades. The fish gather in or near traditional spawning

Red Hind

grounds along sections of the insular shelf during a single one- to two-week period associated with the lunar cycles of January or February. In 1992, token spawning occurred in a Puerto Rican aggregation in the latter part of January, while peak activity took place just prior to, and continued for a week after, the full moon of February. One tagged Red Hind traveled more than 10 miles, at one point crossing water over 600 feet deep and bypassing other aggregations, to spawn at a particular site.

Aggregation areas studied are quite large, typically encompassing over a square mile of patch-reef sea floor, in depths beginning at 60 feet. These extensive tracts normally accommodate multiple congregations of fish, spaced 300 to 400 feet apart. One site was estimated to contain nearly 3,000 fish when at peak density on the day of the full moon. The females average 9 inches and males 11 inches. For the first two days individuals are widely scattered, but, as the full moon approaches, the fish begin to gather in isolated groups of two to six females accompanied by a lone male. A few groups do not have a male, but none contains more than one. It appears that females establish clusters, and males affiliate with one or more of the gatherings. It is believed that individuals stay with the same group and remain at the spawning site throughout the entire aggregation period.

Because of their pallid complexions and distinctly swollen abdomens, females are easily distinguished from males during reproductive periods. When courting, males typically exhibit dusky, mottled patterns with two or three dark, one- to two-inch bars above the anal fin and pale heads with dark areas on the sides of their lips. While the females rest near bottom shelter, the males swim a few feet higher within approximately 50 feet of their females. Now and then, the suitors settle to the bottom and remain stationary for a few minutes. Approaching divers easily disturb the males, causing them to swim a short distance away until the observers leave. The males patrol their territories and occasionally engage in border confrontations with neighboring males. During the brief encounters the rivals raise their dorsal fins and rush each other with open mouths.

A male will sometimes execute a series of short, slow arching rises or hover just off the bottom in a head-up position in what are believed to be courtship displays. Actual spawns, which have been observed rarely, are rather uneventful. A single female lifts a foot or two off the bottom, where a male promptly joins her. Together, the pair rises slightly and releases a cloud of gametes before settling back to the sea floor. From inspection of gonadal tissue, it has been determined that females spawn more than once during each annual gathering. The number of eggs a female is able to produce increases substantially with size. A 12-inch female can produce approximately 100,000 eggs, whereas a 16-inch individual can yield 1.5 million eggs.

Following sustained fishing pressure on Red Hinds throughout the 1980s, a 16-square-mile area of shelf ledge south of St. Thomas, encompassing the species' annual spawning aggregation, was closed from December through February in 1990, then permanently in 1999. By 2004, five years after the permanent closure, the biomass of Red Hinds increased by 60 percent. During the same time the fishing improved outside aggregation periods.

Researchers monitoring the site reported that the aggregation had the most fish present during January and spawning peaked two days before the full moon. After the first spawning period 80 percent or more of the fish leave the area for three weeks before reforming for a second spawning period. Those remaining consisted of the largest females and males. Three years after implementing a year-round no-take policy at the site the number of aggregating fish more than doubled. Not only did the size and density increase substantially during the spawning season, but the sex ratio at the end of the 1980s, which was estimated to be 15 females for each male shifted dramatically to three to five females for each male.

As much as 20 percent of a Red Hind spawning aggregation can be made up of Rock Hind, *Epinephelus adscensionis*. The females of the modest-sized grouper are also obviously full of eggs, while the males display black blotches over a white background. The spawning behavior for this species has yet to be reported.

Graysby with captured Creole Wrasse.

Graysby
Cephalopholis cruentata

Graysbys are normally the most common grouper, inhabiting shallow, high-profile reef systems from Bermuda to the northern shores of South America and throughout the Caribbean and Gulf of Mexico. The seemingly ill-tempered little grouper, which can grow to just over one foot, live within limited territories often located among the nooks and crannies of star and lettuce corals. Although occasionally found at depths to 90 feet, the rather reclusive fish generally haunt the shadowy edges of overhangs or crevices at depths from 20 to 30 feet. Their bodies and fins, ranging from light brown to deep chocolate, are evenly spread with dark spots. Three to five distinctive white spots, spaced along each side of the dorsal fin base, change from white to black when Graysbys become disturbed. Larger individuals occasionally display an indistinct pattern of body bars.

The tenacious predators hunt throughout the day, with distinct peaks at dawn and dusk. Fish, particularly Brown Chromis, *Chromis multilineata,* that move to and from the reef at sunrise and sunset during the Graysbys' favorite hunting period, make up the bulk of their diet. The stomach of one specimen contained four Brown Chromis, all in the same state of digestion, suggesting that they were consumed within a narrow time frame. Other small fishes, including Cherubfish, *Centropyge argi,* cardinalfishes, Bicolor Damselfish, *Stegastes partitus,* Blue Chromis, *Chromis cyanea,* and Yellow Goatfish, *Mulloidichthys martinicus,* are also common prey. Shrimp, crabs, mantis shrimp and a few other crustaceans comprise the remaining 20 percent of their food. Most invertebrates are taken just after sunset as they emerge from their daytime hiding places. During the day Graysbys often hunt with roving Goldentail Moray, *Gymnothorax miliaris,* and Spotted Moray, *G. moringa,* as the eels slip from crevice to crevice in pursuit of concealed prey. The hunters stay close, hoping the morays' surprise attacks will flush a fish into the open. As many as four Graysbys accompany a single eel. The fish routinely joust among themselves to gain an advantageous position near openings where the eel is foraging. During the prolonged hunts, the typically shy fish are so intent on their business that they can easily be approached by divers.

Graysbys are protogynous hermaphrodites that generally change from females to males during their fourth or fifth years, when they reach a length of about eight inches. The transition, which usually occurs after the species' annual spawning season, is followed by a quick growth spurt. There is no visible sexual dimorphism in the species; but virtually all individuals over 10 inches are males. Most females mature sexually during their third year, at a length of six inches, and by their fifth year all females have changed into males.

A single dominant male and several somewhat smaller females share a territory, averaging 200 square feet, where resources and mates are aggressively defended. Ritualistic, but intense, turf wars occasionally break out between two similar-sized individuals, possibly territorial males. We watched a 15-minute scrap between two 10-inch Graysbys late one afternoon only to realize years later that this aggressive behavior was a possible prelude to spawning. When first observed, the pair was glaring at one another across a white sand shelf. Without warning, both simultaneously attacked with open mouths, crashing together so hard they spun in a complete turn, causing one to retreat a foot or two and pause, facing in the opposite direction. We thought the battle was over, but, in less than 10 seconds the retreating fish turned and cautiously began inching toward its rival until the pair's heads, from snouts to gill covers, touched. The fish imperceptibly pushed against its opponent until both were lying on their sides. They abruptly broke, pivoted and paralleled one another, facing the same direction. Within seconds, they were racing together back and forth through four sharp turns before stopping, as suddenly as they started, in a second extended face-off. In little more than a minute they attacked each other again. This time the power of their rush was so forceful that the fish, with jaws locked, sent up a storm of sand as they tumbled over the bottom. The series of events was repeated several more times in succession before one fish, finally losing heart, disappeared inside a reef pocket.

Courtship and spawning event begin with two rival Graysbys poised in a lengthy face-to-face standoff.

Following much shorter mouth-to-mouth altercations between males, which typically take place on sand channels between coral mounds, males pair spawn within their territories with each member of their small harems. After witnessing our first Graysby spawning bout we realized why it had taken so long to finally observe the behavior. It had a lot to do with our inaccurate search image. We had been looking for spawning to take place around dusk, the typical time for most groupers. Instead the episodes took place in mid-afternoon, and in both of the events we witnessed, following mouth-to-mouth altercations between males—much like the episode described above. Also, other expected cues were missing. There was limited courtship and the males didn't alter their

Attacks between rival male Graysbys are frequent and vicious.

The fish lie unmoving on the bottom for several seconds with their heads pressed together.

One fish pushes ever so slowly against the forebody of his rival until both lie flat.

The Graysbys suddenly spring into action, parallel one another and race back and forth through four sharp turns before stopping in another face-off and beginning the series of events once again.

Male Graysby briefly courts two of the three females in his harem.

colors or patterns. Finally, the abdomens of the females did not appear swollen and spawning rises only took the male/female pairs up a matter of inches above a coral. Other than the males' brief skirmishes the courtship and spawning rises were undramatic and uneventful.

The reproductive season in the southern Caribbean occurs sometime between May and September but has been reported to take place during other

Pair of Graysbys during spawning rise. Smaller female (left).

Larger male Graysby escorts female on a spawning rise of only a few inches.

months elsewhere. The sex ratio in an unfished population of Graysbys is approximately two-and-a-half females for each male; but on one heavily fished reef system, the ratio was six females for every male.

Juvenile Graysbys are extremely shy. They are distinguished by a yellowish wash covering the upper head and back and a conspicuous white band running from the lower lip to the leading edge of their dorsal fin. One- to two-inch youngsters often find sanctuary in the convoluted outer surface of Giant Barrel Sponges, *Xestospongia muta*. With growth, they seek shelter beneath low ledges, away from high-profile reef structures. The skittish juveniles appear in the open only when making short dashes between nearby hiding places. Juveniles, less than five inches, feed almost exclusively on shrimp, but occasionally consume tiny fishes. Young Graysbys grow about two inches each year, slowing to less than an inch after six years.

Coney displaying the common bicolor phase.

Coney

Cephalopholis fulva

Unlike the closely related and similar-sized Graysby, *Cephalopholis cruentata,* that live in complex coral structures, Coneys prefer isolated patch reefs surrounded by sand in depths between 15 and 60 feet. Adults spend 95 percent of their time within a 15-foot radius of a central coral patch, where they live in harems composed of a male and three to four females. The number of cleaning stations in an area appears to have a direct bearing on choice of home sites. Adults reach a maximum length of 16 inches, but individuals over 10 inches are rare in heavily fished waters. The protogynous hermaphrodites reach sexual maturity at just over six inches, and become males between 6 and 10 inches. The species, which lives in the tropical western Atlantic from Bermuda to Brazil, including the Gulf of Mexico is believed to live for at least nine years.

Adult and juvenile Coneys display one of two distinct color patterns, both highlighted with a sprinkle of bright blue dots. The most common pattern, a solid tan to reddish brown body coat, can quickly change into a bicolor pattern that leaves the back dark while the lower body and tail turn white. The bicolor pattern tends to be displayed more in early morning and late afternoon, and possibly when the fish is excited or hunting. A blotched resting pattern occasionally appears when the fish settles on the bottom for an extended period or perches inside a vase sponge—a common daytime behavior that is not yet understood. A few Coneys, both juveniles and adults, are golden except for small blue dots. Golden adults, believed to be the product of a single recessive gene, generally reside in the deeper end of the species' depth range.

The Coney's diet is equally divided between small fishes and crustaceans, typically shrimp, crabs and mantis shrimp. These bottom hunters occasionally stalk prey in pairs and are relentless once in pursuit of a victim. They frequently shadow-feed with morays and octopuses.

In Bimini, Bahamas, seven days after the full moon of May, we observed two adjacent harems of Coneys courting and spawning during the last hour before sunset. Their activities were centered on two coral mounds 20 feet apart, probably the core sites of each harem's territory. The white bellies of the five bicolored females—two with one harem and three with the other—were dramatically distended with eggs. When the two slightly larger males were on the move or near their females, both displayed a distinct spawning pattern. Their brown backs, from snout to just past the foredorsal, were lightly blotched, and a tan streak ran from tail to pectoral fins.

The females shuttled their bulging bellies about the coral mounds or rested heavily on the bottom, often near one another. When stationary, their bicolored bodies were blotched. The males rapidly swam about the area just off the

Male Coney displaying courtship pattern.

A pair of Coneys during a spawning rise. Male continues to display his courtship pattern during spawning rise. (foreground).

Three hydrated female Coneys hover above a coral head waiting to pair spawn.

bottom, now and again settling next to a low-profile structure or inside a vase sponge where they developed a blotched pattern. Once swimming again, their spawning colors instantly returned. These jaunts, which took them 30 to 60 feet from their territories, kept them apart from the females most of the time, but once back, the males courted with enthusiasm. Returning males swam up to females, twitched repeatedly, performed two or three short upward dashes and even wedged their snouts beneath the females' bellies, attempting to coax them off the bottom. Occasionally their persistence caused the females to flee, with the males in close pursuit.

At 7:15 p.m., 45 minutes before sunset, the first observed spawn occurred. A female rose above a coral mound and paused. In an instant, a male dashed up from below, brushing against her abdomen as he passed, looped back sharply and positioned himself next to her side. Wildly thrashing tails carried the pair straight up 8 to 10 feet, where they released a visible cloud of gametes. Waiting Yellowtail Snapper, *Ocyurus chrysurus,* swarmed the dispersing eggs. Ten minutes later, we watched the same male spawn with the second of his two females. Through the dim light we spotted two other swirling aggregations of feeding Yellowtails, probably marking the sites of other Coney spawns.

On the following evening during the same time period, members of the two harems courted and spawned once again. This time, however, the two neighboring males engaged in a skirmish on a sand patch midway between their territories. The altercation began with two regularly occurring color morphs facing off and ended with a violent attack. During the third evening both males showed their spawning patterns, but we did not see any ripe females or observe courtship or spawning activities.

Hamlets

Genus *Hypoplectrus*

At first glance, the slim, deep-bodied hamlets look more like members of the damselfish family than the typically more robust sea basses. However, on closer inspection the three- to five-inch carnivores' swimming style, large mouths and triangular pupils verify their family status. For decades debate among marine taxonomists has swirled around the species status of the many hamlets that make up the species flock of genus *Hypoplectrus*.

At the heart of the controversy is the essence of taxonomy: What defines a species? The intriguing dilemma centers on the small, brightly colored fishes' almost complete lack of anatomical differences and similar breeding habits; yet, for an unexplained reason, the otherwise identical fish display a variety of different color patterns. At least 18 distinct and regularly occurring color morphs have been documented. Adding even more spice to hamlet soup is the

Barred Hamlet—a common species.

Butter Hamlet—a common species.

Masked Hamlet—a rare genus member.

Golden Hamlet—a rare genus member.

A Shy Hamlet (top left) a possible mimic of a young Rock Beauty.

The Blue Hamlet is a possible mimic of the Blue Chromis.

Blue Chromis—a planktivore.

Courtship between a pair of Butter Hamlets. The female displayer (top left) assuming a snout up pose.

smorgasbord of hybrid color variations that occur in nature.

The stately, little, solitary hunters move about just above the bottom within home ranges that are defended against other hamlets. Their day is spent in a never-ending search for a quick meal of shrimp, crabs, fishes (typically small

Butter Hamlets' three-second spawning clasp.

recently settled gobies, blennies, surgeonfishes and damselfishes), mysids (tiny crustaceans) and an occasional mantis shrimp. The Indigo Hamlet, *H. indigo,* one of the largest species across regions, feeds almost exclusively on fishes.

The genus members, native to the waters of the tropical western Atlantic tend to prefer complex reef structures in moderate depths. The Barred Hamlet,

Courtship between a Shy Hamlet (bottom) and a probable hybrid resulted in four spawning clasps.

Hybrid hamlet—possible cross between a Yellowbelly Hamlet and a Shy Hamlet.

Hybrid hamlet—possible cross between a Blue Hamlet and an Indigo Hamlet.

H. puella, Butter Hamlet, *H. unicolor,* and Black Hamlet, *H. nigricans,* are the most common and widely distributed of the color morphs. The three are well represented throughout the entire region, with the Barred extending into the Gulf of Mexico and up to Bermuda. The distribution of the remaining morphs is more restricted, with several radiating out from distinct population centers, which indicates historic geographic isolation, a factor that possibly allowed the color morphs to differentiate. For example, the Blue Hamlet, *H. gemma,* is almost exclusively found in South Florida and the Keys, while the rare Golden Hamlet, *H. gummigutta,* and Masked Hamlet, *H. providencianus,* primarily live on the submerged banks between Central America and the Greater Antilles. In some areas as many as eight morphs can inhabit the same reef tract.

Like a few other small sea basses, hamlets are simultaneous hermaphrodites that take turns releasing both eggs and milt each evening of the year—a behavior known as egg trading. The strategy is rare among fishes occurring in only 40 of 34,000 species. Fortunately for fish watchers, the little sea basses are so preoccupied with each other during courtship that neither seems disturbed by a diver's presence, allowing a discreet approach to within three or four feet. Unfortunately, light beams tend to distract the courting couples. As early as an hour before sunset, mature hamlets begin acting giddy. At some point they swim rapidly out of their territories in search of mates for the evening. These trips can take them over 100 yards from their permanent territories. It is speculated that because the fish are site-oriented, they are well aware of neighboring hamlets. On their way to traditional spawning sites to mate with established partners, hamlets occasionally happen upon other courting pairs. If they are the same color morph, the cruising hamlets might try to break in, but are generally driven away. When they rendezvous with their regular mates extended courtship begins. For up to an hour before the first spawn, the pairs chase about the bottom, occasionally stopping to feed, all the while keeping a lookout for interfering rivals that are vigorously chased away when detected.

Finally, one of the hamlets assumes the female role and acts as the displayer striking a rigid snout-up pose with fins flared while intermittently snapping its head from side to side. The interim male, whose color is distinctly blanched, mills nervously about the bottom a few feet away. Gradually the two make their way up into the water column, usually just above a prominent feature such as a coral mound or gorgonian bush. Suddenly, the displayer darts to within inches of her mate and pauses. In a blink, the fish clasp like encircling hands with their pectoral fins and tails outstretched and beating fast. The acting male with its head pointed down, belly puckered, fins spread wide and mouth opened into an oval, is enveloped by the arching body of its mate. One one-thousand, two one-thousand, three one-thousand and the fish pop apart with an inconspicuous release of gametes. With color renewed, the male darts a short distance away with its mate in chase. The dramatic, three-second spawning clasps generally begin 10 to 15 minutes before sunset and reoccur every few minutes until the last light fades, with partners changing sexual roles between each bout. The number of clasps per spawning pairs we have observed varied from as few as 2 to a record 10 by a potent pair of Indigo Hamlets. The affairs end when one of the spent partners discreetly vanishes into the night. On two occasions we observed spawning bouts ending on an odd number, which means that the last eggs released weren't reciprocated. Both affairs concluded with aggressive chases. Fertile hybrids (inter-specific hybrids between species in the same genus are generally sterile), the product of crosses between color morphs, have long been used as evidence against granting separate species status to the members of the genus. The argument goes: if the different color morphs are indeed individual species, then there should be extremely limited inter-morph spawning. Of the 28 spawns we witnessed during a two-week period, all but one occurred between like hamlets. The one exception was a brownish colored hybrid (the area's most common hybrid coloration) spawning with a Shy Hamlet, *H. guttavarius.* Hamlets use sight to choose a mate and, if given a choice, nearly always choose a like color pattern. During our dive, when the cross-spawn was observed, we didn't see any other Shy or brown hybrids in the vicinity.

Aquarium breeding research has documented that if a fish is given no option other than to spawn with a different morph, it will do so nearly 50 percent of the time. Field data recorded only 7 of 189 pairings between unlike color morphs. Although limited, this is still a surprising amount of interbreeding between species and might be a reason why the different hamlets are so genetically similar.

In 1980, protein gel electrophoresis performed on nine different color morphs detected virtually no interspecific differences. However, this evidence, which relies on negative data, is considered inconclusive by many. One interesting conclusion that can be drawn from this work is that *Hypoplectrus* has differentiated recently. The genus may not have evolved until after the closing of the Isthmus of Panama, approximately 2.8 million years ago. This date might seem ancient, but, on an evolutionary time scale, hamlets are the new kids on the block. The different color morphs may have been segregated later into population centers by the rise and fall of sea level.

Further muddling the matter is the enticing hypothesis that aggressive mimicry is the mechanism driving speciation in *Hypoplectrus.* There are eight hamlet morphs suspected of having color patterns resembling non-predatory reef fishes that gain a distinct advantage procuring food from the association. For example: the Blue Hamlet closely resembles the plankton-feeding Blue Chromis, *Chromis cyanea,* the Butter Hamlet looks much like the Foureye Butterflyfish, *Chaetodon capistratus,* that dines on Christmas Tree Worm, *Spirobranchus giganteus,* and feather duster tentacles, the Black Hamlet seemingly mimics the algae-eating Dusky Damselfish, *Stegastes adustus,* and Longfin Damselfish, *S. diencaeus,* and the Shy Hamlet has a marked resemblance to the sponge-eating Rock Beauty, *Holacanthus tricolor.* In theory, aggressive mimics should not only look like their models, but should also be less abundant; spend more time with their model than expected by chance; hunt more actively when associated with the model and evolve similar behavioral characteristics. The supposed hamlet mimics meet the first four criteria, but failed to display the behavior traits of those they mimic. Possibly the short evolutionary history of hamlets has not yet allowed them to develop complete mimicry. On the other hand, the similarities could well be nothing more than coincidence.

All this pushed the hamlet question to the point where it became necessary to interbreed color morphs in captivity before re-evaluating their taxonomic status. A laboratory breeding project did just that in the early 1990s. The research established that crosses between color morphs (in this case Blue Hamlets and Butter Hamlets) produced hybrids intermediate to both parents in color and markings. The research also found that self-fertilization of a Shy Hamlet resulted in offspring that resembled the parent species. This helped confirm that color pattern is genetically determinate in *Hypoplectrus.* Early genetic data, from 1980, was unable to distinguish between color morphs, however more recent studies have found a small but inconsistent degree of genetic difference between geographically separated populations. These findings, combined with field observations documenting a strong spawning fidelity between like hamlets as well as verifying a low frequency of hybrid hamlets, have tilted the debate in the direction of granting full species rank to the regularly occurring color morphs.

Of course, everyone's opinion pivots on a preferred definition of species, and there seem to be as many as there are taxonomists. In fact, entire graduate-level courses are devoted to this one perplexing question. Michael L. Domeier, the researcher who interbred hamlets writes, "That there is no agreement of the precise nature of the species is a reflection of the nature of the dynamic process at work and thus we should not anticipate future agreement." His thoughts share good company. After discovering astounding differences in every genus of barnacles he studied, Darwin wrote, "After describing a set of forms as distinct species; tearing up my MS. and making them one species, tearing that up and making them separate, and then making them one again, I have gnashed my teeth, cursed species, and asked what sin I had committed to be so punished."

Harlequin Bass

Serranus tigrinus

Monogamy is not a widely practiced mating strategy on the reef; however, the common Harlequin Bass not only share eggs exclusively with permanent partners, but also remain with them throughout the day, hunting cooperatively and defending jointly held territories for feeding, spawning and sleeping. Pairs generally inhabit one of several large, adjoining territories spread across rubble fields bordering reef structures. Territories held by mated pairs average 600 square feet, while those of solitary adults are only half as large. Juveniles, relegated to territorial outskirts, stay close to their hiding holes in small, weakly defended areas. Occasionally, Harlequin Bass are chased by larger food competitors, such as Graysby, *Cephalopholis cruentata,* and Coney, *C. fulva.* In turn, they attack hamlets and other small, gleaning carnivores that encroach on their territory. When Harlequins confront their neighbors at boundaries, the two momentarily pause and face off with their dorsal and ventrals flared. Occasionally, one will lean to the side with its dorsal spines aimed in the direction of its rival. At other times, the fish cautiously circle each other before ending the extended affair with a back-and-forth chase accentuated by a well-placed nip or two.

Harlequin Bass hover together briefly before spawning.

Throughout the day pairs move about together, just off the bottom within their territories, feeding primarily on shrimp, with a few fishes, mantis shrimp and crabs rounding out their diet. Frequently, one of the hunters veers off from the other to poke around a crevice, but the pair seldom remains apart for long and generally keeps in eye contact. Abrupt feeding strikes are directed at the bottom. The species is unique among pair-bonded fish for its cooperative stalking behavior. The two hunters approach concealed prey from opposite sides of a crevice and attack simultaneously, driving their startled victim toward one mouth or the other.

Pairs of the two-and-a-half- to four-inch adults are usually closely size-related in order to profit from maximum egg productivity. If given the opportunity, however, mates will abandon smaller partners to form new pair associations with larger solitary individuals. Before venturing into a new territory, a visitor must be signaled by an extended fin display from the resident. If welcomed, the guest follows the occupant around its territory. The following day, the newly formed pair actively patrols, hunts and spawns together. If a paired fish disappears, its partner makes courtship overtures the same evening and usually forms long-term bonds with a similar-sized solitary neighbor by the next day. The two live together within their joint territories, unless a portion is confiscated by a bordering pair during the changeover.

The simultaneous hermaphroditic Harlequins mate throughout the year, with increased activity during the week before and after the full moon. Spawning events begin 20 minutes before sunset and can last until 10 minutes after the sun goes down. As with other small egg-trading sea basses, the goal is to spawn as males. The partner taking the female role controls the tempo of reproduction. Courtship starts with the pairs blanching slightly as the acting females lead their mates on a series of short runs interspersed with feeding. Now and again, the females raise their heads and twitch. Finally, the females swim to the top of low-profile features in their territories and signal readiness by flexing their bodies. Acting males move near and, for a brief moment, the pairs hover inches apart. Without warning and faster than the eye can follow, the duos dart straight up several inches, spawn and return separately to the bottom. In the single spawn, each acting female releases her entire payload of eggs. On many evenings, especially those close to the full moon, the partners immediately reverse sexual roles and spawn a second time about two minutes later. However, spawning is not always so egalitarian. Encounters often end with only a single spawn. On other evenings a series of false starts may occur without spawning taking place. At times, both partners simultaneously present the curved-body pose or, in a few instances, neither signals before spawning.

In their quest to fertilize eggs, members of permanent pairs regularly go outside their territories to spawn as males with neighboring solitary adults, but never with members of other established pairs. These encounters are not reciprocated, indicating the secondary social status of loners. This is probably the reason single adults quickly replace missing partners. When pair members spawn with solitary individuals, the encounters occur prior to egg-trading with their partners. In some instances, both partners take part in simultaneous spawning rushes with solitary neighbors. However, things don't go so smoothly if a partner, after spawning as a male, heads for the neighbor's territory without trading eggs with its mate. The spurned companion attempts to block its partner's route, but often relents and joins the tryst.

Lantern Bass

Serranus baldwini

Lantern Bass spend their days scooting about coral rubble and sand bottoms hunting for tiny bits of animal matter. Now and then, the one-and-one-half- to two-inch sea bass dart up into the water column to snag floating specks of plankton. Populations are separated into harems consisting of from one to seven small simultaneous hermaphrodites living within small, overlapping ranges inside the patrolled territorial boundaries of single large dominant males. In the animal kingdom, this social system is unique to Lantern Bass and a second genus member, *Serranus psittacinus,* from the Eastern Pacific. The smaller hermaphrodites have functioning ovarian and testicular tissue with separate ducts for the transfer of eggs and sperm; their visually identical harem-masters are sex-changed males that have lost their ovaries.

The males boldly move around their territories, guarding against the intrusion of neighboring males and chasing subservient hermaphrodites that are much more passive, seldom venturing far from a restricted area and often remaining hidden. Although hermaphrodites occasionally have altercations among themselves, it remains unclear if such conflicts are meant to protect territories or to maintain an individual's position in a hierarchical social system. Lantern Bass frequently survey their surroundings while perched on rocks or hovering in a head-up position just off the bottom.

During the last hour and a half of daylight, when courtship and spawning occur, the hermaphrodites become more settled while males pick up the pace of chasing, surveying and patrolling. Unlike other genus members, males control reproduction, spawning once each evening with their harems of hermaphrodites. Smaller harem members rarely spawn with each other or streak in and mix their sperm with a spawning pair. Courtship begins with a male approaching a hermaphrodite and signaling his intentions by jerking forward in a series of short, rapid movements or bobbing up

Courting Lantern Bass.

and down from a horizontal hovering position. The male approaches from behind and rests his head on the rear of his partner's body or nudges its side repeatedly. If uninterested, the hermaphrodite scampers away and is either chased or ignored by the male. When responsive, the hermaphrodite lifts off the bottom, raises its head and bends laterally into an S-curve. The fish dart up one or two feet and spawn before returning separately to the bottom, where they dodge under cover. At times a hermaphrodite performs the S-curve without being courted, which immediately brings the male flying to its side.

Interesting questions remain unanswered concerning the curious mating behavior of Lantern Bass. It is still a mystery why hermaphrodites have both sexual functions if they spawn only as females; and what mechanism keeps the hermaphrodites from spawning among themselves? And why do males lose their female capabilities, which, if left in place, would greatly increase their productivity? Possibly, pure males have evolved only recently in the species, and there has not been sufficient time for the hermaphrodites to lose their testicular tissue.

Tobaccofish

Serranus tabacarius

Solitary, three- to seven-inch Tobaccofish hunt throughout the day for small fishes and shrimp within their home ranges, which encompass between 40 to 500 square yards of sandy, low-lying rock and patch-reef plains. Several weakly defended territories broadly overlap one another.

During one-hour periods between midafternoon and dusk, the simultaneous hermaphrodites seek out partners at locations within their hunting grounds and engage in entertaining courtship and spawning displays. The prolific egg traders typically spawn over a dozen times each afternoon, with from one to nine different partners. The number of matings and mates varies greatly each day, with a tendency for increased activity around the time of the full moon. Individuals change spawning partners every 10 minutes on average, with reciprocal spawns between partners occurring approximately a minute apart.

Courtship usually begins when one or both partners rise a few feet off the bottom, arch their tails upward and lift their heads. Often one of the fish,

Courting Tobaccofish display blotched backs and black masks.

generally the larger, chases the other briskly across the bottom. When ready to spawn, the first to assume the male's role develops a distinctive black mask, collapses his fins and settles flat and still on the sand. The temporary female, wearing a pale, blotched pattern, flexes laterally and fidgets nearby while repeatedly fluttering her rear dorsal and ventral fins. With an arched body, she begins swimming back and forth, brushing the vent on her white belly across her mate's darkened head. After a few passes the female swims up and away, with the male tagging close behind. As the pair rises several feet above the bottom, the acting male may repeatedly flick his dorsal fin and move to her side in a faltering start-and-stop-motion. But more often he slips up from below and bumps her vent with a partially open mouth. The female abruptly turns, to face her partner, and the pair immediately dashes up together, leaving a gamete cloud a foot above their starting point. Possibly, because home ranges are large and the pairs spawn in random locations, streaking rarely occurs.

Larger Tobaccofish seem to attain a preferred status. They control more extensive home ranges, are seldom chased, typically spawn first and last in the male role and frequently refuse to reciprocate or mate consecutively as females. However, they never develop sufficient dominance to function exclusively as males.

Chalk Bass

Serranus tortugarum

Unlike their brawny relatives that constantly probe the bottom for crustaceans, dainty, baby blue Chalk Bass gather in loose groups and pick plankton, just above rock piles on sandy sea floors. Juveniles, which look much like adults, and sexually mature individuals, over one and a quarter inches, inhabit the shallower end of the species' 30- to 300-foot depth range. Larger adults, which can grow to nearly three inches, generally prefer depths below 100 feet.

Late in the day, the egg traders leave their feeding grounds and travel to an area containing a cluster of small spawning territories, where they hover a foot or two above the bottom with a customary partner of similar size. Unpaired loners patrol the peripheries. Courtship displays are subtle, and with good reason: Chalk Basses are accomplished sneakers that keep a keen eye on nearby pairs, waiting for a chance to dart in at the last moment and add their sperm to the pair's gamete clouds.

The first partner to take the female role glides in front of her mate, cautiously lifts her dorsal fin, and, in slow motion, rises an inch or two. While the pair pauses, members of neighboring couples and loners inch in their direction. When approached too closely, the couple drives the intruders away. Spawning rises are short, quick, and frequently joined by one to five streakers shedding sperm. Often, in an effort to disguise their timing, neither partner signals in any way before their rush; or the couples make false starts, which bring freeloaders flying in their direction. Mates usually spawn more than a dozen times during an evening, with the partners trading eggs with one another 80 percent of the time.

Plankton-picking Chalk Bass hover above open bottom.

Courting pairs located more than five feet from other spawners or isolated by bottom structures can afford to display more conspicuously. In these situations a partner might bob forward and down slightly once or twice. Occasionally her mate answers with a similar signal just before spawning. The same signal is also given when individuals leave protective cover or while heading for night shelter. A fear of predation is added reason for courting Chalk Bass to be cautious—lizardfishes often station themselves near active spawning territories to take advantage of preoccupied courters.

Reef Basslets

Genus *Liopropoma*

Sighting any of the four small, reclusive species forming the Caribbean contingent of *Liopropoma* is a worthy challenge for advanced fish watchers. The beautiful, two- to four-inch, red-orange-black-and-yellow-striped, torpedo-shaped fish with electric blue highlights typically inhabit coral caves and dark reef pockets below 80 feet. However, the most prevalent of the four, the Peppermint Basslet, *Liopropoma rubre,* is occasionally found in water as shallow as 20 feet, especially on slopes with shingled layers of plate corals. An underwater light helps to locate the solitary fish deep within the narrow crevices formed by the coral strata. The light beam is also necessary to bring out the exquisite colors of the lovely little fish. At times the shy loners can be seen hovering at the edge of ambient light. Once aware of divers, they slowly retreat into the shadows, but frequently return a few seconds later for a quick peek.

The three remaining species, the Candy Basslet, *L. carmabi,* Cave Basslet, *L. mowbrayi,* and Wrasse Basslet, *L. eukrines,* tend to dwell deeper and are

Right: *The cheek-to-cheek spawning rise of Peppermint Basslets.*

A two-inch Cave Basslet.

A three-inch Wrasse Basslet.

believed to be common on walls below safe diving limits. Unstriped, salmon-colored Cave Basslets prefer small cave pockets only a foot or two across rather than long ledge overhangs. Smaller and rarer, Candy Basslets are sighted between 50 and 200 feet, but more common below 100 feet. While these three species range widely across the region, the slightly larger Wrasse Basslets appears to be restricted to deeper areas off the U.S. East Coast from North Carolina south to Florida and on to the Gulf of Mexico.

Even though widespread and fairly common, little is known about the group. The genus has been placed in the sea bass family, but debate continues about the proper classification. The species appear to be more closely allied with soapfishes than with other sea basses. Both groups share flat, sloping heads, comparable dorsal fins and vertebrae alignment, and have similar-appearing larvae that sport single long, streaming foredorsal fin filaments with a series of pennant-like floats attached. Unlike hermaphroditic sea basses, *Liopropoma* are born, and remain, males or females. Because of their solitary nature the fish are assumed to be territorial, although aggression between individuals has never been recorded. Even their diet has yet to be substantiated; but the small predators are believed to subsist on crustaceans and small fishes. We had the good fortune to observe a pair of Peppermint Basslets spawning just after sunset eight days following the full moon of September. Two similar-sized individuals appeared from a colony of Finger Coral, *Porites porites,* but immediately vanished when illuminated by a light beam. A minute later the pair returned and remained near one another and only an inch or two above the coral. One had its head elevated and both twitched once or twice before moving together until the sides of their heads touched and pointed up. With their bodies angled apart 45 degrees, they rose slowly for a body length before rapidly jetting up and apart in an apparent spawning rush. The couple instantly disappeared back into the coral and failed to reappear.

Soapfishes

Subfamily Grammistinae

Soapfishes are rather strange and probably very recent descendants of the sea bass family. The group's common name comes from a coating of mucus, known as grammistin, secreted from skin pores. When caught or handled, the fish excrete a lather resembling soap suds. The substance, containing a protein

Greater Soapfish, a nocturnal species, generally rest on their side during the day.

toxin that irritates skin, is believed to protect the species from predation. Fishes kept in captivity with a secreting soapfish often die from the poisonous lather. Soapfishes have yet to be found in the stomachs of predators, but the mucus seems to have little or no effect on cleaning gobies that happily pick parasites off soapfishes.

The group is represented by two genera in the Caribbean: *Rypticus,* with four species, and *Pseudogramma,* which includes only a single small, reclusive species. Greater Soapfish, *Rypticus saponaceus,* the species most frequently sighted by divers, behave oddly. During the day, the fish spend much of their time lying on their sides or leaning against a coral head, as if sick or asleep. When approached closely, they promptly right themselves and skulk away to other protected locations. The 8- to 12-inch gray fish, with white mottling, large mouths and projecting lower jaws, regularly inhabit patch reefs in depths from 10 to 60 feet. Soapfishes primarily feed at night on small fishes, shrimp, crabs and mantis shrimp. On one occasion we watched a Greater Soapfish slink about the bottom after dark, engulfing one cardinalfish after the other illuminated by our hand lights. In the Gulf of Mexico, adults of the five- to eight-inch Whitespotted Soapfish, *R. maculatus,* have been observed cleaning Goliath Grouper, *Epinephelus itajara.* Members of both genera are known to be protogynous hermaphrodites.

Courting male Greater Soapfish (rear) repeatedly swam by the side of a female.

The female often made abrupt turns to avoid the courting male.

A single published observation of spawning reported a pair of Greater Soapfish embraced, a half-hour before sunset, in a spawning clasp similar to hamlets, *Hypoplectrus* spp. However, in 2011 we photographically documented the extended courtship and spawning rise of a pair of Greater Soapfish beginning 45 minutes after sunset. The courtship and high spawning rise had few

The pair initiated several false spawning rises.

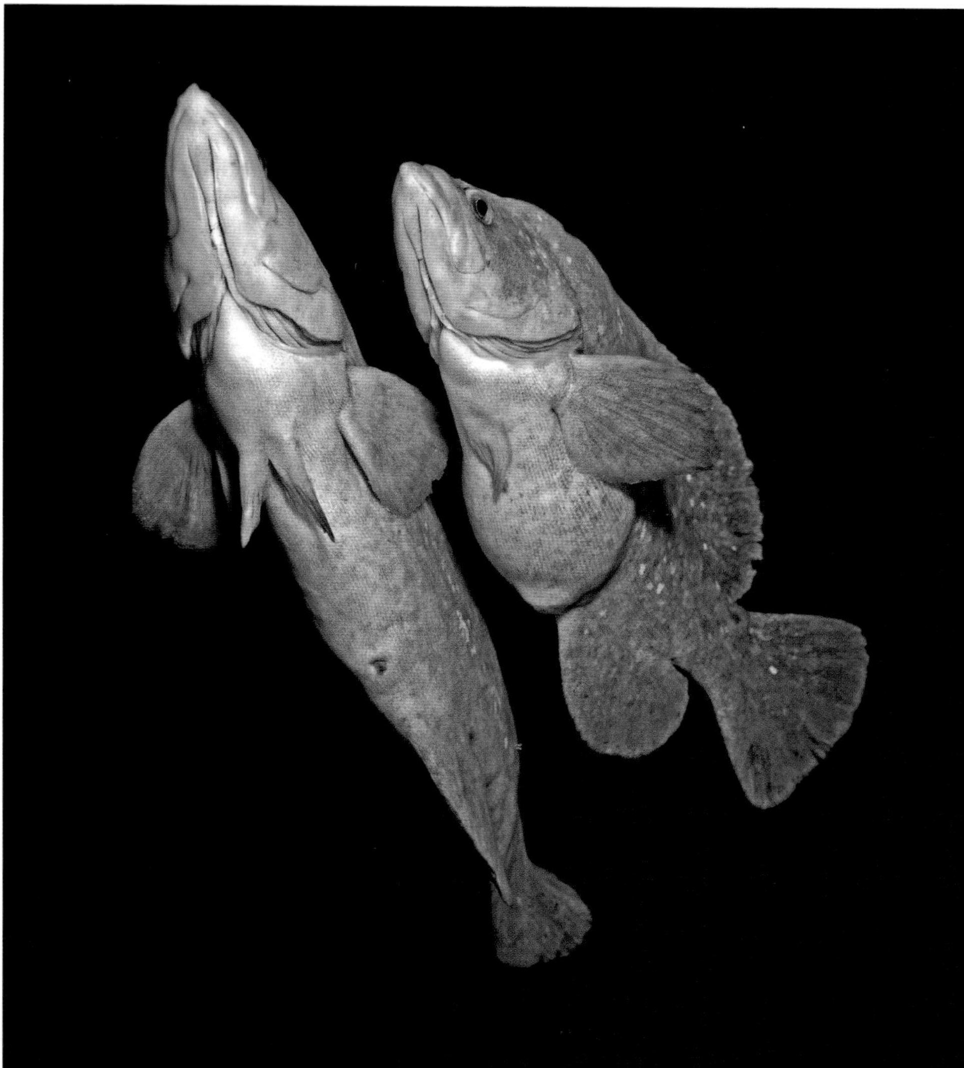

Spawning rise carried the pair 50 feet above coral slope where gametes were released.

similarities to hamlets. The portion of the courtship we observed took place at a depth of 60 feet near or just above the bottom of a coral-studded slope in Bonaire. The fish caught our attention because two normally solitary predators were sighted together on a sand patch between living coral heads. During the event, the approximately 10-inch male soapfish and the somewhat smaller female (evident because of her obviously swollen abdomen) swam parallel rapidly across the bottom, rolled around one another and made three false spawning rises before streaking up side-by-side toward the surface. Gametes were released about 10 feet below the surface. The following evening we discussed our observation with a friend who that night watched a male soapfish spawn with two females minutes apart.

Jawfishes

Family Opistognathidae

Yellowheads hover above their burrows picking zooplankton when currents flow.

Yellowhead Jawfish

Opistognathus aurifrons

Passionate courtship displays, male mouth brooding and burrow building are a few of the entertaining behaviors that make jawfishes so fascinating to observe. Typically, the rather small, elongate, blunt-headed fishes with great bulging eyes and large mouths inhabit sand and rubble plains, where they construct rock-lined burrows in the seafloor.

Of the eight jawfish species inhabiting the region's waters, the Yellowhead Jawfish is by far the most commonly sighted. These pearly, three- to four-inch, fairylike fish burnished with bluish fins and yellowish heads spend much of their day above their burrow entrances plucking zooplankton from the current. Colonies, which normally consist of 2 to 10 individuals, are found on sand aprons encircling coral and rock reefs at depths from 5 to more than 150 feet. The association with reef structures ensures an abundance of buried coral, rock and shell rubble essential for the construction of their masterfully built burrows.

Yellowheads usually allow divers to approach within five or six feet before dropping close to their burrow entrances or slipping inside tail first. Most reappear a minute or two later and cautiously begin feeding once again. They

Right: *Males mouthbrood clutches of eggs for five to seven days.*

Male Yellowhead churning batch of fresh yellow eggs.

seem to be more easily approached when currents are bringing in an increased supply of planktonic food. If you remain motionless, a colony will become used to your presence and entertain you for an entire dive.

From early spring to fall, males incubate clutches of eggs inside their mouths. For fishes, oral brooding is considered an advanced parental behavior. It occurs in eight fish families, the majority from fresh water. In the tropical western Atlantic only two families, jawfishes and cardinalfishes, are known to care for their eggs in this manner. Mouth brooding significantly limits the number of offspring produced, but provides superior protection during incubation and better prepares them for the perilous larval stage.

When searching for brooding males, look for distended cheeks. Males with eggs are quite cautious, usually backing into their burrows when you are 10 or more feet away, but periodically stick their heads above the sand to check your whereabouts before slipping back out of sight. Some brooders refuse to surface with divers nearby, while others seem to lose interest and go about their business even when observers are only a few feet away. If you have the time and patience, it is worth the wait to watch males churn their mouthfuls of eggs.

Churning, as the behavior is known to scientists, occurs when the clutch is partially spit out of the mouth and quickly sucked back in. This action removes waste, aerates the eggs and allows the embryos to mix, so all will have a chance to develop equally. The behavior serves much the same purpose as fin-fanning by benthic egg layers. The egg mass might be agitated two, three or four times in rapid succession. Churning can occur as often as every minute, but at other times, especially if you remain too close, males stubbornly keep their eggs tucked safely away.

Clutches are laid throughout the lunar cycle; however 70 percent arrive between the full and new moon. Incubation lasts between five and seven days. The age of a batch can be roughly estimated by the eggs' color. When fresh and heavy with yolk eggs are yellow, gradually turning dark orange during days two and three. As the yolk is exhausted silver begins to dominate as eye pigment develops. During the final two days, the embryos' eyes can be clearly seen inside the clear casings. The egg ball is firmly held together by a tangle

Male Yellowhead churning mature batch of eggs ready for release.

of adhesive filaments allowing the mass to be churned without breaking apart. The tiny, translucent hatchlings are released from the male's mouth within the hour following sunset. Males are extremely cautious and retreat into their burrows if illuminated by light. Before bedding down Yellowheads traditionally cover their entrances as a protection from intruders. If a burrow opening remains uncovered after sunset, a release is imminent. Again, keep light beams low and turned away from the action. If necessary, filter your beam through closed fingers. At this point the male's head can be seen an inch or so below the entrance. When divers are nearby a male will only extend his head slightly above the entrance when releasing his brood. The father huffs and puffs sending hatchings somersaulting out of his mouth in clusters. From our

Male Yellowhead releasing hatching larvae.

The passionate courtship ballet of Yellowhead Jawfish.

Mated pair of Yellowhead Jawfish share honeymoon burrow.

photographs it appears that the balls of eggs are shrouded inside a clear membrane partitioned into multiple compartments containing larvae. A full release of all the hatchlings takes about a minute to complete. Most mature Yellowhead males incubate two clutches per month, on rare occasions three.

Of course, romance leads to clutches, which leads us to the splendid courting ballet of Yellowheads. The monogamous couples, which live only a foot or two apart and share a third burrow as a honeymoon suite, are artists of passion. Courtship, which usually takes place in the early morning or late afternoon, can be identified by a mated pair of Yellowheads remaining unusually close or sharing the same burrow. The

A courting male Yellowhead Jawfish displays to his partner.

Yellowheads' animated courtship ballet can last for the better part of a dive. If you enjoy a good romance, calmly settle on the sand nearby and enjoy an unforgettable spectacle.

Courtship begins with a male approaching his mate in a series of swoops. Once close, he arches his back, flares his fins and opens his great mouth, as if singing an aria. The pair often separate, or drop out of sight in the third burrow, for minutes at a time before reappearing and continuing their nuptial dance. This ritual culminates with an exchange of eggs and fertilization inside the pair's honeymoon burrow. Later the female resumes her normal feeding behavior. For the next week the male will feed briefly leaving his clutch inside the burrow while he is outside picking zooplankton.

Courting male arching, flaring and opening mouth.

Yellowheads are not only lovers but also marvelous

Yellowhead clearing sand from burrow.

engineers. Their skillfully built burrows are tributes to industry. It is common to see a Yellowhead busily excavating mouthfuls of sand and rocks to maintain their present burrows; however the construction of a new burrow takes place less frequently. The demanding project, which takes nearly eight hours to complete, starts with the excavation of a funnel-shaped pit about eight inches in diameter and six inches deep. Next, rocks and shell shards supporting a vertical tunnel are positioned with the skill of artisans until the passageway is flush with the surface sand. Afterwards, a chamber is excavated below. It is usually tunneled beneath a buried stone that serves as support. Immovable rocks encountered during the dig dictate the room's final size and shape. Completed chambers can be up to 10 inches long and 3 inches high. A blanket of sand serves as a finishing touch—a mighty feat for a four-inch fish.

Housekeeping chores routinely pick up pace late in the afternoon as the colony prepares for the night. Loose sand, pebbles and shells are brought up and blown clear. Within minutes of sunset a rock, shell or tuft of algae is pulled over each entranceway to hide the burrow's location. Yellowheads emerge at first light and begin feeding immediately.

There is good reason for Yellowheads to be wary; sand plains are hunting

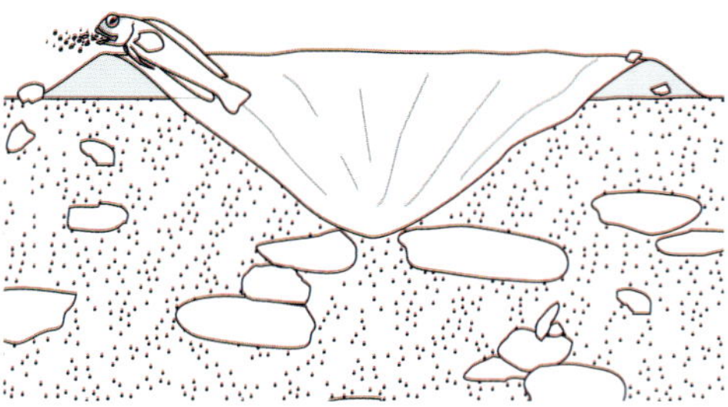

A six-inch-deep pit is dug by mouth.

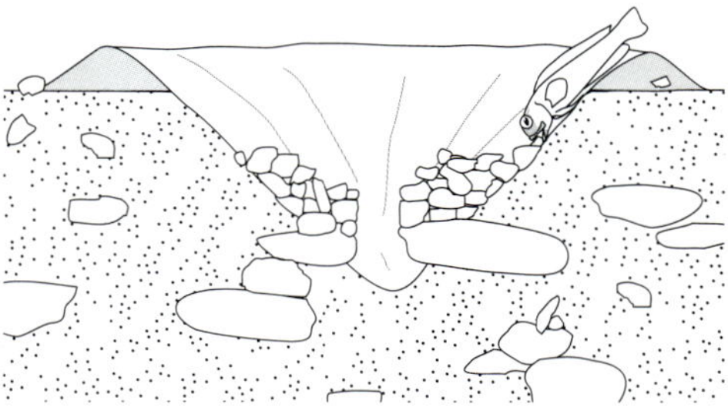

Shell, rock and coral are positioned.

Digging a large hole by mouth is only the beginning of an eight-hour burrow building project.

grounds for several large piscivores. The gravity of approaching danger can be gauged by a jawfishes' reaction, which ranges from slowly moving closer to their burrows to headfirst dives inside. When large prowling predators, such as Hogfish, *Lachnolaimus maximus,* Yellowtail Snapper, *Ocyurus chrysurus,* Margate, *Haemulon album,* or Southern Stingray, *Dasyatis americana,* pass nearby, Yellowheads dive for cover. Occasionally, when directly threatened, they even cap the openings with rocks. We once watched a feeding Yellowhead move away from its burrow to allow a prowling Sharptail Eel, *Myrichthys breviceps,* to slip inside its burrow uncontested. The approaches of smaller, less menacing, sand flat hunters, such as Yellowhead Wrasse, *Halichoeres garnoti,* and Slippery Dick, *H. bivittatus,* usually provoke mild aggression rather than flight. When smaller fishes like Hovering Dartfish, *Ptereleotris helenae,* or Bridled Goby, *Coryphopterus glaucofraenum,* approach, they are energetically driven away by a hostile show of flared fins, open jaws or, occasionally, mouthfuls of sand spit in their direction.

Yellowhead Jawfish regularly feud with neighboring Sand Tilefish, *Malacanthus plumieri,* a larger burrow-builder that constructs substantial, rock-covered burrows within defended territories on sand flats. Small tilefish are readily chased away, but the approach of larger tilefish force Yellowheads

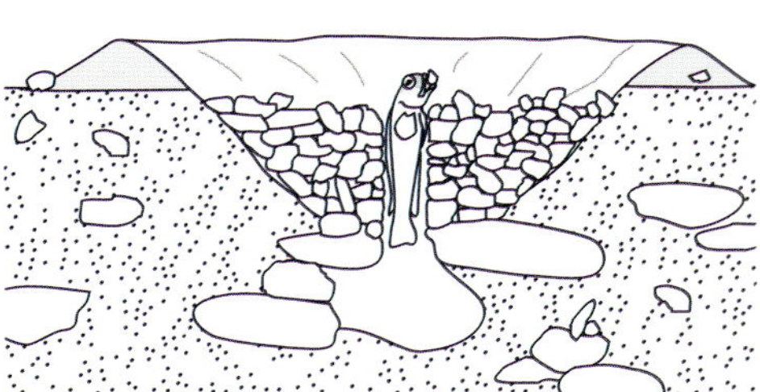

A chamber is excavated below the pit.

A blanket of sand covers the stonework.

Wormfish temporarily share the burrow of a Yellowhead.

For an unknown reason, a Yellowhead flicks its tail under its gill cover.

back inside their burrows. Sand Tilefish have been observed circling an occupied burrow and aggressively darting toward a Yellowhead when it emerges and, in a few cases, actually thrusting their snouts into the opening. These encounters usually end with the tilefish stealing choice rocks ringing the jawfishes' burrows.

Even though small intruders are not tolerated near a burrow, adult Slippery Dick have been seen entering both occupied and unoccupied burrows and remaining for several minutes. And then there is the rather strange association with the Pugjaw Wormfish, *Cerdale floridana,* a rarely sighted, three- to four-inch, eel-like fish that burrows in soft sand. One afternoon we watched three wormfish move in and out of two adjacent burrows with no apparent concern from the resident Yellowheads. In one case, the jawfish moved aside to allow the slender fish to slip in and out. At one point two wormfish and the Yellowhead perched side-by-side for several seconds with their heads projecting from the same burrow. After almost an hour, the wormfish swam off, one at a time, in different directions.

Banded Jawfish

Opistognathus macrognathus

Even though Banded Jawfish share a number of family traits with Yellowheads, including male mouth brooding and burrow building, the family members differ in size, color and feeding behavior. These drab five- to eight-inch burrow dwellers feed on bottom-dwelling invertebrates and small fishes instead of drifting zooplankton. So, instead of flitting above their burrows during the day, Banded Jawfish spend their time peeking over the rims of burrow entrances, encircled by tidy mounds of rocks and shells.

Banded Jawfish build their burrows on sand bottoms dotted with patch reefs in 3 to 35 feet of water. Resting in their burrows with only their heads exposed,

A male Banded Jawfish churns his maturing clutch of eggs.

their mottled bodies blend into their surroundings like ghosts. When a Banded Jawfish is discovered, look around carefully and you will probably see a few others nearby, staring intently in your direction. A cautious diver can usually get to within a few feet of its burrow before the fish disappears from view. After becoming used to having you about, an individual will usually venture into the open and busy itself rearranging rocks or spitting sand.

Colonies of Banded Jawfish typically include from 3 to 11 individuals. Within the colony, mating groups, consisting of a single male and one or, occasionally two females, live in close proximity. Males, slightly larger than females, receive two to four clutches of eggs each month, brooding multiple clutches at times. Research indicated that courtship is synchronized with the full moon. Males hold the egg masses for five days before releasing larvae into the water above their burrows at night. Settling 18 days later, small juveniles, less than an inch long, immediately build miniature burrows near an established colony and feed on zooplankton for the first few weeks before switching to a diet of bottom-dwelling invertebrates. Except for a prominent spot on their foredorsal fins, juveniles look like miniature adults.

Jawfish are not the only burrowers inhabiting the sand plain. One afternoon we watched a Banded Jawfish excavating a new burrow. The fish bolted back to its old burrow as we approached, but was quickly harassed back by a neighboring Sand Tilefish that had laid claim to the abandoned building materials. During the jawfish's absence, we quickly measured the construction. It was five inches wide, seven inches deep and just over a foot in length. The vertical walls of the rectangular hole were smooth indicating that the jawfish had taken advantage of the break in the substrate when selecting its new burrow site.

Finally deciding to ignore us, the jawfish returned and began digging in earnest blowing out mouthfuls of sand and wedging together coral and rock to create a roof. Persistence rather than precision seemed to rule the

The Banded Jawfish courtship display illustrates the origin of the species' common name.

enterprise. If, after several attempts, a rock refused to fit into place, the jawfish simply abandoned the project and returned to blowing sand. Later, it would take up the cause of the fallen rock once again, as if the laws of physics had altered since its earlier effort. In this hit-or-miss manner, an overhead support slowly took shape.

When predators pass near their burrows, jawfish rotate like periscopes to keep their eyes on the intruders. Other than Sand Tilefish that seem to enjoy making off with their neighbors' burrow rocks, relationships with other sand dwellers are more amicable. Slippery Dick, which frequently prowl around colonies, are confronted with a gaping mouth but otherwise ignored. Small Blue Dartfish, *P. calliura,* and Hovering Dartfish that occasionally build their tiny sand burrows near Banded Jawfish appear to be tolerated.

One afternoon we watched a lengthy confrontation between two Yellowhead Jawfish and a small Banded Jawfish that had chosen to build its home two feet away from their adjacent burrows. Every few minutes one of the Yellowheads

Digging a new burrow.

An overflowing mouthful of sand.

would charge in the direction of the interloper with its fins spread and mouth opened. When we visited the site two days later, the Banded Jawfish was nowhere to be found, evidently chased away by the smaller Yellowheads.

Dusky Jawfish

Opistognathus whitehursti

The three- to four-inch Dusky Jawfish are cryptically colored and marked to match the shallow, 4- to 10-foot, sandy rubble and Turtle Grass, *Thalassia testudinum,* bottoms where they build their burrows. Individuals of both sexes live in random clusters within large, widely scattered colonies. These small, active burrow builders appear to construct new residences more frequently than other jawfishes, at times more than 25 feet from their previous homes. If you move rocks a few inches away from their entrances, Dusky Jawfish are not shy about leaving the burrows' protection to retrieve the misplaced stones. Males are believed to be promiscuous, incubating egg clutches for eight- to nine-day periods once or twice each month from spring to late fall with peaks during the days following the full moon.

Male Dusky Jawfish with clutch of eggs.

Lionfishes

Family Scorpaenidae

A Red Lionfish on a Bahamian reef.

Possibly no other fish has been studied so intensely by so many people in such a short period of time as the Red Lionfish, *Pterois volitans,* a native to the western Pacific, and the similar-appearing Devil Firefish, *P. miles,* from the Indian Ocean and Red Sea, since their invasion of the western Atlantic. With no known predators or competitors in their new ocean to check their progress, the beautiful eating machines, with the inherent attributes of consummate invaders, are making a smorgasbord of native reef fishes and crustaceans, and as a result, threatening to irreparably alter marine coral reefs ecosystems for decades to come. DNA analysis in 2007 indicated that 93 percent of Atlantic invaders tested were *P. volitans* and seven percent *P. miles.* (In this chapter the two species will be collectively referred to as "lionfish.")

The source of the invasion will never be known for certain, but most likely originated from aquarium pets released, one by one, over the years into the Intracoastal Waterway in Southeast Florida. The trouble began when the sporadic releases eventually produced the numbers necessary to establish a breeding population. Like the vast majority of bony fishes, lionfish are pelagic spawners that release large numbers of eggs into the open sea. After hatching, many of the larvae are whisked away by the currents—this mechanism is responsible for driving the incredible colonization of lionfish from the Carolinas to the Spanish Main. Weeks later survivors of the oceanic voyage settle to the sea floor as wispy 10-millimeter post-larval/juveniles ready to eat their way into adulthood.

Larval lionfish

One-inch juvenile lionfish

Isolated lionfish had been reported in the tropical waters of Florida's southeastern coast from as early as 1985. The first multiple sighting came in from Palm Beach, Florida in 1994. So it seems somewhat surprising that the first significant population in the western Atlantic took hold off North Carolina where minimum water temperatures traditionally dip to between 50 to 60 degrees F. However, lionfish are tough customers, able to survive the sub-tropical water of southern Korea and Japan in their native range. It is believed that the same warm Gulf Stream currents that carried larvae north along the southeastern U.S. coastline allowed the adult population to survive the area's temperate winters in the warmer currents offshore.

Concerns about lionfish in North Carolina began to grow in 2000 when two adults were sighted. During the following year 17 more lionfish were found at eight different offshore locations in depths from 85 to 260 feet. That same summer two juveniles were discovered near dock pilings off the southern coast of Long Island, N.Y. These and other scattered individuals swept so far north appear unable to survive the region's harsh winters. That same year a juvenile found in a tide pool on the shores of Bermuda, in the middle of the Atlantic 650 miles east of the Carolinas, clearly indicated the lionfish's distribution potential. Additional reports followed from Georgia in 2001, and northeast Florida in 2002.

In 2004, divers made the first confirmed sighting in the central Bahamas. Initially, local dive operators treated the fish as a novelty, but as months passed, others appeared, and soon lionfish were everywhere. Over the next two years lionfish spread south through the island chain in unprecedented numbers. Finally in 2007, larvae made it across the wide, openwater barrier from the Bahamas to the Caribbean. Soon the aliens were being reported along both the north and south coastlines of Cuba. Three years later, the invasion had advanced east to Puerto Rico, west to Yucatan, north into the Keys and up into the Gulf of Mexico, then south past Jamaica before reaching South America and heading east. They reached the shores of Barbados, in the far southeastern Caribbean in November 2011. In 2019 the question remains: Can lionfish make their way across the great freshwater plumes pouring from the Orinoco and Amazon Rivers and invade Brazil?

Regular culling of lionfish has proven helpful controlling local shallowwater populations.

Genetically, only nine haplotypes have been identified in western Atlantic invaders compared to 36 found in the lionfish's historic western Pacific range. The low number indicates that the invading populations probably resulted from a limited number of individuals. This disparity could eventually result in a genetic bottleneck limiting the population explosion, or conversely, may have selected for "superiorly adapted fish."

Like other members of the scorpionfish family, lionfish bear an arsenal of venomous spines on their dorsal, ventral and anal fins that discourage potential predators. Lionfish are seldom aggressive, however they often stand their ground with fins bristled and heads lowered when approached or intimidated. With contact the loose fleshy sheath encasing each venomous spine slides down as the shaft depresses glands at its base forcing venom up through a pair of grooves into the puncture wound. Envenomation is seldom if ever fatal, but intense pain, numbness, swelling, bleeding and blistering are typical. Victims might also experience dizziness, nausea and headaches. In rare cases convulsions and paralysis occur. If pain and swelling continue, or if a broken spine remains embedded, victims should seek medical attention. In the interim, a hot-water soak or compress (not to exceed 110 degrees F.) will help denature the toxin, significantly relieving swelling and pain.

With the unexpected explosion of lionfish in the Bahamas researchers doubled down on their efforts to understand the threat. Initially there was little to go on, other than a couple of feeding studies from the Red Sea and venom research. Lionfish, in their home water, were known to be opportunistic generalists with massive appetites and wide-ranging diets that vary with prey availability. Foremost on everyone's mind was the impact the new predators might have on native fishes. It didn't take long to find out. An early feeding study, published in 2009 sampling the stomach contents of more than 1,000 lionfish, culled from various habitats in the Bahamas, was disturbing. Forty-one fish species from 21 families made up 78 percent of the identified prey samples. Crustaceans, primarily shrimp, and a scattering of mollusks contributed another 14 percent. Invertebrates dominate the diets of juveniles whereas adults tend to prefer fish.

Lionfish spent their nights hunting with peaks in activity at dawn and dusk. They also hunt during the day, but prefer hiding out inside crevices waiting for the advantage of darkness to swing in their favor. The invaders certainly aren't picky eaters. Eight fish families—damselfishes, wrasses, goatfishes, basslets, sea basses, gobies, cardinalfishes, and blennies—made up 48 percent of the lionfishes stomach content by volume. Appearing toward the bottom of the long list of species eaten was an octopus, a seahorse and a male jawfish with eggs still in his mouth, enough to make any fish watcher wince.

Nine years later a multi-regional study encompassing the Bahamas, Belize, Bermuda, Costa Rica, Florida, northern Gulf of Mexico, North Carolina and the U.S. Virgin Islands identified 167 prey items from 78 families, found in the stomachs of 8,175 specimens. A quick glance at their lengthy list of prey items makes it clear that lionfish are targeting plentiful, easy-to-catch prey that fits in their mouths. The study also bolstered the invader's reputation as gluttons; one fish found among the stomach contents measured nearly half the length of the lionfish that had swallowed it whole. Another stomach contained 87 recently eaten items. In a separate incident, a lionfish was seen gulping down 20 fish over a half-hour period. To hold such quantities a lionfish's stomach can

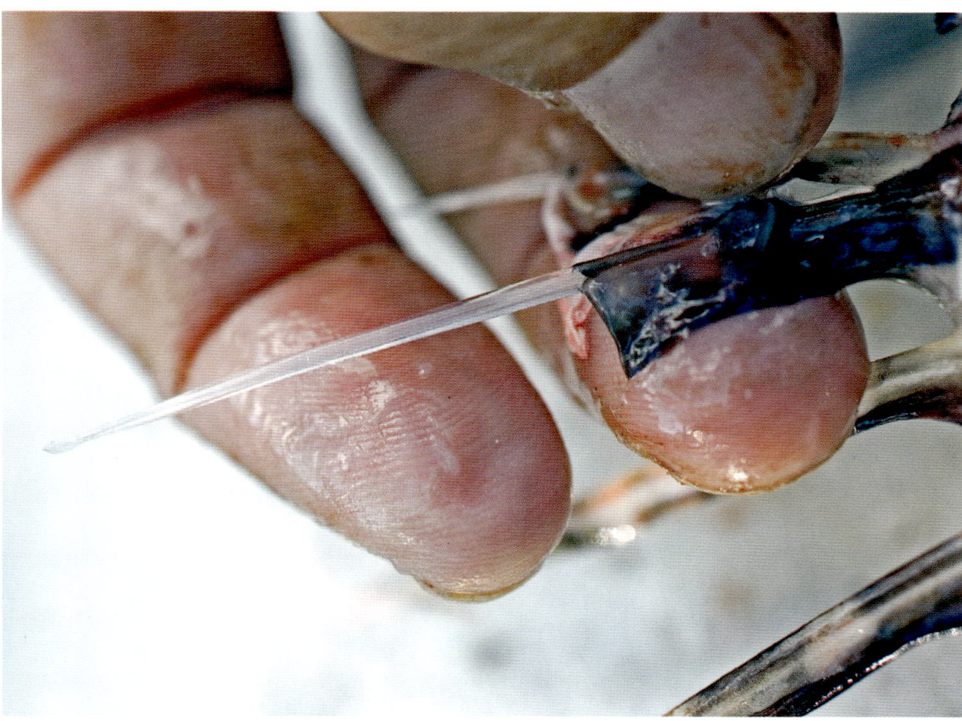

A venomous dorsal fin spine exposed by pulling back its fleshy covering.

Lionfish in Bermuda have always had a strong preference for invertebrates, especially since much of the local population migrated deep and began eating Red Night Shrimp.

expand to 30 times its normal size. For unknown reasons invertebrates—primarily shrimp, a scattering of crabs and a few other crustaceans—are more commonly eaten in the northern half of the range. The first of the lionfish arriving in Bermuda took up residence on shallow reefs, but later, after culling began, migrated to depths, reaching the greatest density below 180 feet where they developed a keen interest in Red Night Shrimp, *Cinetorhynchus rigens,* elevating their already strong preference for crustaceans to 60 percent of their diet, the highest consumption of invertebrates in the invaded range.

Another alarm sounded in 2012 when a two-year study recorded a 65 percent decline in the biomass of fish inhabiting a Bahamian reef. Subsequent studies also documented troubling declines of small reef fishes as well as reduced numbers of larval recruits at sites where lionfish had taken up residence. Even though the new predators do not appear to compete directly with commercially important groupers and snappers they have been documented eating their juveniles. Also worrisome, Atlantic lionfish were reported to be eating young parrotfish and surgeonfish, dominant herbivores crucial for controlling competitive algae on surfaces where coral larvae settle and grow.

On their native reefs lionfish are considered highly successful hunters; however Pacific populations remain balanced. Discovering the mechanism or mechanisms that keep them in check could be an invaluable piece of knowledge. Evidence suggests that their larval stage between spawning and settling is a likely place to investigate. For instance, surveys of larval recruits conducted along the Great Barrier Reef rarely included lionfish even though lionfish are prolific, year-round spawners. It has also been speculated that juvenile lionfish, with pliant, less venomous spines, may be targeted more frequently by predators in the Pacific.

Lionfish make good use of three swim bladders to hover motionlessly, or glide about the reef like winged phantoms—a mirage of bristling spines and fins, and cryptic striping that look, move and smell unlike anything fish and shrimp in the Atlantic are programmed to perceive. During the day the nocturnal hunters customarily take shelter, alone or in small clusters, inside shady protective structures. Although primarily solitary hunters, lionfish occasionally prowl in packs slowly corralling small, abundant bottom dwellers with their heads held low and their oversized pectoral fins spread like wings. However, blowing jets of water to disorient victims, a common technique employed in the Pacific, has been infrequently observed in the Atlantic.

It has long been acknowledged that invaders, detached from traditional ecological constraints such as competition, grow larger and faster in their new surroundings. Lionfish, frequently referred to as "perfect invaders," clearly fit the model. A combination of venomous spines, unique hunting skills, prolific reproductive capacities, life expectancies of 15 years and absence of historic predators and parasites has transformed invading lionfish into "hyper-successful nuisance predators" with the ability to negatively alter ecosystems spanning multiple trophic levels.

Even thought lionfish in both the Pacific and Atlantic Oceans invest equal time hunting, and experience similar kill rates, the advantages bestowed on them in the Atlantic are obvious when compared with those inhabiting home waters. In the Atlantic, they not only grow twice as fast, but attain a larger size reaching 19 inches compared 15 inches in their ocean of origin. They are also far more numerous in the Atlantic. Researchers recorded population densities as high as 400 lionfish per 10,000 square meters in the Atlantic compared to 26.3 in their Pacific range. Once established, the invaders consume a more diverse diet, and substantially larger prey than before. They also eat twice the amount food that similar-sized native predators consume.

A dissected lionfish from a food preference study.

Adding to their prowess, lionfish have the capacity to adapt to a surprising range of habitats including natural and artificial reefs, docks, jetties, bridge pilings, seagrass beds, mangroves and can survive in brackish estuaries with low salinity for extended periods of time. Unfed aquarium specimens survived for three months and lost only 20 percent of their body weight. Lionfish also tolerate a wide span of depth ranges. In the Bahamas, they live from 4 to 450 feet, and most astounding, an adult was sighted from a submarine at 1,000 feet below the surface. Equally impressive, a full-grown lionfish was drawn to the nightlights of a research vessel working in the depths of the Atlantic hundreds of miles offshore.

The distribution of larval lionfish by currents and swimming is believed to be responsible for the invaders' rapid, far-reaching distribution. Once established, adult lionfish generally remain close to home for extended periods. A study of adult lionfish inhabiting both patch reefs and extended reef systems in the Bahamas found that most moved less than 50 yards from where they were tagged while several others were resighted as much as 800 yards away, and one individual turned up 15 days later almost a mile away. Most of this movement appeared to be away from heavily infested patch reefs to less populated areas. A subsequent study found that crowded conditions increase their time away from shelters as well as lengthening foraging trips to seagrass meadows. A North Carolina study employing acoustical tags, demonstrated strong site fidelity and limited movement. A similar pattern held true in a Florida estuary where after several weeks 74 percent of the subjects remained within 10 yards of where they were tagged.

Although lionfish have morphological indicators of hemaphroditism they are classified as gonochoristic, reproducing as male/female pairs. Females in the western Atlantic become reproductively active at a length of about eight inches. Depending on their size females release between 1,800 and 41,945 eggs as often as every two to three days. In some regions they spawn the entire year. As staggering as these numbers seem, the egg production of many large mature reef fishes are more than an order of magnitude greater than lionfish. For instance, Red Grouper, *Epinephelus morio,* can release up to 2.3 million eggs with each spawn and Red Snapper, *Lutjanus campechanus,* 3.4 million. The uncommon adaptation of shrouding eggs in gelatinous rafts, which limits reproductive capacity, must provide some ecological benefit, such as protecting the brood until the larvae hatch.

No one to date has documented lionfish spawning, which, from the presence of ripe gonads late in the day, appears to take place during the night. Supporting this idea, lionfish outfitted with acoustical tags have been recorded rising to the surface at night, indicative of spawning rises. Observations from the Red Sea report courtship beginning shortly after dusk when groups of three to eight lionfish gather. Divers and aquarists describe courting pairs making short, repeated side-by-side dives toward the bottom before facing off and making slow circling ascents. Just beneath the surface the female releases a pair of gelatinous egg masses (one produced by each ovarian lobe), which are fertilized as they float to the surface. Embryos hatch 36 hours later and grow into half-inch larvae that will settle to the seafloor 25 to 40 days later.

A personally related account from Bonaire offers evidence of an early morning spawning event prematurely interrupted by an itchy trigger finger. Spotting a disorganized silhouette of four tightly clustered lionfish heading for the surface, one of a trio of shore-diving lionfish hunters shot a single fish from the tangle causing two others to bolt for the bottom leaving behind a lone female that discharged her rafts of eggs.

As the population explosion in the Bahamas unfolded during the mid-2000s it became evident that lionfish were here to stay, and their spread into the Caribbean inevitable. Worst-case scenarios, hinted at early in the invasion, gained traction as stories of disappearing reef fishes began making headlines. This was a definitive juncture in fisheries biology; no one was sure of what should or could be done. Culling and creating a new seafood market seemed a natural place to begin. The dive community responded with a call to arms; and in the underwater world arms mean spearguns—a form of hunting long criticized by conservationists. Soon slogans, like "Not On My Reef" and "Eat Them to Beat Them," popped up across the Internet as the ideas took shape. In 2009, Reef Environmental Education Foundation (REEF), whose volunteers had been monitoring the Bahamian invasion since 2007, sponsored the first lionfish derby in the heavily infested waters of Abaco in the Bahamas. The weekend event removed 1,408 lionfish. It was an auspicious beginning.

Culling of any form was never expected to eradicate lionfish and hasn't. Nonetheless, organized tournaments have played an effective role controlling populations on reefs near coastal communities and diving resorts where sufficient manpower can be marshaled. A study analyzing the effects of culling estimated that it would require monthly efforts removing between 35 and 65 percent of lionfish annually before prey fish population could rebound. Considering labor alone, culling remote reefs, seagrass beds, mangrove lagoons, and nursery grounds isn't feasible. Tournaments, our best line of defense to date, are still going strong a decade after their inception, continuing to raise awareness, encourage individual removal efforts, collect data and kill lionfish. When combined with cook-offs and lionfish tastings, derbies are a great way to promote a new, homegrown cuisine for local fishing industries.

Early in the invasion, large, native predators, such as groupers and sharks were expected to help minimize rapid population growth. But this hasn't turned out to be the case. To date, native predators haven't been reported regularly consuming lionfish without prompting or training. In 2007, we watched speared, but still living, lionfish being offered to Reef Shark, *Carcharhinus perezii,* during shark feeds in the Bahamas. Incoming sharks typically stopped abruptly inches from the injured fish before darting away. It required persistent prompting to entice hungry sharks into taking the bait. Most of the seized lionfish were spit out, but a few were swallowed. Other native predators including, Green Moray, *Gymnothorax funebris,* Spotted Moray, *G. moringa,* Nassau Grouper, *Epinephelus striatus,* Tiger Grouper, *Mycteroperca tigris,* Nurse Shark, *Ginglymostoma cirratum,* and Yellowtail Snapper, *Ocyurus chrysurus,* are known to consume lionfish. But in most, if not all of these cases, the prey were either juveniles, adults wounded by hunters, or involved predators conditioned by hand feeding to take injured fish—a practice soon discouraged after trained predators turned into aggressive nuisances. In Bonaire, we watched a local lionfish hunter follow a large Yellowtail Snapper that repeatedly stopped and hovered just outside crevices where lionfish were hiding. Oddly, after the speared quarry was left on a sand ledge, the snapper was hesitant to take the bait, and only halfheartedly fed after the fish was cut into pieces. As happens, the Yellowtail turned out to be a pest, making frequent unwelcomed approaches even though no one in our group carried a spear. A similar "pointing" behavior involving a resident Nurse Shark was observed in the Cayman Islands. But in this instance, intact lionfish were swallowed whole. The sharks didn't show any negative effects from the spines or venom.

In Honduras, Bermuda, the Cayman Islands and other locations the largest concentration of lionfish live between 160 and 450 feet—well below safe diving limits. These deep-dwelling populations include many of the largest and most prolific breeders that produce the largest eggs and most robust larvae. Technical divers with the training to penetrate such depths are in short supply. Fish traps, or fish pots in Caribbean parlance, have long been the fishing method of choice in the islands. Having fished local reefs bare of market fishes in most locations, fishers began dropping their traps farther offshore. Soon lionfish were being brought up from as much as 400 feet. In Bermuda, where the greatest density of lionfish live below 180 feet, researchers modified lobster traps, (fish pots having been banned in Bermuda since 1990) left them unbaited to minimize bycatch, and went fishing for lionfish. Their efforts substantially increased catch rates when dropped near lionfish inhabited bottoms. It seems that the predators, which routinely seek shelter during the day, are drawn to the traps for refuge or curiosity rather than food.

Convincing the public to eat lionfish has not been a problem. If fact, many are willing to pay a premium on environmental principles. The problem facing the burgeoning lionfish market is supply. In many areas where lionfish filets and sandwiches have been added to menus, culling soon depleted the readily accessible stock from shallow, inshore reefs. This eventuality, in conjunction with populations shifting to deep habitats, has prompted innovators to begin devising traps designed for depth. The devices, primarily meant for commercial operations, must be light, portable, cheap, easy to build, limit bycatch, and most important be sufficiently functional for fishers to make a living. A prototype, of what has been named the "Gittings Trap" after its primary designer, showed promise during 2018 trials in the Gulf of Mexico. The unbaited, half-moon-shaped design is constructed from two curved stainless steel frames, hinged and covered with nylon netting. With bottom contact, the frames, kept closed during the descent, open flat forming a six-foot circle with a lionfish attractor made from plastic, hardware store lattice standing upright in the center. When hoisted the frames snap shut like a clamshell. During tests, at depths from 120 to 300 feet, 232 lionfish and only 22 bycatch were captured.

As the lionfish invasion nears the end of its second decade, researchers and environmentalists, armed with a growing library of data, are taking a second look at the invasion. Without question, lionfish have brought destruction wherever they've settled, hitting small prey fish the hardest as they spread to every corner of the central western Atlantic. Although lionfish invaded the oilrigs and artificial reefs of northern Gulf of Mexico in unexpected numbers reminiscent of the Bahamas, not all is doom and gloom. On the promising side, there have been few signs of lionfish outcompeting groupers, snappers and other market fishes as feared. They have not yet threatened the extinction of any native species, nor have they made a serious dent in herbivore populations. And for now, the explosion that swept the Bahamas appears to be slowing, even in areas unaffected by culling. Most encouraging, lionfish do not appear to be creating as much havoc in southern Caribbean as they did in the northern regions of the range. Nonetheless, there is no reason to be complacent; this remains an invasion in progress and researchers are far from certain what the lionfish's next move might be.

An early version of the "Gittings Trap" designed to be used in commercial ventures for culling lionfish from deepwater.

Parrotfishes

Family Scaridae

A terminal phase Blue Parrotfish eliminates a chalky plume of excavated waste.

Virtually every reef in the tropical western Atlantic is home to a bustling community of parrotfishes. These robust assemblies are typified by foraging herds of nondescript, two- to six-inch females, and large, solitary, brightly colored males methodically taking bite after bite of algae from the bottom. Parrotfishes share many traits with their forerunners, the wrasses. Members of both families swim primarily with their pectoral fins, change sex from females to males (protogynous hermaphroditism), generally exhibit two or more color patterns within the same species (dichromatism), and have complex social systems and reproductive strategies. However, unlike most wrasses, which use conspicuous canines to capture hard-shelled crustaceans, the teeth of parrotfishes have fused into powerful beaks capable of rasping algae from the porous skeletons of dead coral rock. In the process, large quantities of the reef's structure—calcium carbonate—are routinely consumed. On average, nearly 75 percent of the gut contents of parrotfishes is composed of inorganic sediment. Broad, bony tooth plates, known as the pharyngeal mill, grind the grit into tiny bits, and a specialized alimentary tract extracts food, leaving the remains with nowhere to go but back to the reef as sand. In fact, many divers first notice parrotfishes because of their habit of voiding long plumes of excavated waste as they swim. The family's copious

Right: *When feeding on algae covered rock, Parrotfishes excavate large amounts of calcium carbonate, which is returned to the reef as sand.*

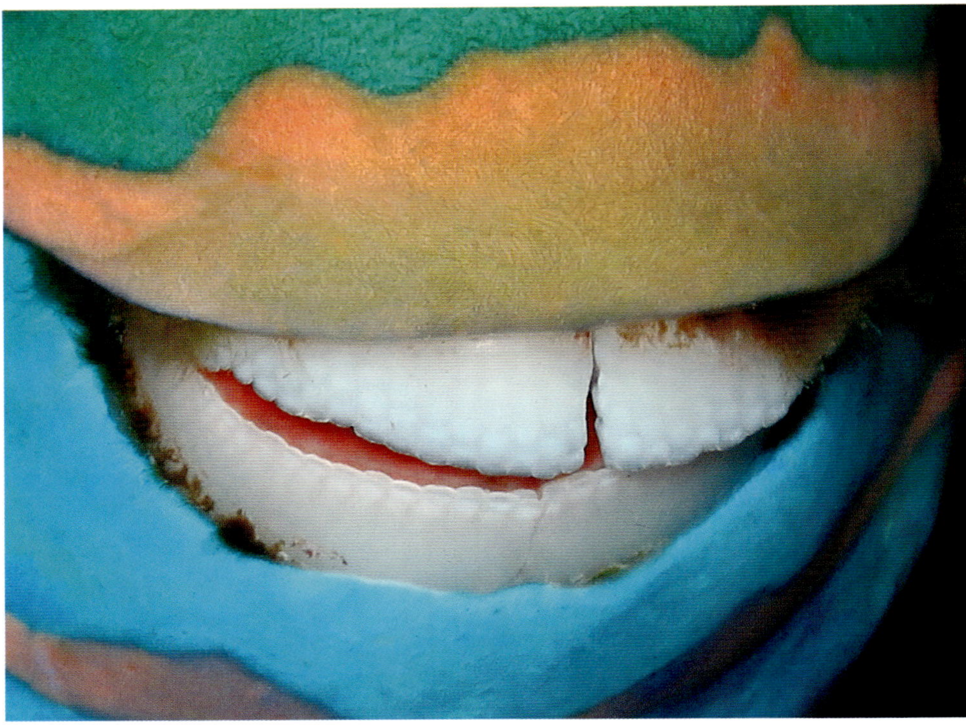

The fused, algae-scraping teeth of a Queen Parrotfish.

processing system generates much of the sand associated with tropical reefs and beaches. Caribbean parrotfishes seldom eat living coral, even though a small number of polyps are inadvertently consumed during their relentless foraging. However, two species, Stoplight Parrotfish, *Sparisoma viride,* and Queen Parrotfish, *Scarus vetula,* have been found to regularly take gouging bites from living coral, especially from the lobes of Boulder Star Coral, *Orbicella annularis.* Besides filamentous algae, the average parrotfish diet includes the tender, uncalcified tips of algal bushes, seagrass blades, occasional crustaceans and, now and again, a bite or two of sponge. Like many herbivores, parrotfishes, when given the opportunity, will supplement their diets by taking bites of flesh from dead fish. Contrary to the territorial nature of many reef fishes, that partition and defend limited food supplies, parrotfishes specialize in specific segments of a plentiful and rapidly renewable plant food, allowing several species to share overlapping feeding grounds harmoniously.

The region's 16 species, from the worldwide catalog of 100, vary in size from the three-inch Bucktooth Parrotfish, *Sparisoma radians,* to the massive four-foot Rainbow Parrotfish, *Scarus guacamaia.* A typical population includes a few large, flamboyantly colored, dominant males, known as terminal phase males, numerous small, rather drab, mostly female members in the initial phase, and sexually immature juveniles.

The origins of the western Atlantic's two primary parrotfish genera, *Scarus* and *Sparisoma,* have been traced to two primitive wrasselike species: the

Emerald Parrotfish, *Nicholsina usta,* a rarely sighted inhabitant of South Florida and northern Gulf of Mexico seagrass beds, and the Bluelip Parrotfish, *Cryptotomus roseus,* a slender, tropical seagrass dweller. Both of these transitional species, the only representatives of their genera, have fused beaks with serrated edges— remnants of individual wrasselike teeth.

Of the 52 species in the genus *Scarus,* six make their home in the Caribbean. Genus members can be identified visually by upper beaks that overlap their lower jaws. *Scarus* share a complex life history—females that live long enough eventually change sex and color patterns. The genus has two types of males: primary males, those born male, and secondary males, those that were once females. If this sounds complicated, brace yourself for the even more perplexing social organization of parrotfishes in genus *Sparisoma*. Males are all secondary and occasionally remain in the initial phase instead of making the more conventional transformation into the terminal phase. At least three *Sparisoma* species including the Stoplight Parrotfish develop testes from ovaries without first functioning as females. The females of some *Sparisoma* are often as large or larger than males in either phase and, adding to the muddle, in some species, many females do not change sex at all. This odd little genus of eight, with upper beaks that slip inside the lower, is restricted to the waters of the tropical western Atlantic. Members of these two genera have distinct grazing behaviors. *Scarus* species scrape filamentous algae from the surface at a bite rate two- to

Like many herbivores, Stoplight Parrotfish occasionally supplement their diets with bites of flesh from dead fish.

A Queen Parrotfish sleeps cloaked inside a mucus cocoon secreted from the mouth.

four-times more frequent than *Sparisoma* species that excavate into the rock leaving visible grazing scars.

Strict, size-related hierarchies that allow only the largest individuals in the initial phase to transform into terminal males govern many species of parrotfishes. These two- to three-week metamorphoses, usually occurring after the disappearance of dominant terminal males, reorder the gonads from ovaries to testes and confer a bright new coat on the recently transformed males. The colors and markings differentiating the phases do not always remain stable. Bright terminal male Bucktooth Parrotfish and Bluelip Parrotfish, both inhabitants of seagrass meadows, fire up their most intense colors when courting, but tone down markedly when traveling outside the safety of their normal habitat. Initial phase Redband Parrotfish, *Sparisoma aurofrenatum,* can also alter color patterns to camouflage with their surroundings. Occasionally subordinates in the terminal phase signal submission to a dominant territorial male by suppressing their colors and markings. Two of the large *Scarus,* the Midnight Parrotfish, *S. coelestinus,* and Blue Parrotfish, *S. coeruleus,* exhibit the same basic color schemes throughout life.

At night, family members utilize a variety of sleeping strategies. Most small parrotfishes seek safe haven deep inside reef pockets. Bluelip Parrotfish, like many wrasses, bury themselves under sand. Each evening at dusk, large Rainbow Parrotfish and Midnight Parrotfish return from distant feeding grounds to sleep inside a customary cave or overhang. A few parrotfishes bed down on open sand or next to a coral structure and change into gaudy night colors. The sleeping fish are often in such deep trances that they can be lifted while still slumbering. However, this is not recommended; in many cases, a disturbed fish suddenly awakens and may bolt blindly into the reef. Without question, the award for the most unusual sleeping habits goes to a few species of *Scarus,* notably the Queen Parrotfish that occasionally cloak themselves in mucus nightgowns secreted from their mouths. The transparent cocoons were once believed to mask their scent from night-prowling predators such as morays, but recent research from Australia has proven that their mucus cocoons actually shield the sleeping fishes from parasites.

Parrotfishes are broadcast spawners, releasing their gametes with single or multiple mates in open water. Most species become reproductively active for an hour either in the morning or the afternoon throughout the year. For a few species, including Redband Parrotfish and the Greenblotch Parrotfish, *Sparisoma atomarium,* terminal males control permanent territorial harems that include from two to seven females, with which they pair spawn exclusively. The terminal males of other species, such as Yellowtail Parrotfish, *S. rubripinne,* and Redtail Parrotfish, *S. chrysopterum,* defend spawning sites only during reproductive periods. Spawning sites are usually near prominent structures on the downcurrent outer edge of reefs, where egg-laden females arrive at an appointed hour.

Pair-spawning parrotfishes all participate in similar courtship and spawning routines. Females, obviously swollen with eggs, gather at traditional sites and calmly forage while excited males, with fins collapsed and tails arched upward, zip in and out of the area. When ready to spawn, females slowly rise several feet off the bottom, where they pause. Returning males temporarily slow, spread their fins, jerk their bodies and often eliminate streams of waste as they assess the status of their partners. When one of the females begins to rise at an angle, the males dart in, positioning themselves close to the females' bodies and, with tail fins providing thrust, blast up at a steep angle with their partners. Eggs released at the apex are immediately enveloped in clouds of sperm. The

A pair of Redband Parrotfish near the peak of a spawning rise.

speed and height of spawning rushes are dictated by a species' size. Large species rise slower and higher than smaller species, which are more prone to predation. Rival males venturing into spawning grounds are vigorously chased. With fins erect, the competing fish zigzag across the bottom at breakneck speed. Challengers must often be chased repeatedly before they relent.

A few species, with numerous initial phase males, such as the Striped Parrotfish, *Scarus iseri,* and Yellowtail Parrotfish, pair spawn and group spawn. On reefs with large populations, initial phase Striped Parrotfish migrate every afternoon along established routes from their shallow inshore feeding grounds to customary group spawning sites on outside reef shelves. There, dozens, or even hundreds, of individuals mill about the bottom before forming into small spawning units composed of multiple males and a single female. The subgroups dart back and forth in parallel formations before shooting up to spawn. These spawning rushes are often contagious; as soon as one group makes its move, other groups follow in rapid succession. Occasionally, these same initial phase males become streakers keeping low profiles as they slink around the spawning sites of terminal males constantly watching for opportunities to dart in and add their sperm to a pair's gamete cloud. Young interlopers, known as sneakers, spawn with females waiting to spawn with territorial males.

Fertilized eggs drift with the currents for about 25 hours before hatching into tiny yolk-sac larvae. After three days, with their onboard supply of fat depleted, the tiny bits of life must fend for themselves in the open ocean or starve. Settlement back to the reef occurs a few weeks later and, in some cases, many miles from their point of origin. The solitary juveniles begin their benthic existence hiding among algae, where they forage for tiny protein-rich crustaceans and single-celled plants, called diatoms.

Striped Parrotfish

Scarus iseri

Loosely organized feeding herds of small, two- to five-inch, initial phase Striped Parrotfish are common in most rock bottom and reef habitats from 3 to nearly 100 feet. The constantly foraging herbivores primarily scrape filamentous algae from rock and dead coral pavement, and also browse on thin mats of organic matter that collect in the bottoms of shallow sand depressions. Both the black-and-white-striped initial phase and seven- to eight-inch, greenish blue terminal males resemble the closely related and similar-sized Princess Parrotfish, *Scarus taeniopterus.*

Although females predominate, the initial phase also includes a substantial percentage of young primary males. Those in the colorful, all-male terminal phase are derived from initial phase males or initial phase females that change sex. Depending on the local environment and population size, those in the initial phase join a defended feeding territory, a nonterritorial stationary herd or large roving herds.

Groups living in defended feeding territories, averaging 150 square yards of sea floor, are most common in regions with large populations. Territories occur most frequently in areas of sparse coral cover in 30 to 35 feet. On heavily populated reefs the entire bottom may be occupied by defended territories, forcing surplus fish to join roving bands. Territorial groups are composed of a single dominant female and from two to eight smaller subservient females that actively defend their personal feeding grounds from territorial neighbors and roving bands of initial phase trespassers. Territorial females can be distinguished by yellow ventral fins extended during confrontational displays. Defense is strictly size-related, with females chasing only initial phase intruders of equal or smaller size. From time to time dominant females engage in border disputes with neighboring matriarchs. These brief tiffs begin when one of the females feeds or hovers near a common boundary. The neighboring dominant promptly approaches to within a foot and extends her yellow ventral fins. The challenged female either swims away or abruptly turns toward her adversary and strikes the same defiant pose. Face-to-face and with mouths agape, the rivals charge repeatedly, but in nearly every case stop just short of contact.

Terminal males generally associate with one or several adjoining territories. They aggressively defend their turf from smaller, usually nonterritorial terminal males, but ignore larger males. Occasionally large initial phase males also defend territories. Most of the territorial males' day is spent swimming energetically about their domains, with brief pauses to forage with groups of females.

It has been documented that the species' system of shared territorial defense

Several male, initial phase Striped Parrotfish break away from the main spawning aggregation to follow a single ripe female.

allows larger group members to feed at a faster and steadier rate than solitary foraging individuals or members of roving groups. On small reefs, where population levels are too low to support roving groups, nearly all feeding areas are controlled by either damselfishes or territorial parrotfishes. In these situations small, non-territorial individuals feed at a much slower rate, making group living, even at the cost of subservience, advantageous. However, on large heavily populated reefs, small territorial females harvest less than their similar-sized counterparts living outside territories. An underling's contribution to group defense gives larger females the opportunity to mature at an accelerated rate, which, in turn, allows the smaller members to advance more rapidly through the group's hierarchy. Most females spend approximately two years as members of a feeding territory before sex reversal, which takes about 10 days to complete.

As light levels fall each evening, Striped Parrotfish, except for dominant females that sleep inside their territories, migrate along traditional routes that follow protective reef structure to deeper water. Initial phase fish leave first, followed by terminal males. Night is spent deep within reef crevices. Although it has not been observed in the wild, individuals in aquariums occasionally sleep within mucus bubbles. At daybreak the procession reverses, with terminal males taking the lead back along the same path to feeding areas. On arrival, females present themselves submissively before dominant females by extending blanched ventral fins and listing to their sides, exposing white bellies.

At many reef sites, roving bands comprised of initial and terminal phase males and a few initial phase females swim randomly about just above the bottom, descending together at intervals to rasp algae from rocks. These groups, often numbering in the dozens and regularly including a scattering of Doctorfish, *Acanthurus chirurgus,* Blue Tang, *A. coeruleus,* and goatfishes, remain stable over a period of months. The third type of social organization, nonterritorial stationary groups, consists of both primary and secondary males and small females.

Both pair and group spawning occur every afternoon throughout the year. Reproduction has been reported to be more intense in the winter in some areas of the Caribbean and during the summer at others. Territorial terminal males pair spawn with members of their feeding groups and occasionally with stray females. Within territory boundaries males, with fins erect, repeatedly visit ripe females feeding on the bottom before rapidly swimming out of the area. Eventually, one or more females rise up to 10 feet off the sea floor and calmly await the males' return. Once back, the males circle their hovering females, often jerking and occasionally discharging plumes of waste. When the females begin to rise, males quickly join them. The pairs, with the males stationed against the females' lower backs, rapidly ascend several feet before releasing visible gamete clouds. Often males spawn in rapid succession with several partners before jetting off in a great circle that eventually brings them back to their territories and more waiting females.

Group spawning involving hundreds of initial phase fish occurs in regions with large populations. At midafternoon, roving bands migrate to a traditional site in deeper water near an outer reef shelf. Fish arriving over an extended period join groups already feeding on the bottom. After a period, tightly knit groups of 20 to 30 individuals rise about 10 feet off the bottom and dart about before splinter groups suddenly break away and release spawn several feet above the others. The first group to spawn appears to spur others into action. A researcher observing a large spawning aggregation counted 35 spawning rushes within one minute.

Princess Parrotfish

Scarus taeniopterus

The body shape, size and markings of the Princess Parrotfish are confusingly similar to those of the Striped Parrotfish, *Scarus iseri.* The two species can be differentiated by a solid upper and lower borders on the tails of both phases of the Princess Parrotfish. This marking is absent on the Striped Parrotfish. Princess Parrotfish typically inhabit areas of heavy coral growth in depths from 15 to 50 feet. Few primary males occur in the initial phase, and only a scattering of primary males has been documented to transform into terminal phase males.

Terminal and initial male Princess Parrotfish compete to spawn with a single ripe female.

Parrotfishes attend cleaning stations frequently throughout the day, but seldom remain for more than a few seconds.

Most terminal males dominate from three to five initial phase females that forage together as groups within the boundaries of permanent territories varying from 100 to nearly 500 square yards. Populations also include wandering bands of initial phase fish. These groups swim together a few feet off the bottom before dispersing as they descend to feed on filamentous algae. After a flurry of bites the groups reform and are off again to new locations. The species has also been reported to consume the mucus covering of sponges and the thin, greenish brown layers of detritus and diatoms that accumulate in bottom sand depressions. Although terminal males are generally solitary foragers, they occasionally join groups of feeding females. Most territorial males defend their borders from other males and spend a great amount of energy chasing both harem members and wanderers.

As daylight begins to fade, Princess Parrotfish assume a novel head-up stance for up to five minutes. Positioned with their tails on the bottom, the fish darken across their backs, rotate slowly in a circle and rapidly open and close their mouths. The posture is quite similar to that of fishes being cleaned, although no cleaning organisms are in the vicinity. This interesting behavior has yet to be explained.

Possibly because safe sleeping sanctuaries are limited within the species' low-profile feeding grounds, only a few adults remain to sleep in their territories while the others migrate, often up to 300 yards, each evening to find night refuge in complex reef structure. Once situated within protective pockets, mucus bubbles are extruded from their mouths to insulate the fish from parasites. At first light they return along routes followed the evening before. Upon arrival at their feeding grounds, terminal males often find their territories occupied by other males that are quickly chased away.

Although Princess Parrotfish and Striped Parrotfish are similar in social organization, spawning strategies, size, habitat, colors and markings, hybrids between the species have never been reported. Striped Parrotfish spawn during mid-to-late afternoon, while Princess Parrotfish become reproductively active from early to midmorning. This difference between spawning periods appears to be the main variable keeping the genes of the two species from mixing.

Terminal males court and spawn year-round within their territories in typical parrotfish fashion. The males swim rapidly around their territories with fins compressed for an average of 20 minutes before spawning. From three to nine females change from a striped to a solid pale brown and rise six or seven feet off the sea floor, indicating their readiness. Males wearing subdued colors circle the hovering females before darting to their sides. Together, the pairs make quick spawning rushes of only a few feet before shedding gametes and parting. No streaking or sneaking has been observed by terminal or initial phase males.

Queen Parrotfish

Scarus vetula

One- to two-foot Queen Parrotfish feed by scraping algae from the surface of dead coral skeletons within large shallow-water territories typified by high-profile coral structures. A survey taken in Puerto Rico found three fish in the initial phase for every terminal phase male. Although rare, there are a few primary males in both the initial and terminal phases, but no sex-changed secondary males in the initial phase.

Queen Parrotfish typically join harems consisting of three or four initial phase females and a single terminal male. Unattached Queens are vigorously chased away from jointly defended grazing territories that often encompass over 1,000 square yards of reef. Individuals tend to feed separately within territory boundaries, but, occasionally, two to three initial phase fish forage in close proximity. This species is one of the few parrotfishes that is known to occasionally sleep in the open inside transparent mucus envelopes.

Pair spawning occurs within territorial boundaries during a two-hour midmorning period. The spawning rush is spectacular, carrying the pairs over 10 feet into the water column before gametes are released in large visible clouds. Researchers in Puerto Rico observed a spawning assembly on an outer reef in 65 feet of water during the mornings of cooler winter months. Terminal phase males, outnumbering those in the initial phase six-to-one, milled about the bottom within small defended territories. Excited males courted females passing through their areas but with little success. Only 12 of 50 courtship attempts ended in a spawning rush and, of these, four were joined by a nearby male.

Blue Parrotfish
Scarus coeruleus

Midnight Parrotfish
Scarus coelestinus

Rainbow Parrotfish
Scarus guacamaia

Little is known about the three largest members of the genus *Scarus,* and there has yet to be a published observation of their spawning behaviors. Blue Parrotfish are solitary feeders that frequently graze over the open sand surrounding reef structures. Although reported to reach four feet in length, we seldom observe members of the species larger than two feet. The species is monochromatic, with both initial and terminal phase individuals displaying a similar powder-blue wash. Larger terminal males, develop conspicuous squared-off snouts. Juvenile Blue Parrotfish inhabit shallow, protected reef habitats. Smaller juveniles are light blue to grayish with a yellow wash extending from snout to dorsal fin. This coloring slowly disappears as the fish mature.

Although Midnight Parrotfish and Rainbow Parrotfish are obviously different in both color and size, they were once considered to be the same species because of significant anatomical similarities. The Midnight Parrotfish is the only family member in the western Atlantic exhibiting a single color pattern throughout life. This adaptation may be due to the species' habit of foraging with swarming schools of similarly colored Blue Tang, *Acanthurus coeruleus.* Both species share a diet of filamentous algae scraped from reef rock. By force of numbers these feeding aggregations overwhelm the defenses of farming damselfishes that have no trouble protecting their personal algal gardens from solitary herbivores, but have little chance fending off a swarm of feeding herbivores.

In Key Largo, Florida fish surveyors from the Reef Environmental Education Foundation (REEF) observed a large group of Midnight Parrotfish streaming down the side of a coral ridge and entering overhangs at the formation's base.

Midnight and Rainbow Parrotfish (right & above) gather in large numbers at dusk before bedding down inside deep crevices and beneath overhangs for the night.

Terminal phase Rainbow Parrotfish—the largest parrotfish inhabiting the Caribbean.

Close observation revealed distinct white bite marks on several egg patches of Sergeant Major, *Abudefduf saxatilis.* The roving group repeated the same behavior further down the reef line.

Rainbows, the largest parrotfish in the Caribbean, tend to spend their day feeding in the shallows near shore. A gut content analysis found Manatee Grass, *Syringodium filiforme,* and algae scraped from coral pavement to be a major component of their diet. Juveniles and greenish blue initial phase adults also frequent shallow habitats and mangrove forests. It is believed that the large herbivores migrate to shallow feeding grounds during the day and return to the reef late each evening for shelter.

At dusk on three consecutive evenings in June, we observed 50 or more large Rainbows and Midnights suddenly converge on a reef site with an abundance of caverns and ledge undercuts near Bimini, Bahamas. The fishes arrived about a half hour before dark in mixed groups that paraded past, high in the water column, excreting great streams of fecal matter. Neither of the species had been seen at the site before the group's appearance. Within minutes water visibility dropped substantially from the suspended silt. Their processions were dramatic and, to all appearances, ritualistic. Most of the fish displayed erect fins. Shortly after their arrival, the largest individuals of both species aggressively charged oncoming fishes with their mouths opened so wide that their snouts flared back, however we never observed contact. The fish repeatedly dived to the bottom and disappeared inside deep undercuts for several seconds before flying out and rejoining the processions. Often five or six individuals entered the same undercut. When observed with a hand light most were restlessly moving about, while others were wedged far back beneath the overhangs. Years later we witnessed the same behavior on the *Benwood* wreck site off Key Largo. Once again, a group of Midnight and Rainbow Parrotfish appeared in the same manner and in about the same numbers at the deep end of the ship's broken metal framework 20 minutes before sunset. An hour later, during a night dive, we observed the fishes bedded down, but still restless, deep inside the collapsed wreckage.

A territorial confrontation between two terminal male Redband Parrotfish.

Redband Parrotfish

Sparisoma aurofrenatum

Redband Parrotfish inhabit the lower slopes and sand halos of reefs with substantial living coral structures from the shallows to over 70 feet. All males are secondary, having changed from females. Most transform into the terminal phase, but a few retain their initial phase markings. The colorful 5- to 12-inch terminal males are solitary browsers that roam widely inside their large territories. Terminal males vigorously defend their sections of the reef and pair spawn with a single or possibly multiple harems of initial phase females around midday at traditional sites along outer reef slopes. The 3- to 6-inch females display various patterns, including green bodies highlighted with bright red fins, a lackluster coat of black and white stripes, as well as a mottled appearance. As the unassuming little initial phase fish move slowly about the bottom taking deliberate bites, they often switch to camouflage coats. Fortunately for identification purposes, a small white saddle blotch on the tail base is always present on members of the initial phase.

Of all the Caribbean parrotfish species, Redbands have the most varied diet, including not only seagrasses, leaf and filamentous algae, but surprisingly, sponges, crabs, brittle stars and urchins. Individuals do not migrate to night shelter; instead, they bed down next to sponges or in shallow depressions.

Pair spawning occurs daily for one-hour periods in late afternoon. Courtship begins when terminal males, with fins spread, busily cajole harem members as they feed on the bottom. Between visits, the males swim rapidly out of the area for several minutes at a time. Eventually females rise several feet into the water column, where they wait for the males' return. Females ascending at the same time tend to remain relatively close together. If males are slow to reappear, the females occasionally drop back to the bottom and continue feeding. The males generally return along sweeping circular routes as if scrutinizing the situation from a distance before approaching. Once near, they spread their fins, twitch, rapidly beat their tails from side to side and excrete plumes of waste. Females often respond by voiding fecal matter before initiating the spawning rush by slowly ascending. Males quickly position themselves behind the rising females. Pairs, propelled by the males' thrashing tails, climb several feet at a steep angle before releasing gametes.

Terminal phase Redband Parrotfish (right) courting a female in his harem.

A female Redband hovers in the water column waiting to spawn.

A courting male Redband dashes about before joining a female waiting to spawn.

Stoplight Parrotfish

Sparisoma viride

One- to two-foot Stoplight Parrotfish, in both the reddish brown initial phase and bright green terminal phase, are among the most common and noticeable members of the Caribbean reef fish community. Studies in Belize, Grand Turk, and Bonaire determined that the protogynous hermaphrodites have an extremely flexible social structure much like the Striped Parrotfish. An extensive five-year research project conducted in Bonaire concluded that Stoplights adapt their feeding, mating and social behavior to take advantage of different environments and food supplies.

For example, on the island's shallow-water sites with abundant, fast-growing algal crops, which could support dense herbivore populations, most terminal phase males and initial phase females live together in nonterritorial groups in less than 10 feet of water. At the same time, on the deeper reefs just offshore a much smaller number of Stoplights control three-quarters of the reef as territorial harems. Also, roving groups of terminal phase males and separate groups of initial phase males and females range freely up and down the coastline for hundreds of yards. However, on reefs with limited food resources, Stoplights live exclusively in territorial harems composed of a single terminal male and from 1 to 14 initial phase females.

Ratios between the female-dominated initial phase and the all-male terminal phase were found to range from nearly two-to-one to over four-to-one, indicating a significant variation across the Caribbean and from reef to reef. On the other hand, initial phase males were generally rare and never made up more than 10 percent of any population studied. Researchers also determined that females reached sexual maturity between six and seven inches, but seldom produced ripe eggs until sometime later indicating a period of rapid growth between maturity and reproduction. It was also discovered that the change

In a recently documented behavior, Stoplight Parrotfish take repeated bites from living coral tissue. It has been speculated that the conspicuous white scars might serve as territorial markers.

from the initial phase to the terminal phase occurred over a wide range of ages and sizes.

Stoplights settle to the reef as three-eighths-inch immature females. The juveniles typically inhabit complex bottom structures in depths from 12 to 38 feet. When young females reach sexual maturity on reefs with large populations, they either join shallow-water groups or become members of harems. Unlike many protogynous species, the largest female living in a harem does not change sex when the dominant male disappears. Instead, initial phase females leave their harems at unpredictable times and join nonterritorial groups where they make the three-week sexual transition into terminal phase males. Even though these terminal phase males were, on average, larger than territorial males, few are sexually active, instead putting their energy into growth.

An initial phase Stoplight Parrotfish takes a bite of living coral.

In Bonaire, Stoplights pair spawned during 90-minute periods each day throughout the year ordinarily between 7:00 and 9:30 a.m., although the time varied at different sites. At the northern site studied, territorial males spawned with members of their harems and with additional females from the shallow-water groups; however spawning never occurred inside shallow feeding areas. Instead, a few males from nonterritorial groups spawned each day in deeper water with females from their groups. On occasions, group males would band together and overrun harems where large territorial males attempted to spawn exclusively with their females. At the site, smaller nonterritorial group terminal males regularly replaced dominant territorial males that disappeared from their harems. A few of the largest group males, termed supermales by the researchers, migrated below 90 feet each morning where they fiercely defended temporary territories and pair spawned with large females that arrived separately and, often, in such numbers that they had to wait to spawn. One supermale was observed spawning 16 times within a 10-minute period. However, most females from nonterritorial groups as well as territorial females spawned twice a day, first with a supermale and then with their usual mates. After the reproductive period, the supermales returned to the shallows to continue feeding.

A second Bonaire study site located to the south, with different offshore topography, and a much smaller Stoplight population had an alternate spawning regimen. The area's surge-swept, bare-rock shallows unfit for heavy grazing, limited competition, allowing terminal males to control the entire population within territorial boundaries. Few juveniles inhabit the southern site, possibly because of the area's lack of branching corals, which provide shelter essential for survival. However, the area hosted numerous initial phase males, made evident by a 14 percent occurrence of streak spawning—an alternate reproductive strategy almost nonexistent among the larger population at the northern study site. Young males, which look identical to females, are able to get away with their covert behavior by hiding in the area's heavy gorgonian cover that grows inside the territories of terminal males.

Instead of feeding by scraping algae from the surface of dead corals and reef pavement, Stoplights use powerful jaws to gouge out the algae growing below a structure's surface. Like other members of genus *Sparisoma,* Stoplights prefer concave rather than flat surfaces and typically take most of their bites from rubble and the dead bases of coral mounds. This excavating method of feeding causes a five times greater consumption of inorganic sediment than that taken in by similar-sized Queen Parrotfish, *Scarus vetula,* which ordinarily consume only surface growth. Stoplight Parrotfish typically forage on upper and lower reef slopes in depths from 5 to 75 feet. However, groups feed on the dense matrix of inshore coral rubble. Deep-water territorial populations have a much easier time excavating the porous coral skeletons that proliferate below 15 feet. However, slow-growing, deepwater algae is limited, and must be diligently guarded from neighboring Stoplights. As a result, territorial terminal males spend nearly a fifth of their time high in the water column patrolling their turf. Inside the deeper territories the thick algal gardens of damselfishes cover much of the reef. Here the story is much the same with the little gardeners, spending

Large groups of Yellowtail Parrotfish gather in the afternoon to spawn.

much of their time chasing trespassers, but for some unknown reason they allow resident Stoplights to take a bite or two from their prize crops before driving them away.

In Bonaire, St. Croix and Puerto Rico researchers have observed both terminal males and initial phase Stoplights taking repeated bites from the living tissue of Boulder Star Coral, *Orbicella annularis,* and Boulder Brain Coral, *Colpophyllia natans,* leaving conspicuous white scars on the coral. The same fish returned to the heads at 2 to 15 minute intervals, removing polyps and skeletal material along the edges of preexisting scars. Seventeen bites were taken from one coral head in just over an hour. Later, another group of Stoplights began grazing on an adjacent, previously unscarred lobe. The majority of damaged colonies were located on reefs with large populations of territorial Stoplight Parrotfish leading to the speculation that the white lesions might serve as territorial markers.

Yellowtail Parrotfish

Sparisoma rubripinne

A combination of the Yellowtail Parrotfish's lackluster colors, scattered populations and preference for shallow-water surge zones makes the species one of the least-noticed family members in the Caribbean. Yellowtail Parrotfish in both phases feed in scattered groups of up to 40 individuals. Adults primarily scrape filamentous algae from dead corals in the shallows, but they also consume algal bushes from reef slopes and seagrass blades. Males in both phases are sex-changed females. The size of both sexes in the initial phase often equals or surpasses that of individuals in the terminal phase. There are roughly four times more Yellowtails in the initial phase than the terminal phase, but because such a large number of males are in the initial phase, the sex ratio for the species is nearly equal.

Even though the species does not establish or defend permanent territories, we have seen Yellowtails involved in a great deal of chasing throughout the day. This constant aggression is possibly due to the high percentage of males in the initial phase. Pair spawning, beginning early to midafternoon, occurs at customary sites on the reef's outer shelf. The species follows the typical parrotfish pair-spawning scenario. Pairs slowly spiral upward before releasing a cloud of gametes high in the water column. We have watched Yellowtail Snapper, *Ocyurus chrysurus,* closely follow their spawning rises and swarm released eggs. And like other large parrotfishes, close observation by divers may cause delayed spawning or force the groups to move down the reef line.

As might be expected with so many young males, initial phase group spawning also takes place. Researchers have observed group spawns on numerous occasions along the seaward edge of a fringing reef in Puerto Rico. Beginning just before noon and continuing over the next hour, hundreds of initial phase Yellowtails arrived from shallow inshore feeding areas. The assembled fish, approximately 80 percent males, milled about the bottom for about a half-hour before small groups, ranging from 4 to 13 individuals, formed into subunits that rose together in the water column and began bouts of rapid, erratic parallel swimming. Just before a spawning rush, the group paused briefly with heads pointed upward, then shot up about 10 feet, releasing gametes at the peak of their rise.

Redtail Parrotfish

Sparisoma chrysopterum

Redtail Parrotfish in both the initial and terminal phase feed together in small to large groups over seagrass meadows bordering reefs, along lower reef slopes and in rocky shallows. Populations consist of equal numbers of both phases. Those in the initial phase can be males (generally the smaller individuals) or females that are often as large or larger than males in the terminal phase. Males in both phases are secondary, having changed from females. During midafternoon, brightly colored terminal males defend temporary territories where they display to, and pair spawn with, females passing through their areas. No group spawning, sneaking or streaking by initial phase males has been reported for the species.

Greenblotch Parrotfish

Sparisoma atomarium

Three- to four-inch Greenblotch Parrotfish typically inhabit coral-encrusted slopes below 50 feet; but we occasionally encounter the species on reefs as shallow as 35 feet and in seagrass beds less than 20 feet deep. Because of their small size, divers seldom notice these strikingly colored little fish. Both phases are distinctly colored and easily identified. The larger males sport multicolored coats of reds, yellows, greens, blues and tan that intensify during reproductive activity. Large females can turn bright red with yellowish anal fins; but they generally display a dull yellowish brown mottled pattern. Smaller one- to two-inch females have reddish brown backs and white bellies with a series of thin brown stripes.

Terminal male Greenblotch Parrotfish (above) Initial Phase female (below).

Those in the female-dominated initial phase outnumber terminal males nearly seven-to-one. The few males in the initial phase and all terminals are reported to be secondary, having changed from females. On the edge of reef slopes, feisty terminal males aggressively protect small harems from rival males. Rapid spawning rushes that carry the pairs only a foot or two above the bottom take place in midafternoon throughout the year.

Bucktooth Parrotfish

Sparisoma radians

A three-inch male Bucktooth Parrotfish displaying his courtship colors.

Averaging just over three inches in length, Bucktooth Parrotfish are the smallest genus members and one of the smallest herbivorous fish. Individuals in both color phases are the same general size and share similar drab, nondescript color patterns except during courtship and spawning when terminal males switch on a few bright red and blue markings around their heads and acquire broad black margins on their tails. Members of both phases can quickly alter their body patterns to blend with their surroundings. The few males in the initial phase and those in the terminal phase are secondary, having changed from females.

Populations almost exclusively reside within thick tangles of Turtle Grass, *Thalassia testudinum,* or Manatee Grass, *Syringodium filiforme,* where the little parrotfish feed from dawn to dusk on the seagrass and clusters of tiny epiphytes clinging to the blades. Many seagrass meadows are alive with large populations numbering in the hundreds. When piscivores such as Bar Jack, *Caranx ruber,* prowl, we have observed large groups of Bucktooth Parrotfish rise in unison, seemingly from nowhere, and flee across the top of the seagrass en masse. If directly threatened, individuals dive deep into the blades, where they instantly change into their camouflage coats. Random groups occasionally forage in fields of benthic *Sargassum* surrounding seagrass meadows, where they nibble on leaf algae, *Halimeda* spp., or bristle brush algae, *Penicillus* spp. At night Bucktooths sleep on the bottom, nestled deep within the seagrass cover.

Many terminal males defend territories containing harems with up to 11 females that spawn once each day. During 30 to 60 minute afternoon reproductive periods, the excited little territorial terminal males buzz back and forth above the seagrass tops like attack helicopters, scouting aggregations of females foraging below. When females are sighted, the males swoop down displaying spread fins and their brightest colors. Nonterritorial terminal and initial phase males roam across the territories often following bands of unattached females, with whom they pair spawn and occasionally group spawn. But their favorite trick is to hide out in the thick seagrass waiting for a chance to streak spawn with territorial terminal males and their mates. Turnabout is fair play—territorial males regularly participate in group spawns with nonterritorial individuals, and readily interfere with the spawning attempts of neighboring territorial terminal males. Males discharge more sperm when spawning with large females, which release more eggs than small females. These large more productive females also make the highest spawning rises and attract more streakers.

Bluelip Parrotfish

Cryptotomus roseus

Because of their shape, slender, one- to three-and-a-half-inch Bluelip Parrotfish, look more like wrasse than a parrotfish, and with good reason. The species, the only representative of its genus and one of the least common parrotfishes in the western Atlantic, is believed to be a primitive link between the two closely related families. Males in both phases are secondary, having changed from females.

The active fish almost exclusively inhabit seagrass beds in depths from 10 to over 60 feet, where small groups including both phases feed together within the forest of blades. Fishes in the somber brown-to-reddish-striped initial phase are mostly females and outnumber those in the terminal phase seven-to-one.

A terminal male Bluelip Parrotfish displays his bright nuptial colors only during courtship and spawning.

A terminal male Bluelip Parrotfish courting a small female.

Although terminal males are generally larger and have distinctly bulbous heads, they otherwise closely resemble individuals in the initial phase except during courtship, when terminal males intermittently display one of the most brilliant color patterns of any fish in the Caribbean. During courtship the larger terminal males swim rapidly over broad areas, periodically flashing their neon-intense yellow, red, blue and green colors to impress prospective mates. Only pair spawning between bright males and smaller initial phase females has been reported.

We have monitored pair spawning bouts of Bluelip Parrotfish during the slack high tide in midmorning and late afternoons, beneath the spans of Blue Heron Bridge that crosses the Intracoastal Waterway on the southeast coast of Florida. The ability to switch vibrant colors on and off allows the fish to remain less conspicuous to predators for much of the day. At night, like several wrasse species, Bluelips bury in the sand and have been reported to secrete protective mucus envelopes.

Spawning rise of Bluelip Parrotfish carry the pair a few inches off the bottom.

Sharks & Rays

Shark families Carcharhinidae, Sphyrnidae & Ginglymostomatidae – Stingray Family Dasyatidae

A Tiger Shark attracted to a fish-feed in the Bahamas.

Unfortunately, divers in the Bahamas and Caribbean seldom sight sharks in the wild. Although not uncommon, the primarily nocturnal apex predators rarely prowl the reefs during the day, and at night are too wary to approach noisy, light-toting visitors. However, shark action, and plenty of it, is guaranteed at a half dozen locations in the Bahamas and at scattered sites across the Caribbean. Several Bahamian operators have been developing their local shark populations for years, and are happy to take dozens of visitors each day to regularly scheduled shark encounters. The sharks have become so accustomed to the handouts that the arrival of a dive boat is sufficient to summon dozens of the magnificent animals to the dinner table. At Great Exuma, where we dived a few years back, a frozen block of fish scraps molded inside a five-gallon bucket, known as a "chumsicle," attracted several dozen Reef Shark, *Carcharhinus perezii,* within minutes. Adding even more novelty, the event takes place within a marine protected area where plenty of big groupers hang around. The instant the chumsicle was lowered underwater the sharks became edgy, tightened their circles and picked up their pace as a cloud of Yellowtail Snapper, *Ocyurus chrysurus,* descended on the block stirring up an oily cloud of thawing scraps. The sharks flew into action snapping up loose bits while others chomped down on the frozen block and spun until flayed carcasses floated free. Early on a wily Nassau Grouper, *Epinephelus striatus,* named Gary, a local favorite that never missed a feed, settled behind a sea fan just outside the sand arena waiting for the feeding frenzy to reach its pitch. Uninterested in the bony offerings, it rushed into the free-for-all and picked off a distracted Yellowtail. Less subtle, a four-foot Black Grouper, *Mycteroperca bonaci,* patiently biding its time hovered high above the swirl of darting fishes for a quarter hour before racing down and gulping up the still frozen remains of the chum ball, ending the show.

Divers have mixed opinions about shark feeds. A few on our liveaboard skipped the dive citing the contrived nature of the event; others left the water exhilarated. There is sound reasoning in both camps. Dive operators sponsoring shark feeds emphasize how personal encounters with the celebrated apex predators instill and strengthen a participant's commitment to shark preservation at a pivotal time when market forces are rapidly decimating their numbers around the globe. Countering from the opposite side, a number of shark researchers characterize the feeds as little more than for-profit sideshows, much like circus acts, leading to altered behavior and ecological disruption of the predators' natural dynamics. Let your conscience be your guide.

And then there is the matter of safety. For those brought up in the wake of Jaws-mania, it is a sobering experience to watch the large predators ripping mouth-sized chunks of fish flesh from bait balls a short distance from where they kneel on the sand. As these encounters demonstrate, the likelihood of reef sharks attacking divers has been greatly exaggerated. Typically, Caribbean sharks tend to turn tail and flee at the sight of divers. This is not to say that sharks cannot be dangerous. Whenever sharks are in the vicinity, they should be monitored closely. If their presence makes you uncomfortable, or if a shark makes repeated passes, circles closely or acts aggressive, it would be wise to calmly leave the water. The Great White Shark, *Carcharodon carcharias,* Tiger Shark, *Galeocerdo cuvier,* and Bull Shark, *Carcharhinus leucas,* are believed to be the most dangerous species in the region. Fortunately, Great Whites and Tigers seldom frequent coral reefs, and Bull Sharks are rarely seen where recreational divers commonly visit. In the clear waters of the Bahamas, 25 shark attacks were documented during a 10-year span from 2005 through 2015. The majority involved spearfishing. The single fatality occurred during a shark feed intended to attract aggressive species for photographers.

While humans are seldom at risk from sharks, many species of these magnificent predators are under severe threat from humans. As stocks of traditional market fishes continue to decline, shark fishing has increased substantially to fill the void. Adding to the slaughter, shark fins, considered an aphrodisiac in many parts of Asia and traditionally used for making shark fin soup in China, still bring top prices even after two decades of worldwide protest. In some cases shark fishermen remove nothing more than the fins, dumping the carcasses overboard to rot. In the late 1990s, more than 50 of the sharks from a popular shark feed near Walker's Cay, Bahamas were slaughtered within a few days by a single long-line fishing vessel. The incident prompted the government to prohibit commercial shark fishing from Bahamian waters.

There has been substantial research analyzing the purpose of the wing-shaped head of the eight species of hammerhead sharks in family Sphyrnidae (four inhabit the waters of the western Atlantic), but as of yet there has not been a definitive

A hammerhead devouring a Southern Stingray.

resolution to the adaptation's function. From the studies it appears that the cephalofoil, as the wide flattened head is properly known, has multiple functions including improved depth perception; enhanced electrosensory reception and smell; refined hydrodynamics for sharper and faster turns; and help in manipulating large prey. This last hypothesis was introduced in a paper documenting the feeding strategy of a Great Hammerhead in Bimini, Bahamas.

In 1988 stingray researchers documented a Great Hammerhead, *Sphyrna mokarran,* capture and devour a Southern Stingray, *Dasyatis americana,* near a submerged barge resting in 18 feet of water seven miles east of Bimini. At mid-morning, when the small team first entered the water, three stingrays were lying in the sand near the wreck. An hour later, two more stingrays were noticed speeding toward the barge from the surrounding seagrass flat. Within minutes a stream of apparently panicked rays followed. Uncharacteristically, 7 of the 15 assembled rays entered the wreckage, while others swam rapidly in tight circles and figure eights near the hull. Minutes later, a 10-foot Great Hammerhead, followed closely by a five-foot Blacktip Shark, *Carcharhinus limbatus,* sailed into visibility, paralleled the barge and disappeared. Within minutes the sharks reappeared, made a second pass and left the area once again.

Eight hours later, two researchers snorkeling over the seagrass 300 yards from the wreck noticed a lone stingray with a three-foot wingspan sprinting across the bottom in the direction of the barge, followed by a Great Hammerhead swimming on the surface. As it passed the snorkelers the shark accelerated, caught up with

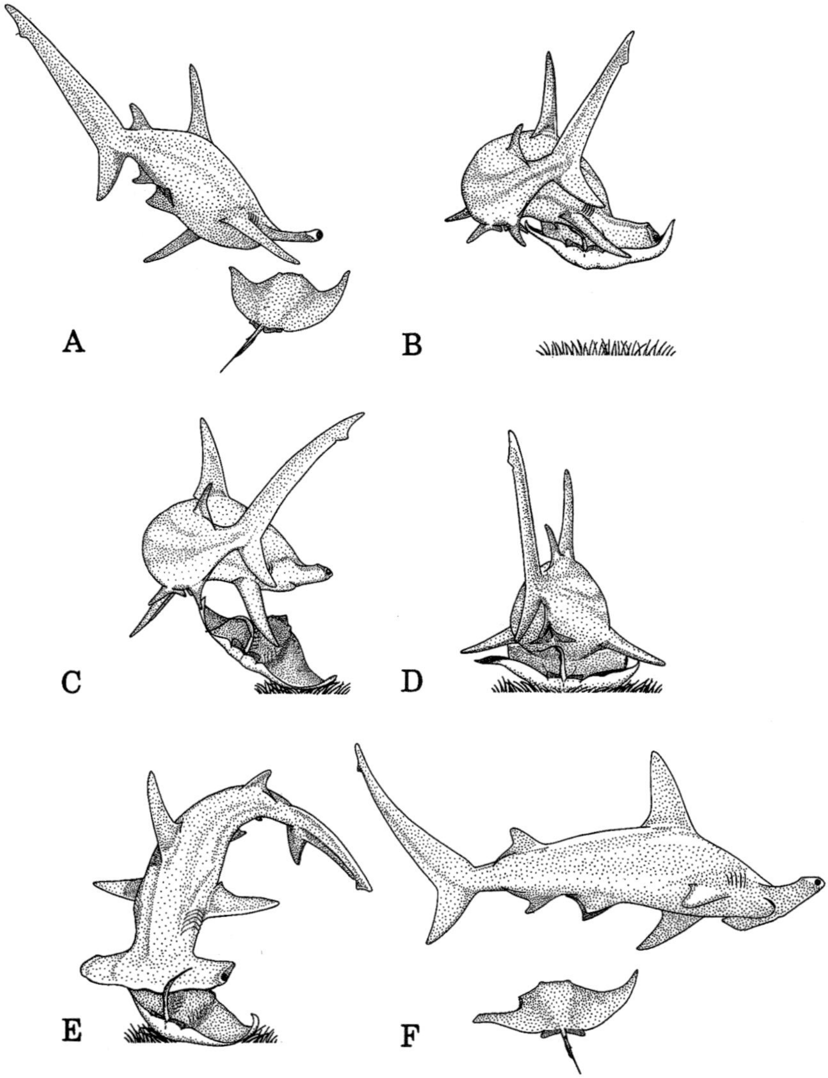

A. *The Great Hammerhead approaches quickly.* ***B.*** *Using its head, the shark strikes the Southern Stingray hard from above.* ***C.*** *The ray bounces off the bottom as the shark brakes overhead.* ***D.*** *A second blow forces the ray back to the bottom where it is held in place by the shark's flattened head.* ***E.*** *The shark pivots and takes the first bite.* ***F.*** *The shark circles before making a second attack.*

the ray and hammered its head into the swimming ray's back from above with such force that it slammed into the bottom. Braking, the shark turned sharply and plowed into its victim a second time, forcing it back down to the sand where it was held in place by the hammerhead's flattened head. The shark pivoted and with a shake removed a crescent-shaped bite from the ray's left front pectoral fin. As the shark circled, the crippled ray fled for the wreck. Unhurried, the shark followed. Thirty feet from the hull it attacked again, striking the ray even harder

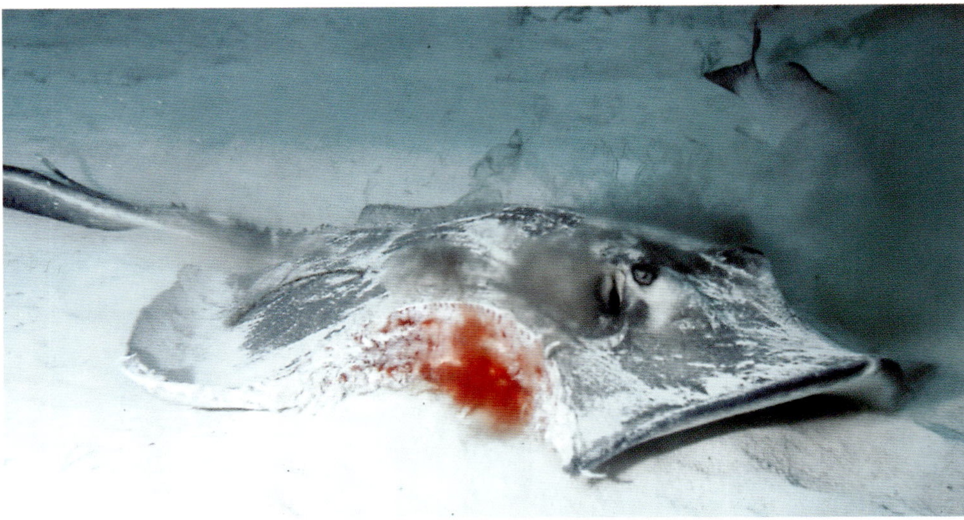

A Southern Stingray immobilized by bites to its pectoral fins.

and once again driving it to the bottom, where the shark took a bite from the right rear pectoral fin. Inexplicably, the stingray did not attempt to use its venomous tail spine during the incident. The stunned ray lay motionless, except for occasionally curling its wing tips. The shark, probably distracted by the onlookers, circled a short distance away, while sharksuckers, *Echeneis* spp., Gray Snapper, *Lutjanus griseus,* Bar Jack, *Caranx ruber,* and Yellowtail Snapper freely scavenged bits of the torn flesh. Twenty minutes later the hammerhead grabbed the limp ray by the head and swam 75 feet from the wreck, stopped and shook the carcass until the rostrum severed and the body fell to the bottom.

The wing-shaped head of a Scalloped Hammerhead.

The same pattern of grabbing, shaking and circling continued until the Blacktip arrived. While the larger shark circled, it darted in and grabbed a scrap of meat. Before the Blacktip could turn to make a second pass, the hammerhead charged with open jaws, sending the Blacktip in retreat. The hammerhead immediately picked up the wingless ray and gulped down the remains headfirst, leaving the long tail dangling from its mouth as it vanished across the seagrass flat just before a second Blacktip and three Atlantic Sharpnose Shark, *Rhizoprionodon terraenovae,* arrived at the scene.

Although pointed-nose sharks are also known to feed on stingrays, rays represent only a small portion of their diets. In contrast, the large number of

A male Nurse Shark bites the pectoral fin of a female during copulation.

A mixture of sperm, seminal fluid and sea water...

stingray spines found in the stomachs of the Great Hammerhead and Smooth Hammerhead, *Sphyrna zygaena,* (96 spines were found inside a single Great Hammerhead) indicates the importance of rays in the diets of both species.

Observations of the reproductive behavior of sharks are quite rare. For five weeks each summer, adult Nurse Shark, *Ginglymostoma cirratum,* gather at traditional inshore breeding grounds across the region. In 1992, researchers monitored a mating congregation of approximately 20 individuals in the Florida Keys. Females, not in the least cooperative, fought off eager advances by thrashing wildly when approached. To avoid the smaller suitors, females occasionally moved

overflows from the cloaca of an impregnated female.

into water too shallow to mate, where they rolled on their sides and dug their pectoral fins into the sand. Males engulfed the females' pectoral fins in their mouth and, with great effort, dragged the struggling females toward deeper water. Six small males were observed working as a tag-team, taking turns nipping and wrestling with a reluctant female that arched and twisted to keep her fins free. Less than 10 percent of mating attempts end in copulation, with only the largest and most tenacious males successful. While the drawn out courtships can last for hours, copulation takes only a minute or two to complete. A successful male must cling fiercely to a female's fin with his mouth, roll her over, wedge his tail under

A two-foot Lemon Shark emerges tailfirst at birth.

her body and twist like a contortionist to insert an erect clasper into a female's cloaca. Prior to insertion, water is drawn into a siphon sac associated with the male's reproductive organs and, along with sperm and seminal fluid, ejected in an overflowing stream through a skin fold in the clasper. After mating, the male collapses to the bottom from fatigue. Months later eggs hatch inside the mother, which gives birth to 20 to 30 pups.

In the shallow waters of the Bimini Lagoon, shark researchers witnessed the first live birth of a shark in its natural habitat when an eight-and-a-half-foot Lemon Shark, *Negaprion brevirostris,* tethered to their boat gave birth. Each of the 10 two-foot pups emerged tailfirst, trailing afterbirth that was quickly devoured by remoras. The youngsters spent their first months at the fringes of the shallow mangrove-lined lagoon before migrating to open water.

Stingrays
Family Dasyatidae

Mating Southern Stingrays make quite a fuss. In one encounter at Stingray City at Grand Cayman Island, seven or eight considerably smaller males vigorously pursued and attempted to pin a lone female to the bottom. One of the males, gripping her wing margin firmly in his mouth, flipped upside-down beneath her body and inserted a clasper. During the five-minute copulation period, the writhing female was partially restrained by the other males, which pressed down on her body as she swam over the sand. When the first male finished, a second male flipped into place and successfully mated with the same female. After he finished the group dispersed.

Numerous smaller male Southern Stingrays attempt to pin a female to the bottom.

Gripping the rear margin of a female's dorsal fin in his mouth, a smaller male flips upside-down and underneath to insert a clasper.

A video sequence of mating Yellow Stingray, *Urobatis jamaicensis,* shows a lone male, similar in size to his mate, mounted head-to-head on her back and twisted awkwardly to insert a clasper. While mounted, the female swam rapidly across the bottom.

Surgeonfishes

Family Acanthuridae

A feeding aggregation of Blue Tang.

Each of the three members of the surgeonfish family inhabiting the tropical western Atlantic—Blue Tang, *Acanthurus coeruleus,* Ocean Surgeonfish, *A. tractus,* and Doctorfish, *A. chirurgus*—are common inhabitants of Caribbean reef fish communities. Surgeonfishes can be identified by the presence of a scalpel-like spine encased in a groove on each side of the tail base. Blue Tangs are easily distinguished by their flat, oval-shaped, powdery blue to dark purple bodies and yellow or white spine grooves. Differentiating between the similar-appearing Ocean Surgeonfish and Doctorfish is a bit trickier. Both species range from bluish gray to dark brown and can pale or darken dramatically, but most are silvery gray. Doctorfish can be recognized by 10 to 12 thin body bars that are often quite faint. At times, both species display pale bands at the base of their tail fins.

Surgeonfishes, which feed continually throughout the day on a wide variety of plants, play a key role inhibiting the growth of filamentous and fleshy turf algae on the reef and contributing large amounts of detritus to the habitat. Blue Tangs are specialized algae-browsers that feed without ingesting sediment and have the long intestines and thin-walled stomachs characteristic of herbivores. On the other hand, Ocean Surgeonfish and Doctorfish, both generalized grazers, have long intestines and thick-walled, gizzard-like stomachs for handling the large volume of detritus and sediment (up to 40 percent) ingested as they forage. Because these materials contain a significant

Ocean Surgeonfish

Doctorfish

number of microinvertebrates, mainly copepods and hydroids, both species can be considered omnivores. Unlike Blue Tangs and Doctorfish, which forage on the reef's structure, Ocean Surgeonfish customarily graze over sand and seagrass beds up to 50 feet from the reef's fringes.

Where large populations occur, Blue Tangs spend much of their day feeding in aggregations that can number more than 100. These swarming assemblies, which regularly include other surgeonfishes and a scattering of Midnight Parrotfish, *Scarus coelestinus,* allow the herbivores to breach the defenses of algae-farming damselfishes. Interestingly, Midnights, which share the same blue coloration, also associate with Blue Tangs outside feeding aggregations, even traveling with them to mass spawning events. The feeding throngs swim along the reef line, descending periodically onto a damselfish garden that they

Mixed feeding aggregation of Doctorfish, Ocean Surgeonfish and Blue Tang.

ravage for 10 to 15 seconds with a flurry of bites before moving a short distance to their next plunder. The nipping attacks of the four- to five-inch damselfishes, which easily repel solitary foragers, are no match for a horde of invaders. Blue Tangs living in small populations, without the advantage of group feeding, spend significantly more time foraging in the water column than their more numerous counterparts.

Large populations of Blue Tangs reproduce in pairs and in resident spawning aggregations. Only pair spawning occurs in small populations. In Bimini, Bahamas, we observed Blue Tangs pair spawning from May through August. Late each afternoon, two to five large males defended personal areas of the water column above an isolated rock reef. Trespassing males were aggressively chased. When approached by smaller, rapidly swimming females, the courting males attempted to position themselves next to their sides. During these brief encounters, the heads and forebodies of the males lightened markedly. After lengthy courtship, pairs completed a few rapid 10- to 15-foot spawning rushes leaving visible gamete clouds a few feet below the surface.

We observed the resident group spawning activities of what we estimated to be 1,000 Blue Tangs and Ocean Surgeonfish for three consecutive afternoons beginning the seventh day after the full moon of March. The spawning area, which covered approximately 5,000 square yards of coral and sand bottom in 45 feet of water, was just inside a dropoff paralleling the seaward shoreline of Turneffe Atoll, Belize. Prior to the gathering, large aggregations of Blue Tangs swam rapidly along the dropoff. An hour before sunset several aggregations converged at the spawning grounds, accompanied by their entourages of Midnight Parrotfish. The fish milled about just above the bottom for 5 to 10 minutes before forming into three to five compact schools, each numbering in the hundreds. The dark clouds of fish repeatedly ascended high into the water column, dropped back to the bottom, briefly dispersed, reassembled and ascended again. Subgroups, numbering from 5 to 25 fish, many displaying white tails (presumably males pursuing single females), periodically broke from the main group and zigzagged about erratically. Eventually a single subgroup dashed toward the surface and expelled gametes, prompting a dozen or more groups to follow suit. Afterwards, subgroups rejoined one of the constantly moving aggregations before breaking free, chasing and spawning once again. The water was a haze of gametes by the time group spawning ceased, approximately 20 minutes before sunset. Isolated pair spawning continued until dark.

As the frenzy of the Blue Tangs reached a fevered pitch, several hundred Doctorfish arrived at the scene and spread across the seafloor in a loose group. When the Blue Tang's reproductive activities subsided, the Doctorfish joined into small subgroups and began to spawn in groups and pairs in a manner similar to the Blue Tangs. Possibly because there were fewer participants, the Doctorfish never formed into compact clouds.

We have observed Ocean Surgeonfish pair spawn during the late afternoon in Bimini. Males encroaching into defended territories often were chased in tight circles, as if attempting to use their sharp tail spines as weapons. Occasionally, territory holders and intruders swam rapidly side-by-side across

Blue Tangs gather at a traditional spawning site an hour before sunset.

Subgroups consisting of a female and several males break away from the gathering.

Pair spawning male Blue Tangs display a white-face pattern.

A territorial confrontation ...

between Ocean Surgeonfish at dusk.

the bottom or briefly hovered next to one another in a nose-down posture.

For nearly a year, researchers in Puerto Rico documented the behavior of a resident spawning aggregation estimated to include more than 20,000 Ocean Surgeonfish and 6,000 Blue Tangs at its zenith. The fishes assembled in the late afternoons during 21-day periods each month from December until March, beginning the day before the full moon until five days after the new moon. Peak activity occurred three to eight days after the full moon in the coolest winter months. From April through November the aggregations were smaller and less intense and spawned only from the third to the ninth day after the full moon. At the height of activity, researchers counted nearly 50 group-spawning rushes in less than one minute. Even though the two surgeonfish species spawn in the same locations at approximately the same times, only one possible hybrid has been reported. Large, ripe male and female Doctorfish were in the vicinity of the Puerto Rico spawning site in the winter months, but were never observed spawning.

The surgeonfishes' tail-base spines, formed by modified scales and attached posteriorly by ligaments to the spinal column, cannot be voluntarily erected. The forward-pointing tip on the convex surface slips slightly from the groove each time the tail flexes. If the exposed point happens to snag onto something

Postlarval surgeonfish

Juvenile Blue Tang

passing by, the spine extends at a right angle from the body. Although reported to carry toxins, a laboratory experiment using the spines of Ocean Surgeonfish to inflict wounds on various reef fishes proved negative.

Under laboratory conditions, the spherical eggs of Ocean Surgeonfish (0.67 mm in diameter) hatch in 28 hours. Settlement peaks during the darkened nights of the new moon, when the large, inch-and-a-quarter, transparent and silvery larvae drop to the seafloor. Blue Tangs tend to settle onto hard reef structures while the Ocean Surgeonfish and Doctorfish choose low-profile rock and coral patches often in shallow back-reef habitats and in seagrass beds near reefs. We have watched transparent postlarval surgeonfish drop to the sand during the night where they began acquiring opacity within seconds. We also observed a settling postlarval Blue Tang already displaying much of its juvenile coloration. The vulnerable youngsters quickly seek shelter where they continue to metamorphose into juveniles. Larval recruits have been documented to settle primarily near established juveniles of their species and to avoid sites occupied by Beaugregory, *Stegastes leucostictus,* a small, territorially aggressive damselfish.

The smallest juvenile Ocean Surgeonfish, already resembling adults in body shape and color, feed in small groups within a confined area. With growth they become aggressive toward one another, especially at dusk, when defending sleeping sites. At times they forage with schools of juvenile parrotfishes. When they reach nearly three inches, they once again feed in small schools. Individuals become sexually mature and join adult schools after approximately two years, at a length of five to six inches. It is estimated that surgeonfishes can live up to 10 years.

To the contrary, bright yellow juvenile Blue Tangs, strikingly different from blue adults, are highly territorial. As they mature, and the youngsters' bodies turn powder blue, except for yellow tails, they mix with small foraging fishes. By the time they are four inches and completely blue they form small groups with their own species. Later, after developing their final blue coloration, they join large foraging aggregations of adults. Interestingly, on two occasions we have encountered solitary individuals, each over five inches in length, still wearing their yellow juvenile colors.

Intermediate phase Blue Tang

Adult Blue Tang

Tilefishes

Family Malacanthidae

Sand Tilefish spend much of the day maintaining their burrows.

Widely spaced rock, coral and shell mounds marking the homes of Sand Tilefish, *Malacanthus plumieri,* are common fixtures on sand plains fronting Caribbean coral reefs. The two- to six-foot-diameter piles, which add structure to an otherwise barren seascape, also provide sanctuary for numerous crustaceans and small fishes, including Bicolor Damselfish, *Stegastes partitus,* cardinalfishes, gobies and, in deeper water, harems of Cherubfish, *Centropyge argi.* The torpedo-shaped, one- to two-foot Sand Tilefish move about just above the bottom in a start-and-stop fashion. While hovering, their long dorsal and anal fins undulate in graceful waves. Sensing danger, tilefish cautiously move toward their mounds and, if threatened, dive into the single tunnel leading down at a 10- to 20-degree slope to an excavated chamber below. The fish spend considerable energy keeping their burrows in good repair. Pieces of rubble from the surrounding sand are carried periodically by mouth and added to the pile, and existing fragments are regularly shifted and rearranged. If building materials are in short supply, tilefish have been known to steal rocks from jawfishes and neighboring tilefish. The rubble roofs provide a barrier against large predators, such as Mutton Snapper, *Lutjanus analis,* and several species of sharks, which have been found with tilefish remains in their stomachs. To expedite hasty escapes, the sand chutes leading to the entranceway are picked free of debris and regularly loosened by a rapidly fluttering body. When researchers experimentally destroyed a burrow, the homeless female built a temporary refuge a few feet away. During the next four days, she constructed a permanent residence at the old site.

Each female maintains an individual burrow within a home range that generally lies adjacent to and overlaps slightly with those of other females. These defended feeding areas cover from 40 to nearly 300 square yards and average 150 square yards in size. The territories of from one to six females are encompassed within the territory of a single male, that appears similar to the females except for his larger size and tail streamers. Males spawn almost exclusively with the females living within their territorial boundaries but otherwise maintain little contact and seldom attempt to dominate partners physically. Even though all males are protogynous hermaphrodites derived from sex-changed females, there does not appear to be a size-related social hierarchy in tilefish harems. Unlike many protogynous species that live in harems, when male tilefish are experimentally removed, the largest female does not automatically change sex and dominate the group.

Females defend their personal feeding territories from other females regardless of the trespasser's size, but the dominant male is allowed free access. In rare instances, females displace neighbors. Territorial females confront female intruders just inside their borders and either chase or escort them out of their area. If, while pursuing an intruder, the fish cross into the trespasser's turf, the tables are turned and the aggressor is chased back into her own territory. Although most confrontations end without further incident, feuding females will occasionally spread their fins, make

Loose rocks are continually added to the burrow pile.

Stealing burrow rocks from a jawfish.

threatening upward jumps, bite, or closely circle each other with their mouths open and gill covers flared. Prolonged spats are settled by territorial males, whose sudden appearance causes the bickering pairs to flee or lower their heads in submission and swim cautiously away, passing under the approaching male.

Tilefish forage throughout the day within their territorial boundaries for an assortment of small bottom-dwelling invertebrates and fishes. Brittle stars, crabs, mantis shrimp, plumed worms, chitons, small wrasses and sand eels make up the bulk of their diet. Females protect their territories' food supplies by aggressively challenging neighboring females and other gleaning carnivores that enter their personal feeding grounds.

Males' territories can include as much as 900 square yards of sandy plain, but the sizes are not correlated to the occupants' body size. As with the home ranges of females, males' territories often overlap slightly. Males occasionally make extended excursions, crossing the territories of one or two neighboring males. The trespassers, either swimming rapidly or creeping slowly near the bottom, are challenged by both territorial males and females. A difference in size does not seem to matter in these exchanges.

When territorial males disappear, neighboring males begin to feud and encroach into the vacated area within an hour. By the second day, things are settled as a male gains control of the abandoned burrow or incorporates part or all of the adjoining territory into his domain. Within hours, the new male begins to make courtship overtures to the annexed harem members. Occasionally, vacated territories are divided among neighboring males. A few males live in territories without mates, but there do not seem to be any adult females without spawning partners. Because females seldom relocate to improve their spawning possibilities, bachelor males must either establish their territories close to females or be patient and wait for recently matured females to set up housekeeping within their boundaries. Mate-stealing is rare, but when it occurs a rather nasty confrontation ensues.

Reproductive activities occur each evening beginning an hour before, and lasting until 10 minutes after, sunset. A female spawns nearly 80 percent of evenings, usually once, but occasionally two or three times. Males might spawn 10 times during an evening depending on the number of females inside their territories. As the light begins to fade, females remain close to their burrows. Courting males visit each in turn, flaring pelvic fins as they approach in a conspicuous series of long, slow, arching dips. Females not yet prepared to spawn generally shun the early advances by swimming away. Later, as females become more receptive, pairs rub against each other, cross bodies and swim parallel to one another across the sea floor. Variations of the courtship dance continue for minutes before the first spawning rise. When ready to release their eggs, females hover three or four feet above the bottom and arch their bellies downward, dramatically accentuating the profile of their egg-swollen abdomens.

Positioned side-to-side and with the female slightly in the lead, the pair slowly ascends at a 60-degree angle. Early rises are typically aborted without release of gametes. At this point the male might swim off to court elsewhere,

The sand chute leading to a burrow's entranceway is regularly loosened with a fluttering body.

Feeding with a Smooth Trunkfish.

Juvenile Sand Tilefish

Burrows are never completed. Building materials are constantly shifted and added.

The arched back of a female (foreground) indicates her readiness to spawn.

The smaller female initiates the slow spawning rise.

but usually the two fish go through the entire routine once or twice more before finally rising 6 to 12 feet and, with a quiver, simultaneously release a cloud of spawn. They immediately part and swim rapidly to the bottom in opposite directions, where the female occasionally rubs her abdomen on the sand. From start to finish, the entire affair takes from one to eight minutes. As darkness approaches the females become more receptive and the males less formal, with partners often spawning as soon as the males arrive.

On six successive days, we watched the same male spawn with his five harem mates. The male followed a different sequence between partners each evening, always spending extra time with the most willing. When we approached too

The pair release gametes at the peak of their spawning rise.

*As the pair break apart after spawning,
a Yellowtail Snapper dashes toward the gamete cloud.*

closely, the tilefish curtailed courtship. If we remained at least 10 feet away, the couples went about their business as usual. On two occasions, three Yellowtail Snapper, *Ocyurus chrysurus,* escorted the male from female to female. The notorious egg-predators hovered menacingly just inches away from the courting couples. Each time a pair began to rise the Yellowtails followed closely, often causing the tilefish to break apart prematurely. After several disrupted attempts, the harried pairs finally spawned, leaving their precious cargoes at the mercy of the Yellowtails, that crisscrossed through the clouds with open mouths before dashing off after the male as he dipped his way in the direction of another partner.

Wrasses

Family Labridae

Green Razorfish grabs a crab dinner.

Possibly because wrasses are generally small, continuously on the go and so common, they tend to be ignored. However, divers and snorkelers who take the time to observe these energetic fishes closely will find their behavior fascinating and their color patterns quite exquisite. Of the over 500 species worldwide, 18 make their homes in the tropical western Atlantic. The typical Caribbean wrasse is a colorful, three- to six-inch, cigar-shaped fish that buzzes about just above the bottom in a small, loosely knit feeding aggregation, known as a herd. However, there are considerable variations in size, body shape and behavior in the family. The variable family ranges from three-inch, plankton-picking Rainbow Wrasse, *Halichoeres pictus,* to mollusk-crushing Hogfish, *Lachnolaimus maximus,* that approach three feet in length. Adding confusion to wrasse identification, many species have three distinct color phases, and a few species display as many as five color and marking patterns during their life cycles.

Like the closely related parrotfishes, wrasses have large, noticeable scales, swim primarily with their pectoral fins, are often quite colorful and change their sex from female to male when certain age, growth or social criteria are met. But unlike parrotfishes that scrape algae from rocks with fused beaks, most wrasses feed on hard-shelled invertebrates including crabs, shrimp, brittle stars and small gastropods. When discovered, the delicacies are seized or scissored free with prominent sets of protruding canines and crushed with powerful pharyngeal teeth. A few species—Creole Wrasse, *Clepticus parrae*, Rainbow Wrasse, Rosy

Razorfish, *Xyrichtys martinicensis,* and adult Bluehead, *Thalassoma bifasciatum*—primarily rely on diets of zooplankton plucked from the currents. Juvenile Spanish Hogfish, *Bodianus rufus,* and Blueheads obtain a portion of their diets by picking ectoparasites from a variety of fishes.

Wrasses inhabit a wide range of habitats, from shallow inshore seagrass beds to deep coral-encrusted outer banks. Except for the open water plankton-feeding Creole Wrasse, all occupy permanent territories in association with other members of their species. Also like parrotfishes, all wrasses are protogynous hermaphrodites—females have the ability to change into terminal males, the pinnacle of reproductive prosperity. Sorting out the species' social organization is, to say the least, challenging. For starters, the majority of wrasses have three distinct color phases representing their social status: the juvenile phase consists of sexually immature youngsters that nearly always live separately from adults; the initial phase includes sexually mature adults that in the majority of species are exclusively females but can be both males or females, and terminal males, nearly always the largest, brightest, most aggressive and least numerous individuals. Females have a strong tendency to select the largest and most ornate males for mating. Unfortunately, eye-catching colors not only attract mates but also predators. For this reason, the conspicuous terminal males of species living around coral habitats with the benefit of many accessible hiding places show the brightest and most permanent

Initial phase Bluehead

A two-inch male Rainbow Wrasse

A two-foot terminal male Hogfish

dichromatism (two different color patterns). Terminal males of species inhabiting sand flats, seagrass beds or hard bottom, without ready access to shelter, exhibit their most striking colors only briefly when courting or engaged in aggressive behavior.

The size of territories varies considerably among species. The small Rosy Razorfish requires only a few feet of bare sand, while large Hogfish lay claim to areas the size of football fields. The terminal males of many wrasse species patrol and defend territories superimposed over the home ranges of three to six initial phase herds, each made up of three to eight fish. Dominant males visit groups periodically throughout the day zooming in with great fanfare before briefly foraging alongside the roving bands. Border disputes periodically take place between neighboring territory holders. Rather than resorting to physical clashes, ritualistic behavior settles most disputes. Rankled males either assume defiant head-down stances, rapidly swim parallel to each other up and down their common borders or engage in face-to-face standoffs with their mouths opened. This common behavior, known as gaping, also occurs between younger members of a group to establish and sustain their social ranking.

Most wrasses belong to a rigid size-related hierarchy that allows only the largest individuals in the initial phase to become terminal males. Sex-changed males are known as secondary males. Primary males, the few individuals born male, cannot change sex, but with adequate size may one day develop into the terminal phase. The transformation from an initial phase female to terminal phase male is predictable, surprisingly rapid and irreversible. Within hours after researchers experimentally removed the terminal males from a small Bluehead population, the largest initial phase female, that had spawned as a female the day before, began behaving aggressively, developed a slightly darkened head while courting and attempted to spawn as a male. Intermediate color changes were evident by the second day and distinct in four days. Eight days after the initiation of sex change it is physiologically possible for an individual to produce sperm. Within several weeks the full color transition is complete.

All wrasses are broadcast spawners that, as a rule, pair spawn daily at established locations within their permanent feeding territories. However, the terminal males of the ubiquitous Blueheads travel each afternoon to the downcurrent end of a reef line and take temporary control of a prominent structure, such as a large coral head, as a spawning site. Initial phase females soon follow, migrating from their home ranges along established routes.

The different species of cigar-shaped wrasses court and spawn in a similar fashion. With fins spread and displaying their brightest colors, terminal males repeatedly dive toward gatherings of females as they mill about the bottom. Suitors signal their readiness to spawn by executing a series of rapid vertical jumps. Egg-laden females rising in midwater are quickly joined by speeding males. Side-by-side, the pairs rocket up two or three feet where they simultaneously releasing gametes.

Dominant males have no intention of sharing their many females. Young initial phase males, in their quixotic quest to reproduce, rely on stealth to penetrate the terminal males' defenses. Because their color patterns mimic those of females, young males, known as sneakers, are able to hang out undetected near spawning sites and, when the terminal males are temporarily occupied, sneak a quick spawn with waiting females. Streakers, on the other hand, lurk close by and dash in, at just the precise moment, to mix their sperm at the apex of a paired spawning rush. Occasionally, spurious young sneakers carry their masquerade to extremes by pretending to spawn with dominant males. Terminal males must resort to inspecting the vents of initial phase individuals to weed out the impostors that, when discovered, are aggressively chased from the site. These fast-paced pursuits, which often take the rivals 20 to 30 feet from the spawning site, are common occurrences during reproductive periods.

Terminal males have little problem dominating reproductive activity at isolated reef sites where populations are limited; but on large reefs the ratio of initial phase males increases substantially for several species of wrasses. On these reefs the young males congregate upcurrent from the preferred mating sites and mass spawn with passing females. Of course, their highly competitive hit-and-miss tactics have significant disadvantages. During group spawns, an individual's chance of successfully fertilizing eggs is reduced by the number of competing males. Initial phase males also have difficulty recognizing the sex of their look-alike cohorts. It is estimated that nearly half of group spawns do not even involve the release of eggs. To increase their odds, younger males have greater gonad mass and can release more sperm in proportion to body size than pair-spawning terminal males.

At dusk, some species seek night shelter inside nooks and crannies while others, including the Slippery Dick, *Halichoeres bivittatus,* Yellowhead Wrasse, *H. garnoti,* Rainbow Wrasse and Clown Wrasse, *H. maculipinna,* dive beneath the sand where they remain until the first light of dawn. Razorfishes have taken this behavior a step further by diving under the sand's surface, not only at night but whenever threatened. Larger, plankton-feeding Creole Wrasse find night refuge deep within tube sponges or crevices.

Like nearly all reef fishes, wrasses begin life as pelagic larvae that settle to a reef environment after an extended period. Researchers are able to establish the age and date of settlement for most reef fishes by counting concentric markings on three sets of tiny bones, known as otoliths, at the base of the braincase. These markings are especially clear in wrasses. Wide settlement bands indicate that most family members spend, on average, 24 days in the open ocean. Razorfishes, however, can remain adrift for up to three months. Recently settled wrasses are believed to bury in the sand while they complete their metamorphosis into juveniles.

Bluehead

Thalassoma bifasciatum

From reef to reef, Blueheads are one of the most common species in the tropical western Atlantic. Throughout daylight hours countless initial phase herds forage for small crustaceans on reefs, rubble beds and gorgonian fields, from depths of 3 to 80 feet. The relentless, ever-present hunting packs are the scourge of benthic egg layers. By sheer force of numbers they regularly breach the defenses of a

A terminal male Bluehead courts an initial phase female.

nest's guardian and ravenously plunder eggs for minutes at a time before they can be driven away. When currents flow, the same herds switch feeding strategies and move up in the water column to pick plankton or pluck ectoparasites from hovering fishes at cleaning stations. After maturing into terminal phase males, individuals feed almost exclusively on benthic invertebrates. Unlike most wrasse species, Blueheads do not defend territorial feeding grounds or females; instead, herds roam freely in broad overlapping home ranges, where they are periodically visited by displaying terminal males throughout the day.

Blueheads are protogynous hermaphrodites, able to change from females to males. The initial phase is made up of females, non-sex-changing primary males and secondary males that have changed from large females. When terminal males disappear from a reef area, the largest initial phase individuals in the area go through a succession of intermediate color and marking changes while developing, over several weeks, into conspicuous blue-headed terminal males.

Spawning strategies, the number of fish in each color phase and sex ratios vary dramatically between small and large Bluehead populations. On small isolated reefs with fewer than 200 individuals, pair-spawning terminal males control all the preferred downcurrent spawning sites and nearly all mating activities. On these reefs terminal males make up about 10 percent of the population, and the initial phase consists almost entirely of females. When populations hover between 200 and 500, terminal males still occupy the better sites but can no longer expend the energy necessary to exclusively dominate mating. In these populations, group spawning between an aggregation of initial phase males and single females takes place just upcurrent from the spawning sites of terminal males. As populations on larger reefs grow to more than 1,000, less than one percent of individuals are in the terminal phase, while nearly 50 percent of the initial phase is made up of males. In these situations, those individuals in the terminal phase still spawn in pairs, but the competition from initial phase males is so great that their reproductive activities are relegated to less desirable sites at the perimeter of the reef. Even there, young male sneakers and streakers continue to exert competitive pressure. On the largest reefs with over 10,000 fish, initial phase herds, including both males and females, migrate in steady streams from feeding areas as far as a mile away to spawn in large aggregations, while pair-spawning terminal males are virtually non-existent.

On small to medium reefs, during daily, two-hour, late-morning to early-afternoon spawning periods, terminal males occupy customary sites where they wait for females to arrive from their feeding grounds. To draw attention to themselves or possibly to demonstrate that their spawning site is safe from predators, males perform a series of short upward dashes in rapid succession, known as signal jumps. When females approach, the males excitedly swim in tight circles, occasionally spiraling upward. When ready to spawn, one, or in some cases multiple, females rise from below. The males slow, allowing their temporary mates to pass just ahead. Side-by-side and almost touching, they swiftly ascend at a sharp upward angle. In an instant the swollen females turn on their sides, positioning their ventral slits close to the males, and release eggs, which are immediately enveloped by a cloud of sperm. On medium to small reefs, terminal males controlling the prime sites pair spawn as many as 25 times each afternoon, while females, strapped by energy-expensive egg production, spawn once every other day for up to two years of their reproductive life cycles.

Initial phase Blueheads break through the defenses of a Sergeant Major and swarm its nest full of eggs.

Up to 50 initial phase male Blueheads dash into the water column in pursuit of a single, ripe female.

Despite the terminal male's bright coloration, his striking markings appear to have little or no effect on a female's mate selection. When terminal males were experimentally removed from traditional spawning sites, females remained faithful to their original sites that were occupied within two days by new terminal males from nearby sites. Their previous spawning partners did not follow the new males. Possibly females, always vulnerable to predation, prefer accustomed migration routes and spawning areas, where they are familiar with the location of shelter holes. Further research also has documented that spawning sites, used continually for generations, were not intentionally chosen to confer dispersal advantages to pelagic offspring. After removing all the Blueheads from a small isolated patch reef, researchers introduced a fresh population captured miles away. Within days the new residents were engaging in reproductive activities, but they adopted different spawning sites from those used by their predecessors. This study indicates that the use of a particular spawning site is probably a learned behavior.

On large reefs where yellow initial phase Blueheads spawn in groups, their frenzied early afternoon behavior easily attracts the diver's eye. These high-energy episodes take place toward the reef's seaward edge, at traditional spawning sites used daily. Look for several hundred individuals excitedly moving about the substrate in loose aggregations. Every few minutes, from 5 to 50 small males form tightly knit groups rapidly swimming several feet off the bottom in parallel, arrowhead-shaped formations. If you look closely a single

female, swollen with eggs, can be seen leading the dodging and darting pack as it rises four or five feet above the substrate. In a split second, several members break free, rush upward, form into a tight ball and in unison shed a visible cloud of gametes. On large reefs when action is intense, spawning balls form one after the other. Even though large terminal males are believed to be 20 to 50 times more reproductively successful than small initial phase males, it is estimated that, because of their commanding numbers, nearly 75 percent of eggs are fertilized by spawning groups.

Blueheads spend their first 6 to 10 weeks of life as pelagic larvae before settling to shallow water habitats in unpredictable mass recruitment waves. They immediately bury in the sand where they remain while changing into juveniles. After emerging, bright yellow juveniles gather in restricted pockets of the reef or at the edge of seagrass beds where they remain until sexually mature. The few able to survive predation during this perilous period develop a dark midbody stripe and join a local herd as initial phase Blueheads. Individuals do not leave the area where they settle, and remain faithful to the herd they initially join. At dusk, each fish takes refuge inside a personal reef pocket that is defended from cohorts.

A Clown Wrasse races just above the bottom displaying his courtship colors.

Clown Wrasse

Halichoeres maculipinna

Although the small, cigar-shaped Clown Wrasses are seldom noticed, they are usually quite common around the sandy fringes of patch reefs at moderate depths. Initial phase herds, numbering from three to three dozen individuals, often mix with juvenile surgeonfishes, young parrotfishes and other wrasse species. A single, brightly colored terminal phase male controls a permanent territory where he dominates as many as 30 initial phase fish that group into three to six herds. One to three subordinate terminal phase males round out the species' territorial society.

Initial phase Clown Wrasse can be females or primary or secondary sex-changed males. Each phase retains vestiges of their juvenile markings of bold gold, black and white stripes. Herds, composed primarily of females, remain stable, moving about and feeding together inside limited home ranges. Within these feeding groups, there is a rigid size-based pecking order with larger members frequently forcing smaller colleagues aside at food sources. Constantly moving herds spend their day rooting about the bottom for worms and other invertebrates hidden just beneath the sand. Even though they do not wander far, feeding grounds often overlap those of their neighbors. When paths cross, the herds intermingle briefly without incident.

Subordinate terminal males can be primary or secondary males. Unlike dominant males, they seldom display intense colors, never patrol territories in the water column and do not court or spawn. When briefly chased by a dominant male, which frequently happens, they collapse fins, turn tail and flee. However, subordinates systematically visit each initial phase herd within their territories.

Socially superior territorial males can be recognized by their intense colors and a dark mid-body blotch prominently displayed as they patrol a few feet off the bottom, confront neighboring dominants, chase subordinates, inspect herds, court and spawn. When not engaged in these behaviors their colors fade, and it is difficult to distinguish them from subordinates. As in most wrasse species, when a dominant male disappears, an extended power struggle ensues to gain control of his territory. Not only do local subordinates compete, but neighboring subordinates and dominants cross boundaries to vie for the vacant position. After several hours of chasing and displaying, one of the challengers establishes authority and immediately begins displaying to his initial phase herds and patrolling boundaries while proudly displaying his brightest colors.

The large territories of Clown Wrasse, ranging between 500 and 1,200 square yards and contiguous to other territories, are vigilantly patrolled and defended by both types of terminal phase males. When males in adjacent territories confront one another, they intensify markings, face off with their mouths open and canines exposed or, abreast, race back and forth, only inches apart, each fish prudently remaining just inside his own boundaries. Such encounters are the only instances when subordinates intensify colors.

Clown Wrasse pair spawn year-round for short periods from late morning to mid-afternoon. Dominant males court during the day by streaking toward a foraging herd with fins flared and colors bright. As spawning episodes begin at various locations within their territories, dominant males increase displays and pursue subordinate males with heightened vigor. When ready to spawn, they advertise their intent with a sequence of quick signal jumps. Ripe females rising slightly off the bottom are joined by males in short, fast spawning rushes. In some Clown Wrasse populations, possibly those with a large number of initial phase males, one or two small males sometimes streak in from the periphery and shed their sperm with the spawning pairs. In these same large populations, isolated initial phase pair spawning and group spawning also take place. Subordinate males, changing into terminal males, have yet to be observed participating in reproductive activities. It is believed that dominant males easily detect their attempts to spawn because they no longer resemble females.

A terminal male Yellowhead Wrasse displaying a normal color pattern.

Yellowhead Wrasse

Halichoeres garnoti

Contrary to the social tendencies of cigar-shaped wrasses, Yellowhead Wrasse are less gregarious. Solitary terminal males and small groups in the initial phase feed within large, broadly overlapping home ranges with little conflict. Their ranges normally encompass hard coral habitats on deeper portions of the reef. Even the tiny yellow juveniles with electric blue streaks hang out alone or in twos or threes near small coral pockets or along the fringes of seagrass beds.

Instead of constantly searching sand flats for food, individual Yellowheads spend the majority of their feeding time poking inside narrow crevices and under rocks for a variety of invertebrates, such as crabs and brittle stars. These

A terminal male Yellowhead Wrasse in his permanent "Redback" color pattern.

opportunistic carnivores frequently join loose feeding aggregations consisting of other wrasses, porgies, Yellowtail Snapper, *Ocyurus chrysurus,* Bar Jack, *Caranx ruber,* and Spotted Goatfish, *Pseudupeneus maculatus.* It is generally believed that Yellowhead Wrasse do not defend feeding territories because their food supply is too widely distributed to favor group foraging. It would be logistically impractical for males to keep tabs on groups of females constantly moving about such large areas.

A typical population consists of slightly more females than males and slightly more initial phase than terminal phase individuals. Though unproved, it appears that all males are secondary sex-changed females, whether in the terminal or initial phase. Many individuals in a population exhibit a variety of intermediate color patterns between the initial and terminal phases. It has yet to be determined if these Yellowheads are in the initial, terminal or a transitional phase, or even if they are all actually males. Complicating the matter further, large Yellowhead Wrasse can brighten dramatically, fade to white and temporarily lose their prominent black midbody bar. There is a striking variation of the Yellowhead Wrasse in the Caribbean, known as the "Redback" because of its bright red upper rear body and midbody bar. The variation was once only found in Bermuda, where it is the exclusive color pattern of the Yellowhead Wrasse, but around 2014 they began to be spotted in the southern and western Caribbean.

For three successive afternoons we watched the same Yellowhead terminal male shuttle back and forth along a low reef line between three initial phase groups about 40 feet apart. The largest

A terminal male visits his sleeping site,

noses around for 10 minutes

and dives below the sand for the night.

gathering consisted of approximately a dozen females. The groups moved calmly about the bottom, occasionally feeding while the male was out of the vicinity. When the male approached, he performed one or two rapid, eye-catching signal jumps that carried him at least five feet from his starting point. In response, several females would rise together five or six feet off the sea floor and hover. When a single female began to rise again the male darted to a position with his belly next to her back and his head even with her pectoral fins. Together they shot up in a blur and released gametes several feet above the starting point. After one or two pair spawns, the male typically returned to the bottom and rapidly swam out of the area. We once watched a male spawn with 10 females in succession with only seconds between.

Approximately an hour before sunset up to a dozen terminal phase Yellowheads, displaying bright colors, gather together and begin their nightly bedding routine by moving as groups across broad areas of the bottom. The individuals do not feed or show aggression toward one another. After a short period, the big males break into smaller bands that are joined by a few Yellowheads in the initial phase. Together they nose about sand patches, repeatedly going back and forth between several spots. The large individuals begin to show preferences for particular locations by returning to the same areas repeatedly. Ten to 15 minutes later one of the bright males suddenly dives into a patch of sand he has been frequenting and drives his body below the surface by rapidly beating his tail. The others nose about where their colleague disappeared. Minutes later a second terminal male disappears nearby in the same manner. Well before sunset all have buried for the night. Preferred sleeping areas tend to be in loose pockets of sand next to low-profile rock ledges.

We observed the same terminal male bury himself at the same site on successive evenings. When we shoveled our hands into the sand where a fish disappeared, it came flying out and immediately began nosing about once more invariably choosing the same site to dive back into 5 or 10 minutes later.

Slippery Dick

Halichoeres bivittatus

All sizes and color phases of cigar-shaped Slippery Dicks feed much of the day in small, loosely knit, opportunistic hunting packs alertly patrolling the bottom for a wide variety of animal matter. When a diver's fin disturbs the sand or a rock is overturned, exposing brittle stars, crabs or shrimp, these tenacious herds are first on the scene. Often, even before a horrified diver can reposition the rock, its hidden fauna are picked clean. Although herds occasionally forage about reefs, they are most common on the sand aprons surrounding widely scattered patch corals and on hard bottom in less than 15 feet of water. Small initial phase primary males and females dominate population numbers. There are also quite a few primary males in the terminal phase. The skin tones of Slippery Dicks change dramatically to blend with their habitat. Those feeding over white sand pale considerably with only the black mid-body stripe remaining distinct; individuals inhabiting shadowy algae-covered terrain display a greenish or dark reddish brown coloration.

A pair of young male Slippery Dicks challenge one another for dominance.

Even though there is a noticeable difference between initial and terminal phase color patterns, the species is considered only partially sexually dichromatic because random females occasionally display terminal colors. However, individuals with the most pronounced terminal patterns are invariably males. Terminal color patterns are brightened to their full intensity only during courtship and spawning; otherwise they fade, often blurring the visual distinction between the sexes.

It is believed that terminal males do not form harems or maintain permanent feeding territories, but instead temporarily defend specific spawning sites along the outside edge of shallow reefs near their feeding grounds. Reproductive periods, beginning in late morning and extending into early afternoon, coincide with those of Clown Wrasse, and Yellowhead Wrasse, indicating that all three species are reacting to a similar set of environmental cues. Much like Blueheads, pair spawning occurs at these traditional sites, with occasional initial phase males streaking in to add their sperm at the apex of rapid, three- to four-foot spawning rises. Typical of their hyperactive nature, terminal males jet about just above the bottom between clusters of females as courtship intensifies. Adding to the fast-paced action, smaller males are constantly chased. Terminal males frequently leave the spawning area for minutes at a time. When they return, one to three egg-swollen females, ready to release their hydrated eggs, rise off the bottom. The approaching terminal males dash into the group and, in a blur, pair spawn with one hovering female after another before zooming off once again. Although we have never observed this species spawning in initial phase groups, this behavior has been reported.

A terminal male Rainbow Wrasse courts with fins spread and colors bright.

Rainbow Wrasse
Halichoeres pictus

Rainbow Wrasse, the smallest of the eight cigar-shaped wrasse species in the Caribbean, feed on zooplankton in the water column on the outer slopes of shallow patch reefs and gorgonian thickets. Loosely structured herds of 20 to 40 fish regularly mix with equal numbers of similarly colored initial phase Blueheads. Because of the initial phases' close resemblance to plentiful young Blueheads, these slender, two- to three-inch Rainbow Wrasse are frequently overlooked. The species is distinctly sexually dichromatic, with terminal phase males bearing a striking blue, green and yellow coat that radiates brightly during their frequent courtship displays. As the lively males streak away from aggregations of females their color fades, only to intensify each time they approach another group of females. A large black blotch on each side of the tail base becomes prominent during these encounters. To date, only pair spawning has been reported for this species. At dusk, terminal males dive beneath the sand for the night.

A second before spawning.

Both color phases include primary and secondary males, with secondary males more abundant in the terminal phase. A typical population contains slightly more females than males. Small groups of tiny one-inch, golden juveniles with black-and-white-striped heads can typically be found along deep walls, where they often mix with Masked Goby, *Coryphopterus personatus.*

Creole Wrasse

Clepticus parrae

After watching the smaller, cigar-shaped wrasses going through their bottom-oriented routines, it is difficult to imagine that the robust-bodied, semi-pelagic Creole Wrasse are even remotely related. Although this abundant species has gone through considerable modification to adapt successfully to life as an openwater feeder, it still relies on the reef structure for cleaning stations, spawning sites and night sanctuary. Their small protrusible mouths draw in transient plankton including pelagic tunicates, small jellyfishes, larval invertebrates and pteropods (free-swimming sea slugs and snails). In mid to late afternoon, rapidly moving columns of Creole Wrasse, often numbering in the hundreds, sweep along the outer reef, propelled by wildly beating pectoral fins. As they fly by, small groups of a dozen or more peel off and assume stationary, snout-up or snout-down positions at cleaning stations. On the outer reefs this species is, by far, the cleaners' best customer.

Juvenile Creole Wrasse settle on reef habitats, where the tiny, translucent, one-inch fish with lavender overtones gather in small clusters near the tops of coral heads. The initial phase is made up exclusively of females, that change into terminal males when they reach a length of six to seven inches. As they change sex, their anal and dorsal fins develop elongate filaments, and the posterior half of their purplish bodies gradually transforms into an artist's palette of yellows, oranges and magenta.

The fast-paced courtship of Creole Wrasse is quite a performance. Spawning occurs year-round in mid to late afternoon but, although frequent, does not take place daily. There seems to be no territory defense or streaking, but larger males

A courting pair of Creole Wrasse (male above).

regularly and aggressively displace smaller males attempting to solicit a mate. A population's courtship, often lasting for hours, tends to occur in waves of 10 to 20 fish that pair off at random locations within customary spawning areas. The show begins with males shadowing females from above, while the females cut and dive erratically in an attempt to lose their single-minded suitors. During these prolonged chases that can last for 15 minutes, the males intensify colors, extend their white mouth membranes into pronounced puckers and spread their dorsal and anal fins. Pairs are so preoccupied that they regularly rocket past divers without notice. Once in the throes of passion, males become so determined to spawn that they may attempt to court juveniles or even other species.

When finally ready to comply, the females swim to the bottom and slow, allowing the males to position beneath them. With the backs of their heads, the males begin to push their partners toward the surface. At this point the females often have a change of heart and attempt to flee, forcing the males to scramble wildly to maintain position. When uncooperative females are pacified, the males press their snouts against the listing females' flared gill covers and, with beating tails, slowly push their mates upward for several feet before the females release eggs. The males immediately move through the cloud, releasing sperm before setting out to court new partners.

Rosy Razorfish

Xyrichtys martinicensis

Three wrasse species in the circumtropical genus *Xyrichtys,* commonly known as razorfishes, inhabit open sand flats in the tropical western Atlantic. The group's thin bodies and bony, hatchet-shaped heads that allow the slender fishes to dive into the sandy bottom when threatened are the source of the group's common name. Without their ability to hide beneath the sand it would be impossible for the little fishes to survive on the barren sand plains where there are few other options to avoid predators. The Pearly Razorfish, *X. novacula,* the largest and the least common of the three species, feed solely on benthic

A male Rosy Razorfish (left) makes up to nine visits with each female in his harem during afternoon courtship.

invertebrates, with a preference for small, buried mollusks and crustaceans. Green Razorfish. *X. splendens,* with males topping out at six inches, dine on both benthic invertebrates as well as planktonic food. Rosy Razorfish, the smallest and most abundant of the group almost exclusively pick zooplankton from open water. In some locales the three species inhabit adjoining colonies covering several hundred square yards of sand.

Like their two genus mates, Rosy Razorfish are protogynous hermaphrodites born as females that can later, when size, age and social requirements are met change into terminal phase males. These transformations can be relatively quick. A researcher documented the largest initial phase female in a harem change into and spawn as a terminal phase male within two weeks even though a total color change had not been completed. Rosies live within territories measuring a few feet across. Mosaics of adjoining territories form colonies containing from a dozen to more than a hundred fish, generally in depths from 12 to 20 feet. However, colonies have been observed below 100 feet. Well established colonies have been reported to contain densities of up to five fish per square yard. Territories toward the center of colonies contain harems consisting of a single dominant male and typically three to seven smaller females, with which the male attempts to exclusively mate. In what must be an extremely unusual, and most demanding endeavor, one male was reported to retain 35 females within his domain. Territories at the fringes of colonies generally contain single males always on the lookout to acquire females and form harems of their own. Haremic males kept neighboring rivals from encroaching into their personal realms by challenging them at the boundaries where the pairs race side-by-side. At times both large and small colonies migrate en masse to nearby locations or coalesce with neighboring groups. It hasn't been established if these moves are gradual or sudden.

When living in open spaces, having multiple eyes constantly on the lookout for danger is a lifesaver. Males and females, forever cautious and highly site-attached, seldom travel more than three feet from home ranges and feed no more than a few feet off the bottom. To escape rapidly approaching predators, razorfishes maintain one, or in some cases up to three, personal sand patches inside their territories where they prefer to bury in the sand. Owners frequently move in and out of each private patch to keep the sand loose as well as removing buried pieces of coral and rock that could impede their dives. When predators are sighted a fright reaction spreads throughout a colony. Rosie's vary their response depending on the level of the threat. At the first indication of danger, feeding fish drop to the bottom, list to their sides and acquire a banded pattern as camouflage. If the threat persists, Rosies slowly move toward their maintained sites where they prefer to dive beneath the surface. Once buried, individuals can remain hidden for up to an hour or tunnel away beneath the sand for some distance, but most fish pop their heads out of the sand after only a few minutes at the exact spot where they disappeared. Although razorfishes decidedly prefer their maintained sites, they will, as a last resort, bury beneath the sand wherever they happen to be.

Living on the open sand is extremely dangerous. As an example, on a daily basis Rosies experience three times the number of strong reactions to predators

Male Rosy Razorfish emerges from the sand during courtship.

as Blueheads a similar-sized wrasse species that inhabit complex bottoms with a number of readily accessible hiding holes. Rapidly approaching Bar Jacks, barracudas, Yellowtail Snapper and lizardfishes generate the most frequent and strongest fright responses. Although razorfishes are completely concealed, a few large piscivores—rays, dolphins and hammerhead sharks—have the ability to detect their hiding places. Random, cone-shaped depressions pock the bottom where dolphins recently hunted buried razorfishes.

Colony members primarily pick zooplankton from the water column when currents flow. They also intermittently rummage through the sand for small invertebrates and occasionally nab small crustaceans uncovered while tidying their dive patches. If competitive carnivores such as crustacean-eating goatfishes or Yellowfin Mojarra, *Gerres cinereus,* Slippery Dick or Eyed Flounder, *Bothus ocellatus,* feed nearby, male razorfishes, although smaller, do not hesitate to defend their feeding grounds by giving chase.

With multiple females vying to become males, there is little peace and harmony within harems. Terminal males frequently attack large females vying to change into terminal males while larger females help sustain their social position by continually harassing smaller females. Males also periodically intervene in squabbles between females.

Males typically brighten their colors and spread their fins when passing females throughout the day. However, the pace picks up each afternoon during lengthy hour-and-a-half courtship affairs when males begin visiting each female in his harems circling, nudging and showing off brightened colors. As courtship continues, males spend additional time with each prospective mate repeatedly slipping back and forth along their sides before hovering head down only an inch off the bottom. At the height of their advances males ritualistically dive beneath the sand. Pausing seconds later with their bodies halfway above the surface the males quiver briefly before emerging completely and assuming a rigid head-up stance with their bodies trembling and fins taunt. At times during courtship females also bury in a similar fashion. As courtship intensifies females showing obvious abdominal swelling turn the table by making advances toward the males. The same courtship routine is typically repeated multiple times before the females, one at a time, indicate their readiness to spawn by slowly rising a foot or so above the bottom. Males rush to their sides and together the pairs make rather slow rises that take two to four seconds before gametes are released in visible puffs about four feet above the bottom. Males average slightly more than three spawns each day, while the majority of females spawn once. No sneaking or streaking or group spawning has been documented. After spawning ends, the fish remain close to the bottom and become progressively inactive until sunset, when they finally bed down for the night by diving into one of their personal patches of sand.

A most interesting reproductive characteristic of the Rosy's anatomy is the females' unique white abdomens highlighted by a series of vertical reddish orange lines. On each of a suitor's multiple courtship visits, females curl their bodies into an exaggerated S-shape conspicuously exposing their expanding bellies. Dissection has revealed that the tissue covering the area functions as a transparent window believed to enable males to evaluate the female's swelling ovaries. However, this ploy does not always guarantee success. Field research has found that males do

The swollen belly of the female Rosy Razorfish indicates her readiness to spawn.

A pair of Rosy Razorfish simultaneously release a cloud of gametes at the apex of their three- to four-foot spawning rise.

not spawn with a quarter of the females they court on a given afternoon. This limitation appears to be related to their unusually small testes compared with other similar-sized wrasse species. Scientists suspect that the females' curious red and white abdominal anatomy might have evolved as a means for female Rosy's to compete with harem mates for fertilization.

In a rare behavior among spawning fishes, the males of both Rosy and Green Razorfish occasionally end spawning rushes by taking bites out of the females' clouds of expelled eggs. Video analysis has shown the males modifying their normal spawning rush when they consume eggs, leading to the idea that males anticipate cannibalizing their mates nutrient-rich eggs and probably don't release sperm during these efforts. One set of observations documented cannibalism three percent of the time during 150 Rosy Razorfish spawning rises. A second study recorded a 40 percent rate of cannibalism. An egg consumption rate of 25 percent has been recorded for Green Razorfish. A similar cannibalistic behavior has not been observed for any other broadcast spawning reef fish in the Caribbean, although it has been reported for Scalefin Anthias, *Anthias squamipinnis,* a small, plankton-feeding sea bass from the Indo-Pacific. Pearly Razorfish, exclusive benthic carnivores, haven't been observed feeding on their mates' eggs.

Spanish Hogfish

Bodianus rufus

From juveniles to the terminal phase, Spanish Hogfish exhibit the same distinctive purple and golden color pattern and basic body shape. Juveniles from less than one inch to just over three inches are active cleaners that busily swim about coral structures in ones, twos or threes. Client fishes swim near and pause, waiting to have ectoparasites picked from their bodies and occasionally from inside their opened mouths and flared gill covers. When the fish become sexually mature females, at about four inches, they join a harem and switch to a diet that includes a broad assortment of benthic invertebrates. Their primary foods include crabs, brittle stars, sea urchins and a variety of gastropods. After reaching six to seven inches, females make a final weeklong transformation into secondary terminal phase males that can eventually reach a maximum length of 16 inches.

Courtship and pair spawning between females and their terminal male harem-masters occur throughout the year, beginning approximately an hour before sunset. Preceding spawning, which takes place shortly after sunset, loose groups of three to four females mill about just above the bottom inside their territories. For brief periods the small groups disappear into the reef structure before suddenly reappearing. The large terminal male that has usually been elsewhere

A male Spanish Hogfish displays a prominent set of incisors used to capture and crush crustaceans and gastropods.

Signaling her readiness to spawn, a female Spanish Hogfish curves into an S-shape.

The slow spawning rise of Spanish Hogfish.

Juveniles act as cleaners.

returns and makes passes near the females, prompting one or two to rise higher above the reef and signal their willingness to spawn by curving their bodies into an S-shape. The male swims slowly to the side of a female and together the pairs calmly rise three or four feet before shedding gametes and returning to the reef. After one or two spawns the male swims away, only to return several minutes later to spawn with the remaining females. This pattern is usually repeated multiple times, depending on the size of the harem, for a 10- to 15-minute period before spawning is complete.

A terminal male Hogfish attempting to crack a tough crab carapace.

Hogfish
Lachnolaimus maximus

Male Hogfish, capable of reaching three feet in length, are the largest wrasses in the region. Even the largest males, with elongate, concave snouts and reddish brown masks, all start life as females. The smaller females reach sexual maturity in three to five years at 12 to 16 inches in length. Both sexes frequently display a dramatic reddish mottled pattern when feeding near the bottom or settling next to cleaning stations. Unfortunately, this highly desired food fish has little fear of divers and has been overharvested by spearfishing and traditional fishing methods in many areas.

Males establish and patrol long, narrow territories, just over 300 feet in length, occupied by harems of 10 to 15 females. During the day little interaction occurs among harem members as they independently root about the sand for mollusks, including snails, bivalves, and occasionally crabs. Powerful jaws crush the hard-shelled invertebrates.

Courtship and pair spawning take place daily in the cooler winter months, during the last hour and a half before sunset. Males begin courtship by patrolling their territories at an accelerated pace. When prospective mates are sighted, they extend their long foredorsal spines and dive toward the females. If rebuffed, the males continue on to the next potential partner. When ready to spawn, females lift off the bottom and swim slowly upward at an angle. Males, approaching from behind, pick up speed as they switch from pectoral fin sculling to powerful strokes of their wide tails. Side-by-side and streaking toward the surface, the pairs roll to their sides with the males above, and release gametes 20 to 30 feet above the bottom. Splitting apart and looping, they dive toward the bottom. As their paths cross, the males briefly extend their dorsal spines. A single male may take part in dozens of paired spawning rushes in an afternoon, while females are believed to spawn only once. The larvae hatch in 24 hours, begin taking food after two days and settle in shallow water habitats two to three weeks later.

COMMON NAME INDEX

Index Legend: photos - bold, photos and text - bold underlined.

S

T

W

SCIENTIFIC NAME INDEX

Index Legend: photos - bold, photos and text - bold underlined.

430

R

S

T

U

X

Index Legend: photos - bold, photos and text - bold underlined.

References

Reproduction

Appeldoorn, R.S., D.A. Hensley and D.Y. Shapiro. 1994. Egg dispersal in a Caribbean coral reef fish, *Thalassoma bifasciatum*. II. Dispersal off the reef platform. Bull. Mar. Sci. 54: 271-280.

Barlow, G.W. 1981. Patterns of parental investment, dispersal and size among coral reef-fishes. Env. Biol. Fish. 6: 65-85.

Bauer, J.A. and S.E. Bauer. 1981. Reproductive biology of pigmy angelfishes of the genus *Centropyge* (Pomacanthidae). Bull. Mar. Sci. 31(3): 495-513.

Burnett-Herkes, J. 1975. Contribution to the biology of the red hind, *Epinephelus guttatus*, a commercially important serranid fish from the tropical western Atlantic. Ph.D. Dissertation. Univ. Miami, Coral Gables.

Carter, J. and D. Perrine. 1994. A spawning aggregation of dog snapper, *Lutjanus jocu* (Pisces: Lutjanidae), in Belize, Central America. Bull. Mar. Sci. 55: 228-234.

Colin, P.L. 1992. Reproduction of the Nassau Grouper, *Epinephelus striatus*, (Pisces: Serranidae) and its relationship to environmental conditions. Env. Biol. Fish. 34: 357-377.

Colin, P.L. 1995. Surface currents in Exuma Sound, Bahamas and adjacent areas with reference to potential larval transport. Bull. Mar. Sci. 56(1): 48-57.

Colin, P.L. 1996. Longevity of some coral reef fish spawning aggregations. Copeia 1996: 189-192.

Colin, P.L. and I.E. Clavijo. 1988. Spawning activity of fishes producing pelagic eggs on a shelf edge coral reef, southwestern Puerto Rico. Bull. Mar. Sci. 43(2): 249-279.

Doherty, P.J., D.M. Williams and P.F. Sale. 1985. The adaptive significance of larval dispersal in coral reef fishes. Env. Biol. Fish. 12: 81-90.

Domeier, M.L. and M.E Clarke. 1992. A laboratory produced hybrid between *Lutjanus synagris* and *Ocyurus chrysurus* and a probable hybrid between *L. griseus* and *O. chrysurus* (Perciformes: Lutjanidae). Bull. Mar. Sci. 50(3): 501-507.

Domeier, M.L. and P.L. Colin. 1997. Tropical reef fish spawning aggregations: defined and reviewed. Bull. Mar. Sci. 60(3): 698-726.

Hoffman, S.G. and D.R. Robertson. 1983. Foraging and reproduction of two Caribbean reef toadfishes (Batrachoididae). Bull. Mar. Sci. 33: 919-927.

Hoffman, S.G., M.P. Schildhauer and R.R. Warner. 1985. The costs of changing sex and the ontogeny of males under contest competition for mates. Evolution 39: 915-927.

Hourigan, T.F. and C.D. Kelley. 1985. Histology of the gonads and observations on the social behavior of the Caribbean angelfish *Holacanthus tricolor*. Marine Biol. 88: 311-322.

Hubb, C.L. 1955. Hybridization between fish species in nature. Syst. Zool. 4(1): 1-20.

Johannes, R.E. 1978. Reproductive strategies of coastal marine fishes in the tropics. Env. Biol. Fish. 3: 65-84.

Leonard, J.L. 1993. Sexual conflict in simultaneous hermaphrodites: evidence from serranid fishes. Env. Biol. Fish. 36: 135-148.

Lobel, P.S. 1980. Nesting, eggs and larvae of triggerfishes (Balistidae). Envir. Biol. Fish. 5: 251-252.

Randall, J.E. and H.A. Randall. 1963. The spawning and early development of the Atlantic parrotfish, *Sparisoma rubripinne*, with notes on other scarid and labrid fishes. Zoologica 48: 49-60.

Robertson, D.R., C.W. Petersen and J.D. Brawn. 1990. Lunar reproductive cycles of benthic-brooding reef fishes: reflections of larval biology or adult biology? Ecol. Monog. 60(3): 311-329.

Ross, R.M. 1990. The evolution of sex-change mechanisms in fishes. Env. Biol. Fish. 29: 81-93.

Sadovy de Mitcheson, Y. and P.L. Colin (eds) 2012. Reef fish spawning aggregations: biology, research and management, Vol 35: Fish and Fisheries Series. Springer, New York, NY.

Shapiro, D.Y., D.A. Hensley and R.S. Appeldoorn. 1988. Pelagic spawning and egg transport in coral-reef fishes: skeptical overview. Env. Biol. Fish. 22: 3-14.

Thresher, R.E. 1984. "Reproduction in Reef Fishes" Neptune City, NJ: T.F.H. Publications.

Turner, G.F. 1993. Teleost mating behaviour. "Behaviour of Teleost Fishes" 2nd (T.J. Pitcher, ed.) Chapman & Hall.

Warner, R.R. 1987. Female choice of sites versus mates in coral reef fish, *Thalassoma bifasciatum*. Anim. Behav. 35: 1470-1478.

Warner, R.R. 1988a. Traditionality of mating-site preferences in coral reef fish. Nature 335: 719-721.

Warner, R.R. 1988b. Sex change in fishes: hypotheses, evidence, and objections. Env. Biol. Fish. 22(2): 81-90.

Warner, R.R. 1990. Resource assessment versus tradition in mating-site determination. Am. Nat. 135: 205-217.

Warner, R.R. 1995. Large mating aggregations and daily long-distance spawning migrations in the Bluehead Wrasse, *Thalassoma bifasciatum*. Env. Biol. Fish. 44: 337-345.

Warner, R.R. and D.S. Ross. 1978. Sexual patterns in the labroid fishes of the western Caribbean, I. The wrasses (Labridae). Smithsonian Contrib. Zool. 254: 1-27.

Warner, R.R. and S.G. Hoffman 1980. Local population size as a determinant of mating system and sexual composition in two tropical marine fishes (*Thalassoma* spp.). Evolution 34(3): 508-518.

Warner, R.R. and S.E. Swearer. 1991. Social control of sex change in the bluehead wrasse, *Thalassoma bifasciatum* (Pisces: Labridae). Biol. Bull. 181: 199-204.

Life Cycles and Reef Fish Community Structures

Appledoorn, R.S., D.A. Hensley and D.Y. Shapiro. 1994. Egg dispersal in a Caribbean coral reef fish, *Thalassoma bifasciatum*. II. Dispersal off the reef platform. Bull Mar. Sci. 54: 271-280.

Bakun, A. 1986. Local retention of planktonic early life stages in tropical reef/bank demersal systems: the role of vertically-structured hydrodynamic processes. " IOC/FAO Workshop on Recruitment in Tropical Coastal Demersal Communities," (D. Pauly and A. Yanez-Arancibia, eds.), IOC Workshop Rep 44 (suppl.), UNESCO, Paris: 15-32.

Bauer, J.A. Jr. and S.E. Bauer. 1981. Reproductive biology of pigmy angelfishes of the genus *Centropyge* (Pomacanthidae). Bull. Mar. Sci. 31(3): 495-513.

Bohnsack, J.A. 1982. Effects of piscivorous predator removal on coral reef fish community structure. "Gutshop '81: Fish Food Habits Studies" (G.M. Cailliet and C.A. Simenstad, eds.). Seattle: Washington Sea Grant Publ. 258-267.

Booth, D.A. and D.M. Broshan. 1995. The role of recruitment dynamics in rocky shore and coral reef fish communities. "Advances in Ecological Research" (M. Begon and A.H. Fitter, eds.) Academic Press Inc. San Diego, CA. 25: 309-385.

Castro, P. and M.E. Huber. 1992. "Marine Biology" Wm. C. Brown Communications, Dubuque, IA.

Colin, P.L. 1995. Surface currents in Exuma Sound, Bahamas and adjacent areas with reference to potential larval transport. Bull. Mar. Sci. 56(1): 48-57.

Cowen, R.K., C.B. Paris and A. Srinivasan. 2006. Scaling of connectivity in marine populations. Science 311: 522-527.

Doherty, P.J. 1983. Tropical territorial damselfishes: is density limited by aggression or recruitment? Ecology 64: 176-190.

Doherty, P.J. 1985. Predation on juvenile coral reef fishes: an exclusion experiment. Coral Reefs 4: 225-234.

Doherty, P.J. 1991. Spatial and temporal patterns in recruitment. "The Ecology of Fishes on Coral Reefs." (P.F. Sale, ed.) Academic Press, San Diego, CA: 261-293.

Doherty, P.J. and D. McB. Williams. 1988. The replenishment of coral reef fish populations. Oceanogr. Mar. Biol. Annu. Rev. 26: 487-551.

Fisher, R. 2005. Swimming speeds of larval coral reef fishes: impacts on self-recruitment and dispersal. Mar. Ecol. Prog. Ser. 285: 223-232.

Fisher, R. and J.M. Leis 2009. Swimming performance in larval fishes: from escaping predators to the potential for long distance migration. In: Domenici P., Kapoor B.G. (eds) Fish locomotion: an etho-ecological approach. Science Publishers: 333–373.

Hernandez, F.J., J.A. Hare and D.P. Fey 2009. Evaluating diel, ontogenetic and environmental effects on larval fish vertical distribution using generalized additive models for location, scale and shape. Fish Oceanogr. 18:224–236.

Hitchcock, G.L., et al. 2001. Property fields in a Tortugas Eddy in the southern Straits of Florida. Deep Sea Res Part I Oceanogr. Res. Pap. 52(12): 2195-2213.

Hixon, M.A. 2011. 60 Years of coral reef ecology: past, present, future. Bull. Mar. Sci. 87(4): 727-765.

Hixon, M.A., D.W. Johnson and S.M. Sogard. 2013. BOFFFFs: on the importance of conserving old-growth age structure in fishery populations. ICES Journal of Marine Science, (2014), 71(8): 2171-2185.

Hunte, W. and I.M. Côté. 1989. Recruitment in the redlip blenny, *Ophioblennius atlanticus*: is space limiting? Coral Reefs 8: 45-50.

Hunter, J.R. 1981. Feeding ecology and predation of marine fish larvae. "Marine Fish Larvae: Morphology, Ecology and Relation to Fisheries" (R. Lasker, ed.) Seattle: Univ. of Washington Press. 33-77.

Johannes, R.E. 1978. Reproductive strategies of coastal marine fishes in the tropics. Environ. Biol. Fishes 3: 65-84.

Jones, G.P. 1991. Postrecruitment processes in the ecology of coral reef fish populations: a multifactorial perspective. "The Ecology of Fishes on Coral Reefs." (P.F. Sale, ed.) San Diego: Academic Press. 294-328.

Kelley, S. 1995. Preliminary guide to the identification of the early life history stages of Pomacanthid fishes of the Western Central Atlantic and Gulf of Mexico. NOAA Tech. Mem. NMFS-SEFSC-375.

Kingsford, M.J., E. Wolanski and J.H. Choat. 1991. Influence of tidally-induced fronts and langmuir circulations on the distribution and movement of presettlement fishes around a coral reef. Mar. Biol. 109: 167-180.

Lasker, R. 1981. The role of a stable ocean in larval fish survival and subsequent recruitment. "Marine Fish Larvae Morphology, Ecology and Relation to Fisheries" (R. Lasker, ed.) Seattle: Univ. of Washington Press. 80-87.

Leis, J.M. 1991. The pelagic stage of reef fishes: the larval biology of coral reef fishes. "The Ecology of Fishes on Coral Reefs" (P.F. Sale, ed.) San Diego: Academic Press. 183-230.

Leis, J.M. 2006. Are larvae of demersal fishes plankton or nekton? Adv. Mar. Biol. 51: 57-141.

Leis, J.M. 2010. Ontogeny of behaviour in larvae of marine demersal fishes. Ichthyol. Res. 2010. 57: 325–342.

Lindeman, K.L. 1986. Development of larvae of the French grunt, *Haemulon flavolineatum,* and comparative development of twelve species of western Atlantic *Haemulon* (Percoidei, Haemulidae). Bull. Mar. Sci. 39(3): 673-716.

Mannoch, C.S. 1987. Age and growth of snappers and groupers. "Tropical Snappers and Groupers - Biology and Fisheries Management." (J.J. Polovina and S. Ralston, eds.) Boulder, CO: Westview Press. 329-373.

Morse, A.N.C. 1991. How do planktonic larvae know where to settle? American Scientist 79: 154-167.

Moser, H.G. 1981. Morphological and functional aspects of marine fish larvae. "Marine Fish Larvae: Morphology, Ecology and Relation to Fisheries" (R. Lasker, ed.) Seattle: Univ. of Washington Press. 30-88.

Panella, G. 1971. Fish otoliths daily growth layers and periodical patterns. Science 173: 1124-1127.

Panella, G. 1980. Growth patterns in fish sagittae. "Skeletal Growth of Aquatic Organisms: Biological Records of Environmental Change" (D.E. Rhoads and R.A. Lutz, eds.) New York: Plenum. 519-560.

Paris, C.B. and R.K. Cowen. 2004. Direct evidence of a biophysical retention mechanism for coral reef fish larvae. Limnol. Oceanogr., 49(6): 1964-1979.

Paris, C.B., J. Atema, J-O. Irisson, M. Kingsford, G. Gerlach, C. M. Guigand. 2013. Reef Odor: A Wake Up Call for Navigation in Reef Fish Larvae. PLoS ONE 8(8): e72808.

References

Randall, J.E. 1961. A contribution to the biology of the convict surgeonfish of the Hawaiian Islands, *Acanthurus triostegus sandvicensis*. Pac. Sci 15: 215-272.

Richards, W.L. and K.C. Lindeman. 1987. Recruitment dynamics of reef fishes: planktonic processes, settlement and demersal ecologies, and fishery analysis. Bull. Mar. Sci. 41: 392-410.

Robertson, D.R. 1988. Extreme variation in settlement of the Caribbean triggerfish *Balistes vetula* in Panama. Copeia: 698-703.

Robertson, D.R., D.G. Green and B.C. Victor. 1988. Temporal coupling of production and recruitment of larvae of a Caribbean reef fish. Ecology 69: 370-381.

Robertson, D.R., C.W. Petersen and J.D. Brawn. 1990. Lunar reproductive cycles of benthic-brooding reef fishes: reflections of larval biology or adult biology? Ecol. Monog. 60(3): 311-329.

Robertson, D.R. 2001. Population maintenance among tropical reef fishes: inferences from small-island endemics. Proc. Nat'l. Acad. Sci. 98: 5667-5670.

Roughgarden, J. 1986. Dynamics of a metapopulation with space-limited subpopulations. Theor. Pop. Biol. 5: 163-186.

Sadovy de Mitcheson, Y. and Colin P.L. (eds) 2012. Reef fish spawning aggregations: biology, research and management, Vol 35: Fish and Fisheries Series. Springer, New York, NY.

Sale, P.F. 1980. The ecology of fishes on coral reefs. Oceanogr. Mar. Biol. 18: 367-421.

Seong, S.C., M.F. McGowan and W. J. Richards. 1994. Vertical distribution of fish larvae off the Florida Keys, 26 May-5 June 1989. Bull. Mar. Sci. 54(3): 828-842.

Shulman, M.J. and J.C. Ogden. 1987. What controls tropical reef fish populations: recruitment or benthic mortality? An example in the Caribbean reef fish *Haemulon flavolineatum*. Mar. Ecol. Prog. Ser. 39: 233-242.

Soares, M.C., R. Bshary, S.C. Cardoso, et al. 2016. Face your fears: cleaning gobies inspect predators despite being stressed by them. PLoS One. 7(6): e39781. https://doi.org/10.1371/journal.pone.0039781.

Sponaugle, S., R.K. Cowen, A. Shanks, et al. 2002. Predicting self-recruitment in marine populations: biophysical correlates and mechanisms. Bull. Mar. Sci. 70(1) Suppl.: 341-375.

Sullivan, J.M., D.V. Holliday, M. McFarland, M.A. McManus, et al. 2010. Layered organization in the coastal ocean: An introduction to planktonic thin layers and the LOCO project. Continental Shelf Research. Vol. 30 Issue 1: 1-6.

Sweatman, H.P.A. 1988. Field evidence that settling coral reef fish larvae detect resident fishes using dissolved chemical cues. J. Exp. Mar. Biol. Ecol. 124: 163-174.

Thresher, R.E. and E.B. Brothers. 1989. Evidence of intra- and inter-oceanic regional differences in the early life history of reef-associated fishes. Mar. Ecol. Prog. Ser. 57: 187-205.

Victor, B.C. 1982. Daily otolith increments and recruitment in two coral-reef wrasses, *Thalassoma bifasciatum* and *Halichoeres bivittatus*. Marine Biol. 71: 203-208.

Victor, B.C. 1983. Recruitment and population dynamics of a coral reef fish. Science 219: 419-420.

Victor, B.C. 1984. Coral reef fish larvae: Patch size estimation and mixing in the plankton. Limnol. Oceangr. 29(5): 1116-1119.

Victor, B.C. 1986. Larval settlement and juvenile mortality in a recruitment-limited coral reef fish population. Ecol. Monog. 56(2): 145-160.

Victor, B.C. 1986. Duration of the planktonic larval stage of one hundred species of Pacific and Atlantic wrasse (family Labridae). Mar. Biol 90: 317-326.

Victor, B.C. 1991. Settlement strategies and biogeography of reef fishes. "The Ecology of Fishes on Coral Reefs." (P.F. Sale, ed.) San Diego: Academic Press. 231-260.

Williams, D. McB. 1980. Dynamics of the pomacentrid community on small patch reefs in One Tree Lagoon (Great Barrier Reef). Bull. Mar. Sci. 30: 159-170.

Feeding

Adey, W.H. and R.S. Steneck. 1985. Highly productive eastern Caribbean reefs: synergistic effects of biological, chemical, physical and geological factors. "The Ecology of Coral Reefs" (M.L. Reaka, ed.) NOAA Symp. Ser. Undersea Res. Natl. Oceanic Atmos. Adm., Rockville, MD. 3: 163-188.

Aronson, R.B. 1983. Foraging behavior of the West Atlantic Trumpetfish, *Aulostomus maculatus*: use of large, herbivorous reef fishes as camouflage. Bull. Mar. Sci. 33: 166-171.

Aronson, R.B. and S.L. Sanderson. 1987. Benefits of heterospecfic foraging by the Caribbean wrasse, *Halichoeres garnoti* (Pisces: Labridae). Env. Biol. Fish. 47: 129-141.

Bruckner, A.W. and R.J. Bruckner. 1998. Rapid-wasting disease: pathogen or predator. Science. 279: 2023-2025.

Bythell, J.C., E.H. Gladfelter and M. Bythell. 1993. Chronic and catastrophic natural mortality of three common Caribbean reef corals. Coral Reefs 12: 143-152.

Carpenter, R.C. 1986. Partitioning herbivory and its effects on coral reef algal communities. Ecol. Monogr. 56: 345-363.

Carpenter, R.C. 1990a. Mass mortality of *Diadema antillarum*. I. Long-term effects of sea urchin population-dynamics and coral reef algal communities. Mar. Biol. 104: 67-77.

Carpenter, R.C. 1990b. Mass mortality of *Diadema antillarum*. II. Effects on population densities and grazing intensity of parrotfishes and surgeonfishes. Mar. Biol. 104: 79-86.

Choat, J.H. 1991. The biology of herbivorous fishes on coral reefs. "The Ecology of Fishes on Coral Reefs" (P.F. Sale, ed.). San Diego: Academic Press. 120-153.

Davis, W.P. and R.S. Birdsong. 1973. Coral reef fishes which forage in the water column. Helgol. Wiss. Meeresunters. 24: 292-306.

Emery, A.R. 1968. Preliminary observations on coral reef plankton. Limnol. Oceanogr. 13: 293-303.

Gerking, S.D. 1994. "Feeding Ecology of Fish." San Diego: Academic Press.

Hay, M.E. 1991. Fish - seaweed interactions on coral reefs: effects of herbivorous fishes and adaptations of their prey. "The Ecology of Fishes on Coral Reefs." (P.F. Sale, ed.) San Diego: Academic Press. 96-118.

Hay, M.E., V.J. Paul, S.M. Lewis, K. Gustafson, J. Tucker and R.N. Trindell. 1988. Can tropical seaweeds reduce herbivory by growing at night? Diel patterns of growth, nitrogen content, herbivory, and chemical versus morphological defenses. Oecologia. 75: 233-245.

Helfman, G.S 1989. Threat-sensitive predator avoidance in damselfish-trumpetfish interactions. Behav. Ecol. and Sociobiol. 24: 47-58.

Helfman, G.S. 1993. Fish behaviour by day, night and twilight. "Behaviour of Teleost Fishes" 2nd (T.J. Pitcher, ed.) London: Chapman & Hall. 479-512.

Helfman, G.S., J.L. Meyer and W.N. McFarland. 1982. The ontogeny of twilight migration patterns in grunts (Pisces: Haemulidae). Anim.Behav. 30: 317-326.

Hobson, E.S. 1973. Diel feeding migrations of tropical reef fishes. Helgol. Wiss. Meeresunters. 24: 361-370.

Hobson, E.S. 1991. Tropic relationships of fishes specialized to feed on zooplankters above coral reefs. "The Ecology of Fishes on Coral Reefs" (P.F. Sale, ed.) San Diego: Academic Press. 69-93.

Horn, M. 1989. Biology of marine herbivorous fishes. Oceanogr. Mar. Biol. Annu. Rev. 27: 167-272.

Hughes, T.P., D.C. Reed and M.J. Boyle. 1987. Herbivory on coral reefs: community structure following mass mortalities of sea urchins. J. Exp. Mar. Biol. Ecol. 113: 39-59.

Irvine, G.V. 1982. The importance of behavior in plant-herbivore interactions: a case study. "Fish Food Studies" (G.M. Caillet and C.A. Simenstad, eds.) Washington Sea Grant Program. Seattle: Univ. of Washington Press. 240-248.

Karplus, I. 1971. A feeding association between the grouper *Epinephelus fasciatus* and the moray eel *Gymnothorax griseus*. Copeia. 3: 164.

Kaufman, L. 1976. Feeding behavior and functional coloration of the Atlantic trumpetfish, *Aulostomus maculatus*. Copeia. 2: 377-378.

Lewis, S.M. 1986. The role of herbivorous fishes in the organization of a Caribbean reef community. Ecol. Mongr. 56: 183-200.

Lewis, S.M., J.N. Norris and R.B. Searles. 1987. The regulation of morphological plasticity in tropical reef algae by herbivory. Ecology. 68: 636-641.

Lobel, P.S. 1980. Herbivory by damselfishes and their role in coral reef community ecology. Bull. Mar. Sci. 30 (Spec. Issue): 273-289.

McFarland, W.N., J.C. Ogden and J.N. Lythgoe. 1979. The influence of light on the twilight migrations of grunts. Env. Biol. Fishes. 4: 9-22.

Nursall, J.R. 1982. Behavioral interactions between Caribbean reef fish and eels (Muraenidae and Ophichthidae). Copeia. 1: 229-232.

Paul, V.J. and K.L. Van Alstyne. 1988. Chemical defense and chemical variation in some tropical Pacific species of *Halimeda* (Halimedaceae,Chlorophyta). Coral Reefs. 6: 263-270.

Pressley, P.H. 1981. Pair formation and joint territoriality in a simultaneous hermaphrodite: the coral reef fish *Serranus tigrinus*. Z. Tierpsychol. 56: 33-46.

Randall, J.E. 1961. Overgrazing of algae by herbivorous marine fishes. Ecology. 42: 812.

Randall, J.E. 1967. Food habits of reef fishes of the West Indies. Stud. Trop. Ocean. 5: 665-847.

Randall, J.E. and W.D. Hartman. 1968. Sponge-feeding fishes of the West Indies. Mar. Biol. 1: 216-225.

Robertson, D.R. 1987. Responses of two coral reef toadfishes (Batrachoididae) to the demise of their primary prey, the sea urchin *Diadema antillarum*. Copeia. 3: 637-642.

Russ, G.R. and J. St. John. 1988. Diets, growth rates and secondary production of herbivorous coral reef fishes. Proc. Int. Coral Reef Symp. 6th 2: 37-43.

Sazima, C., J.P. Krajewski, R.M. Bonaldo and I. Sazima. 2007. Nuclear-follower foraging associations of reef fishes and other animals at an oceanic archipelago. Environ. Biol. Fish. 80 (4): 351-361.

Schmitt, E.F. 1997. The influence of herbivorous fishes on coral reef communities with low sea urchin abundance: a study among reef community types and seasons in the Florida Keys. Ph.D. Dissertation. Univ. of Miami. Coral Gables, FL.

Steneck, R.S. 1983. Escalating herbivory and resulting adaptive trends in calcareous algal crust. Paleobiology. 9: 44-91.

Steneck, R.S. 1988. Herbivory on coral reef: a synthesis. Proc. Int. Coral Reef Symp. 6th 1: 37-49.

Wulff, J.L. 1997. Parrotfish predation on cryptic sponges of Caribbean coral reefs. Mar. Biol. 129: 41-52.

Zaret, T.M. 1980. "Predation and Freshwater Communities" New Haven, CT: Yale Univ. Press.

Colors & Camouflage

Casazza, T.L. and S.W. Ross. 2008. Fishes associated with pelagic *Sargassum* and open water lacking *Sargassum* in Gulf Stream off North Carolina. Fishery Bulletin, 106(4): 348-363.

Lorenz, K. 1962. The function of colour in coral reef fishes. Proc. Roy. Inst. Gr. Brit. 39: 282-296.

Lorenz, K. 1966. "On Aggression." New York: Harcourt, Brace & World. Inc.

Nemtzov, S.C. 1993. Diel color phase changes in the coney, *Epinephelus fulvus* (Teleostie, Serranidae). Copeia. 3: 883-885.

Neudecker, S. 1989. Eye camouflage and false eyespots: chaetodontid responses to predators. Env. Biol. Fish. 25: 143-157.

Randall, J.E. and H.A. Randall. 1960. Examples of mimicry and protective resemblance in tropical marine fishes. Bull. Mar. Sci., Gulf Carib. 10: 444-480.

References

Trott, T.M., S.A. Mckenna, J.M. Pitt, et al. 2010. Efforts to enhance protection of the Sargasso Sea. Proceedings of the 63rd Gulf and Caribbean Fisheries Institute. 282-288.

Symbiosis

Bohlke, J.E. and C.C.G. Chaplin. 1993. "Fishes of the Bahamas and Adjacent Tropical Waters" 2nd edition. Austin: Univ. of Texas Press.

Brockmann, H.J. and J.P. Hailman. 1976. Fish cleaning symbiosis: notes on juvenile angelfishes (*Pomacanthus*, Chaetodontidae) and comparisons with other species. Z. Tierpsychol. 42: 129-138.

Côté, I.M. 2000. Evolution and ecology of cleaning symbioses in the sea. Oceanog. Mar. Biol: an annual revue. Vol. 38: 311–355.

Côté, I.M. and M.C. Soares. 2011. Gobies as Cleaners. The Biology of Gobies. Eds, Patzner R., J.L. Van Tassell, et al. CRC Press.

Darcy, G.E., E. Maisel, and J.C. Ogden. 1974. Cleaning preferences of the gobies *Gobiosoma evelynae* and *G. prochilos* and the juvenile wrasse *Thalassoma bifasciatum*. Copeia. 1974: 375-379.

Feddern, H.A. 1965. The spawning, growth, and general behavior of the bluehead wrasse, *Thalassoma bifasciatum* (Pisces: Labridae). Bull. Mar. Sci. 15:896-941.

Foster, S.A. 1985. Wound healing: a possible role of cleaning stations. Copeia. 1985(4): 875-880.

Gorlick, D.L., P.D. Atkins and G.S. Losey. 1978. Cleaning stations as water holes, garbage dumps, and sites for the evolution of reciprocal altruism? Am. Nat. 112: 341-353.

Grutter, A.S. 2002. Cleaning symbioses from the parasites' perspective. Parasitology 124: S65-S81.

Grutter, A.S. 2003. Feeding ecology of the fish ectoparasite Gnathia sp. (Crustacea: Isopoda) from the Great Barrier Reef, and its implications from fish cleaning behaviour. Mar. Ecol. Prog. Ser. Vol. 259: 295-302.

Grutter, A.S., M. De Brauwer and R. Bshary, et al. 2017. Parasite infestation increases on coral reefs without cleaner fish. Coral Reefs. DOI 10.1007/s00338-017-1628-z.

Huebner, L.K. and N.E. Chadwick. 2012. Patterns of cleaning behaviour on coral reef fish by the anemoneshrimp *Ancylomenes pedersoni*. J. Mar. Biol. Assoc. U.K.: 92 (7): 1557-1562.

Johnson, W.S. and R. Ruben. 1988. Cleaning behavior of *Bodianus rufus, Thalassoma bifasciatum, Gobiosoma evelynae,* and *Periclimenes pedersoni* along a depth gradient at Salt River Submarine Canyon, St. Croix. Env. Biol. Fish. 23(3): 225-232.

Levy, J.M. 1994. Taxonomic review of ectoparasites of Epinepheline groupers in the Exuma Cays, Bahamas. Unpublished.

Levy, J.M., K.M. Sullivan and M. de Garine-Wichatitsky. 1994. Account of ectoparasites of Epinepheline groupers in the Exuma Cays, Bahamas. Proc. Gulf Carib. Fish. Inst. 43.

Limbaugh, C. 1961. Cleaning symbiosis. Sci. Am. 205: 42-49.

Limbaugh, C., H. Pederson and F.A. Chance Jr. 1961. Shrimps that clean fishes. Bull. Mar. Sci. Gulf & Carib. 11: 237-257.

Losey, G.S. Jr. 1972. The ecological importance of cleaning symbiosis. Copeia. 820-833.

Losey, G.S. Jr. 1987. Cleaning symbiosis. Symbiosis. 4: 229-256.

Losey, G.S. Jr. and L. Margules. 1974. Cleaning symbiosis provides a positive reinforcer for fish. Science. 184: 179-180.

Pagan-Font, F.A. 1967. The study of the commensal relationship between the conchfish, *Astrapogon stellatus,* and the queen conch, *Strombus gigas,* in southwestern Puerto Rico. Masters Thesis, Univ. of Puerto Rico.

Poulin, R. and A.S. Grutter. 1996. Cleaning symbioses: proximate and adaptive explanations. What evolutionary pressures led to the evolution of cleaning symbiosis? BioScience. 46(7): 512-517.

Randall, J.E. and H.A. Randall. 1960. Examples of mimicry and protective resemblance in tropical marine fishes. Bull. Mar. Sci. 10(4): 444-480.

Sazima, I., R.L. Moura and M.C.M. Rodrigues. 1999. A juvenile sharksucker, Echeneis naucrates acting as a station-based cleaner fish. French Icthyological Society–Cybium. (23-4: 377-380.

Sazima, C., A. Grossman and I. Sazima. 2010. Turtle cleaners: reef fishes foraging on epibionts of sea turtles in the tropical Southwestern Atlantic, with a summary of this association type. Neotrop. Ichthyol. 8(1): 187-192.

Sikkel, P.C. 1992. Interspecific feeding associations between the goatfish *Mulloides martinicus* (Mullidae) and a possible aggressive mimic the snapper *Ocyurus chrysurus* (Lutjanidae). Copeia. 3: 914-917.

Sluka, R. and K.M. Sullivan. 1996. Daily activity patterns of groupers in the Exuma Cays Land and Sea Park, Central Bahamas. Bahamas J. of Sci. 2: 17-22.

Snyder, D.B. 1999. Mimicry of initial-phase bluehead wrasse, *Thalassoma bifasciatum* (Labridae) by juvenile tiger grouper, *Mycteroperca tigris* (Serranidae). Revue Francoise d' Aquariologie.

Soares, M.C., R.F. Oliveira, A.F.H. Ros, et al. 2011. Tactile stimulation lowers stress in fish. Nat. Commun. 2: 534.

Thresher, R.E. 1977. Pseudo-cleaning behavior of Florida reef fishes. Copeia. 4: 768-769.

Thresher, R.E. 1979. Possible mucophagy by juvenile *Holacanthus tricolor* (Pisces: Pomacanthidae). Copeia. 1979(1): 160-162.

Titus, B.M., M. Daly and D.A. Exton. 2015. Temporal patterns of Pederson shrimp (*Ancylomenes pedersoni* Chace 1958) cleaning interactions on Caribbean coral reefs. Mar. Biol.

Victor, Benjamin C. 2018. Faculative cleaning behavior in a western Atlantic sponge goby, *Elacatinus xanthiprora* (Teleostei: Gobiidae) Journal of the Ocean Science Foundation, 31, 1-7.

Vaughan, D. B., A. S. Grutter, M. J. Costello, and K. S. Hutson (2016), Cleaner fishes and shrimp diversity and a re-evaluation of cleaning symbioses. Fish and Fisheries, 00: 1–19.

Wicksten, M.K. 1995a. Associations of fishes and their cleaners on coral reefs of Bonaire, Netherlands Antilles,

Copeia. 1995: 477-481.

Wicksten, M.K. 1995b. Behaviour of cleaners and their client fishes at Bonaire, Netherlands Antilles. J. of Natural History. 32: 13-30.

Wicksten, M.K. 2009. Interactions with fishes of five species of *Lysmata* (Decapoda, Caridea, Lysmatidae). Crustaceana, 82 (9): 1213-1223.

Williams, E.H. and L.B. Williams. 1980. Four new species of *Renocila* (Isopoda: Cymothoidae), The first reported from the New World. Proc. Biol. Soc. Wash. 93(3): 573-592.

Williams, L.B. and E.H. Williams. 1981. Nine new species of *Anilocra* (Crustacea: Isopoda: Cymothoidae) external parasites of West Indian coral reef fishes. Proc. Biol. Soc. Wash. 94(4): 1005-1047.

Williams, E.H., L.B. Williams, R.E. Waldner and J.J. Kimmel. 1982. Predisposition of a Pomacentrid fish, *Chromis multilineatus* (Guichenot) to parasitism by a Cymothoid Isopod, Anilocra chromis. J. Parasito. 68(5): 942-945.

Williams, E.H. and L.B. Williams. 1985. Brood pouch release of *Anilocra chromis* (Isopods, Cymothoidae) a parasite of brown chromis, *Chromis multilineatus* (Guichenot) in the Caribbean. Crustaceana 49(1): 92-95.

Williams, L.B. and E.H. Williams. 1989. Hangers-on. Natural History 1: 40.

Williams, L.B. and E.H. Williams 1998. Ability of Pederson cleaner shrimp to remove juveniles of the parasitic cymothoid isopod, *Anilocra haemuli,* from the host. Crustaceana 71 (8): 862-869.

White, J.W., C.J. Grigby and R.R. Warner. 2007. Cleaning behavior is riskier and less profitable than an alternative strategy for a facultative cleaner fish. Coral Reefs. 26: (1): 87-94.

Senses & Sound Communication

Hawkins, A.D. 1981. The hearing ability of fish. "Hearing and Sound Communication in Fishes" (W.N. Tavolga, A.N. Popper and R.R. Fay, eds.) Springer - Verlag.

Kasumyan, A.O. 2003. The lateral line in fish: structure, function, and role in behavior. J. Ichthyol. 43 (Sup. 2): S175-S213.

Kenyon, T.N. 1994. The significance of sound interception to males of the bicolor damselfish, *Pomacentrus partitus,* during courtship. Env. Biol. Fish. 40: 391-405.

Myberg, A.A. 1981. Sound communication and interception in fishes. "Hearing and Sound Communication in Fishes" (W.N. Tavolga, A.N. Popper and R.R. Fay, eds.) Springer - Verlag.

Partridge, B.L. 1981. Lateral line function and the internal dynamics of fish schools. "Hearing and Sound Communication in Fishes" (W.N. Tavolga, A.N. Popper and R.R. Fay, eds.) Springer–Verlag.

Platt, C. and A.N. Popper. 1981. Fine structure and function of the ear. "Hearing and Sound Communication in Fishes" (W.N. Tavolga, A.N. Popper and R.R. Fay, eds.) Springer–Verlag.

Marine Wildlife Management

Alcala, A.C. and E.D. Gomez. 1987. Dynamiting coral reefs; a resource destructive fishing method. "Human Impact on Coral Reefs: Facts and Recommendations" (B. Salvat, ed.) Antenne Museum E.P.H.E., French Polynesia: 51-60.

Alcala, A.C., G.R. Russ. 2006. No-take marine reserves and reef fisheries management in the Philippines. Ambio. 35 (5): 245-254.

Appeldoorn, R.S. and K.C. Lindeman. 2003. A Caribbean-wide survey of marine reserves: Spatial coverage and attributes of effectiveness. Gulf and Caribbean Research. 14(2): 139-154.

Ault, J.S., S.G. Smith, J.A. Browder, et al. 2014. Indicators for assessing the ecological dynamics and sustainability of southern Florida's coral reef and coastal fisheries. Ecological Indicators 44: 164-172.

Ballantine, W.J. 1997. Design principles for systems of "no-take" marine reserves. Paper for workshop on: the design and monitoring of marine reserves. Fisheries Center, Univ. of British Columbia, Vancouver.

Ballantine, B. 2014. Fifty years on: Lessons from marine reserves in New Zealand and principles for a worldwide network. Biological Conservation 176: 297-307.

Birkeland, C. 2017. Working with, not against, coral-reef fisheries. Coral Reefs 36(1): 1-11.

Berumen, M.L., G.R. Almany, S. Planes, et al. 2012. Persistence of self-recruitment and patterns of larval connectivity in a marine protected area network. Ecology and Evolution 2(2): 444-452.

Bohnsack, J.A. 1994. How marine fishery reserves can improve reef fisheries. Proc. Gulf Carib. Fish. Inst. 43: 217-241.

Bohnsack, J.A., A. Eklund and A.M. Szmant. 1997. Artificial reef research: Is there more than the attraction-production issue? Fisheries. 22: 14-16.

Bohnsack, J.A. 1999. Incorporating no-take marine reserves into precautionary management and stock assessment. NOAA Tech. Memo NMFS-F/SPO-40: 8-16.

Bohnsack, J.A. 2000. A comparison of the short term impacts of no-take marine reserves and minimum size limits. Bull. Mar. Sci. 66: 615-650.

Bohnsack, J.A. 2003. Shifting baselines, marine reserves, and Leopold's Biotic ethic. Gulf and Caribbean Research 14(2): 1-7.

Bohnsack, J.A., J.S. Ault and B. Causey. 2004. Why have no-take marine protected areas? American Fishery Society Symposium 42: 185-193.

Bohnsack, J.A., D.E. Harper, D.B. McClellan, et al. 2010. Coral reef fish response to FKNMS management zones: the first ten years (1997-2007). Progress report to the Florida Keys National Marine Sanctuary. PRBD 08/09-10.

Bohnsack, J.A. 2011. Impacts of coastal protected areas on recreational world records for spotted seatrout, red drum, black drum and common snook. Bull. Mar. Sci. 87(4): 939-970.

Bohnsack, J.A. 2015. A comparison of advantages and disadvantages of fisheries management and marine reserves for reef protection along the Gold and Treasure Coasts of southeastern Florida. Presentation to the Coastal Ocean Task Force by J.A. Bohnsack, Ph.D.

Buxton, C.D., K. Hartmann, et al 2014. When is spillover from marine reserves likely to benefit fisheries? PlosOne, doi.org/10.1371/journal.pone.0107032.

Coleman, F.C., C.C. Koenig and L.A. Collins. 1996. Reproductive styles of shallow-water grouper (Pisces: Serranidae) in the eastern Gulf of Mexico and the consequences of fishing spawning aggregations. Env. Biol. Fish. 47: 129-141.

Coleman, F.C., C.C. Koenig, G.R. Huntsman, et al. 2000. American Fsheries Society Policy Statement. Long-lived reef fishes: The grouper-snapper complex. American Fisheries Society Policy Statement. Fisheries 25(3): 14-21.

Cowen, R.K. and S. Sponaugle. 2009. Larval dispersal and marine population connectivity. Annu. Rev. Marine Sci. 1: 443-466.

Dieseldorff, P.G., W.D. Heyman and J. Azueta. 2013. History and co-management of the artisnal mutton snapper (*Lutjanus analis*) spawning aggregation fishery at Gladden Spit, Belize, 1950-2011. Fisheries Research, 147: 213-221.

Emslie, M.J., M. Logan, et al. 2015. Expectations and outcomes of reserve network performance following re-zoning of the Great Barrier Reef Marine Park. Current Biology. 25(8): 983-992.

Emslie, M.J. 2015. Twice the coral trout in Great Barrier Reef protected zones. https://eatlas.org.au/nerp-te/gbr-aims-effects-of-zoning.

Gilmore, R.G., and R.S. Jones. 1992. Color variation and associated behavior in the epinepheline groupers *Mycteroperca microlepis* (Goode and Bean) and *M. phenax* (Jordan and Swain). Bull. Mar. Sci. 51: 83-103.

Green, A. and P. Lokani. 2004. Designing a resilient network of marine protected areas in Kimbe Bay, West New Britain, Papua New Guinea. TNC Pacific Island Countries Report No. 11/04

Green, A., A. White, S. Kilarski (Eds.) 2013. Designing marine protected area networks to achieve fisheries, biodiversity, and climate change objectives in tropical ecosystems: A practitioner guide. The Nature Conservancy, and the USAID Coral Triangle Support Partnership, Cebu City, Philippines. viii + 35 pp.

Halpern, B.S., S.E. Lester and J.B. Kellner. 2010. Spillover from marine reserves and the replenishment of fished stocks. Environmental Conservation 36 (4): 268-276.

Harrison, H.B., D.H. Williams, R.D. Evans, et al. 2012. Larval export from marine reserves and the recruitment benefit for fish and fisheries. Current Biology, Vol. 22. 22(11): R444-R446.

Hixon, M.A. and J.P. Beets. 1993. Predation, prey refuges, and the structure of coral-reef fish assemblages. Ecol Monogr. 63: 77-101.

Hock, K., N.H. Wolff, J.C. Ortiz, S.A.Condie, K.R.N. Anthony, P.G. Blackwell, et al. 2017. Connectivity and systemic resilience of the Great Barrier Reef. PLoS Biol 15(11): e2003355. https://doi.org/10.1371/journal.pbio.2003355

Huntsman, G.R. and W.E. Schaaf. 1994. Simulation of the impact of fishing on reproduction of a protogynous grouper, the graysby. N. Amer. J. of Fish. Management. 14: 41-52.

Jennings, S. and N.V.C. Polunin. 1996. Impacts of fishing on tropical reef ecosystems. Ambio. 25: 44-49.

Lindberg, W.J. and J.L. Loftin. 1998. Effects of artificial reef characteristics and fishing mortality on gag (*Mycteroperca microlepis*) productivity and reef fish community structure. Florida Dept. of Environmental Protection.

Manooch, C.S. III. 1987. Age and growth of snappers and groupers. "Tropical Snappers and Groupers: Biology and Fisheries Management." (J.J. Polovina and S. Ralston, eds.) Boulder, CO: Westview Press.

Marine Reserves in the United States. 2014. National Marine Protected Areas Center. marineprotectedareas.noaa.gov.

McClanahan,T.R. 1988. Coexistence in a sea urchin guild and its implications to coral reef diversity and degradation. Oecolgia. 77: 210-218.

McClanahan, T.R. 1992. Resource utilization, competition and predation: a model and example from coral reef grazers. Ecol. Model. 61: 195-215.

McClanahan, T.R. and N.A. Muthiga. 1988. Changes in Kenyan coral reef community structure and function due to exploitation. Hydrobiology. 166: 269-276.

McCook, L.J., G.R. Almany, M.L. Berumen, et al. 2009. Management under uncertainty: guide-lines for incorporating connectivity into the protection of coral reefs. Coral Reefs 28(2): 353-366.

Ogden, J. C. 1997. Marine managers look upstream for connections. Science. 278: 1414-1415.

Pauly, D., V. Christensen, J. Dalsgaard, R. Froese, and T. Francisco. 1998. Fishing down marine food webs. Science. 279: 860-863.

Perez, V.B. and A. Tewfik. 2016. Brief history of management and conservation of Nassau grouper and their spawning aggregations in Belize: A collaborative approach. Conference: Gulf and Caribbean Fisheries Institute, Vol. 68.

Sale, P.F., H. Van Lavieren, M.C. Ablan Lagman, et al. 2010. Preserving reef connectivity: A handbook for marine protected area managers. Connectivity Working Group, Coral Reef Targeted Research & Capacity Building for Management Program, UNU-INWEH.

Secretariat of the Convention on Biological Diversity. 2004. Technical advice on the establishment and management of a national system of marine and coastal protected areas, SCBD, 40 pages (CBD Technical Series no. 13).

Polovina, J.J. and I. Sakai. 1989. Impacts of artificial reefs on fishery production in Shimarmaki, Japan. Bull Mar. Sci. 44: 997-1,003.

Roberts, C.M. 1997. Connectivity and management of Caribbean coral reefs. Science. 278: 1454-1456.

Roberts, C.M., J.A. Bohnsak, et al. 2001. Effects of marine reserves on adjacent fisheries. Science. 294 (5548): 1920-1923.

Roberts, C.M., J.P. Hawkins and F.R. Gell. 2005. The role of marine reserves in achieving sustainable fisheries. Philos Trans. R. Soc. Lond. B Bilo. Sci. 360(1453): 123-132.

Russ, G.R. and A.C. Alcala. 1998. Natural fishing experiments in marine reserves 1983-1993: community and trophic responses. Coral Reefs. 17: 383-397.

Russ, G.R. and A.C. Alcala. 1998. Natural fishing experiments in marine reserves 1983-1993: roles of life history and fishing intensity in family response. Coral Reef. 17: 399-416.

Sadovy, Y. 1994a. Summary of papers presented at the grouper symposium of the 43rd annual meeting of the Gulf and Caribbean Fisheries Institute in Miami, Florida (Nov. 1990). Proc. Gulf Carib. Fish. Inst. 43: 482-484.

Sadovy, Y. 1994b. Grouper stocks of the western central Atlantic: the need for management and management needs. Proc. Gulf Carib. Fish. Inst. 43: 43-64.

Sadovy, Y. and A. Eklund. 1999. Synopsis of biological information on *Epinephelus striatus* (Bloch, 1792), the Nassau grouper and *E. Itajara* (Lichtenstein, 1822) the jewfish. NOAA Tech. Rep. Ser.

Sadovy de Mitcheson, Y. and P.L. Colin (eds) 2012. Reef fish spawning aggregations: biology, research and management, Fish and Fisheries Series Vol. 35.

Sherman, K.D., C.P. Dahlgren, J.R. Stevens and C.R. Tyler. 2016. Integrating population biology into conservation management for endangered Nassau grouper *Epinephelus striatus*. Mar. Ecol. Prog. Ser. 554: 263-280.

Sluka, R., M. Chiappone, K.M. Sullivan and R. Wright. 1996. Habitat and life in the Exuma Cays, the Bahamas: the status of grouper and coral reefs in the northern cays. The Nature Conservancy, Florida and Caribbean Marine Conservation Science Center. Nassau, Bahamas: Media Publishing Ltd.,

Williams, N. 1998. Overfishing disrupts entire ecosystems. Science. 279: 809.

Thresher, R.E. 1984. "Reproduction in Reef Fishes" Neptune City, NJ: T.F.H.

Angelfishes

Blasiola, G.C. 1976. *Centropyge aurantonotus* Burgess, 1974 (Pisces: Chaetodontidae): range, extension, and redescription. Bull. Mar. Sci. 26: 564-568.

Bruce, R.W. 1980. Protogynous hermaphroditism in two marine angelfishes. Copeia. 2: 353-355.

Clarke, R.D. 1977. Habitat distribution and species diversity of Chaetodontid and Pomacentrid fishes near Bimini, Bahamas. Mar. Biol. 40: 277-289.

Hourigan, T.F. and C.D. Kelley. 1985. Histology of the gonads and observations on the social behavior of the Caribbean angelfish *Holacanthus tricolor.* Mar. Biol. 88: 311-322.

Hourigan, T.F., F.G. Stanton, P.J. Motta, C.D. Kelley and B. Carlson. 1989. The feeding ecology of three species of Caribbean angelfishes (family Pomacanthidae). Env. Biol. of Fish. 24: 105-116.

Moyer, J.T., R.E. Thresher and P.L. Colin. 1983. Courtship, spawning and inferred social organization of American angelfishes (Genera *Pomacnathus, Holacanthus* and *Centropyge;* Pomacanthidae). Env Biol. of Fishes. 9: 25-39.

Neudecker, S. and P.S. Lobel. 1981. Mating systems of Chaetodintid and Pomacanthid fishes at St. Croix. Z. Tierpsychol. 59: 299-318.

Randall, J.E. 1967. Food habits of reef fishes of the West Indies. Stud. Trop. Ocean. 5: 665-847.

Thresher, R.E. 1984. "Reproduction in Reef Fishes" Neptune City, NJ: T.F.H.Publications.

Basslets

Asoh, K. and T. Yoshikawa. 1996. Nesting behavior, male parental care, and embryonic development in the fairy basslet, *Gramma loreto.* Copeia. 1-8.

Asoh, K. and D.Y. Shapiro. 1997. Bisexual juvenile gonad and gonochorism in the fairy basslet, *Gramma loreto.* Copeia. 22-31.

Freeman, S. and W. Alevizon. 1983. Aspects of territorial behavior and habitat distribution of the fairy basslet, *Gramma loreto.* Copeia. 3: 829-832.

Thresher, R.E. 1980. "Reef fish: behavior and ecology on the reef and in the aquarium." St. Petersburg, FL: Palmetto Publ. Co

Blennies

Buchheim, J.R. and M.A. Hixon. 1992. Competition for shelter holes in the coral-reef fish Acanthemblemaria spinosa Metzelaar. J. Exp. Mar. Biol. Ecol. 164: 45-54.

Clarke, R.D. 1994. Habitat partitioning by chaenopsid blennies in Belize and the Virgin Islands. Copeia. 2003(2): 398-405.

Clarke, R.D. and J.C. Tyler. 2003. Differential space utilization by male and female spinyhead blennies, *Acanthemblemaria spinosa* (Teleostei: Chaenopsidae). Copeia 2003(2): 241-247.

Colin, P.L. and M.R. Gomon. 1973. Notes on the behavior, ecology and distribution of *Lucayablennius zingaro* (Pisces: Clinidae). Carib. J. Sci. 13: 56-61.

Côté, I.M. and W. Hunte. 1989. Male and female mate choice in the redlip blenny: why bigger is better. Anim. Behav. 38: 78-88.

Gibran, F.Z., F. Borges Santos, et al. 2004. Courtship behavior and spawning of the hairy blenny *Labrisomus nuchipinnis* (Labrisomidae) in southeastern Brazil. Neot. Ichthyology, 2(3): 163-166.

Greenfield, D.W. and T.A. Greenfield. 1982. Habitat and resource partitioning between two species of *Acanthemblemaria* (Pisces: Chaenopsidae), with comments on the chaos hypothesis. Smithsonian Cont. Mar. Sci. 12: 499-507.

References

Greenfield, D.W. and R.K. Johnson. 1981. The blennioid fishes of Belize and Honduras, Central America, with comments on their systematics, ecology and distribution (Blenniidae, Chaenopsidae, Clinidae, Tripterygiidae). Fieldiana, Zool. N.S. 8: 1-106.

Greenfield, D.W. and R.K. Johnson. 1990. Community structure of western Caribbean blennioid fishes. Copeia. 2: 433-438.

Hunte, W. and I.M. Côté. 1989. Recruitment in the redlip blenny, *Ophioblennius atlanticus:* is space limited? Coral Reefs. 8: 45-50.

Labelle, M. and J.R. Nursall. 1985. Some aspects of the early life history of the redlip blenny, *Ophioblennius atlanticus* (Teleostei: Blenniidae). Copeia. 1: 39-49.

Marraro, C.H. 1978. Some aspects of the life history and behavioral ecology of the redlip blenny (*Ophioblennius atlanticus*) (Pisces: Blenniidae). Master's Thesis, Dept. Zoology, Univ. Alberta, Edmonton, Canada. Unpublished.

Marraro, C.H. and J.R. Nursall. 1983. The reproductive periodicity and behaviour of *Ophioblennius atlanticus* (Pisces: Blenniidae) at Barbados. Can. J. Zool. 61: 317-325.

Nursall, J.R. 1977. Territoriality in redlip blennies (*Ophioblennius atlanticus* - Pisces: Blenniidae).J. Zool. (London). 182: 205-223.

Robins, C.R. and J.E. Randall. 1965. Three new western Atlantic fishes of the Blennioid genus *Chaenopsis,* with notes on the related *Lucayablennius zingaro.* Proc. Acad. Nat. Sci. Philadelphia. 117(6): 213-237.

Smith-Vaniz, W.F. and F.J. Palacio. 1974. Atlantic fishes of the genus *Acanthemblemaria,* with descriptions of three new species and comments on Pacific species (Clinidae: Chaenopsidae). Proc. Acad. Natl. Sci. Philadelphia. 125: 197-224.

Stephens, J.S. Jr., E.S. Hobson and R.K. Johnson. 1966. Notes on distribution, behavior and morphological variation in some chaenopsid fishes from the tropical eastern Pacific, with descriptions of two new species, *Acanthemblemaria castroi* and *Corallozetus springeri.* Copeia. 424-438.

Victor, Benjamin C. 2019. *Enneanectes flavus,* a new endemic species of triplefin blenny from the southeastern Caribbean (Telostei: Tripterygiidae). Journal of the Ocean Science Foundation, 32, 1-16.

Butterflyfishes

Birkeland, C. and S. Neudecker. 1981. Foraging behavior of two Caribbean Chaetondontids: *Chaetodon capistratus* and *C. aculeatus.* Copeia 1981: 169-178.

Clavijo, I.E. 1985. A probable hybrid butterflyfish from the western Atlantic. Copeia. 1985: 235-238.

Colin, P.L. 1989. Aspects of the spawning of western Atlantic butterflyfishes (Pisces: Chaetodontidae). Env. Biol. Fishes. 25: 131-141.

Colin, P.L. and I.E. Clavijo. 1988. Spawning activity of fishes producing pelagic eggs on a shelf edge coral reef, southwestern Puerto Rico. Bull. Mar. Sci. 43: 249-279.

Fricke, H.A. 1986. Pair swimming and mutual partner guarding in monogamous butterflyfish (Pisces, Chaetodontidae): a joint advertisement for territory. Ethology 73: 307-333.

Gore, M.A. 1982. The effect of a flexible spacing system on the social organization of coral reef fish, *Chaetodon capistratus.* Behaviour 85: 118-148.

Lasker, H.R. 1985. Prey preferences and browsing pressure of the butterflyfish *Chaetodon capistratus* on Caribbean gorgonians. Mar. Ecol. Prog. Ser. 21: 213-220.

Neudecker, S. and P.S. Lobel. 1982. Mating systems of chaetodontid and pomacanthid fishes at St. Croix. Z. Tierpsychol. 59: 299-318.

Pitts, P.A. 1991. Comparative use of food and space by three Bahamian butterflyfishes. Bull. Mar. Sci. 48(3): 749-756.

Randall, J.E. 1967. Food habits of reef fishes of the West Indies. Stud. Trop. Ocean. 5: 665-847.

Reese, E.S. 1989. Orientation behavior of butterflyfishes (family Chaetodontidae) on coral reefs: spatial learning of route specific landmarks and cognitive maps. Env. Biol. Fishes. 25: 79-86.

Thresher, R. E. 1984. "Reproduction in Reef Fishes" Neptune City, NJ: T.F.H. Publications.

Damselfishes

Albrecht, H. 1969. Behaviour of four species of Atlantic damselfishes from Colombia, South America (*Abudefduf saxatilis, A. taurus, Chromis multilineata, C. Cyanea;* Pisces, Pomacentridae). Z. Tierpsychol. 26: 662-676.

Axline-Minotti, B.A. 2003. The role of threespot damselfish (Stegastes planifrons) as a keystone species in a Bahamian path reef. Thesis: for Environmental Studies, Ohio University.

Chaves, Laís de Carvalho Teixeira, C. Gustavo A. Ormond, MiGinty, E.S. and B. Padovani Ferrira. 2012. Space partitioning among damselfishes in the Caribbean coast of Panama: the role of habitat preferences. Neotropical Ichthyology, 10(3): 633-642.

Cummings, W.C. 1968. Reproduction habits of the sergeant major, *Abudefduf saxatilis* (Pisces, Pomacentridae), with comparative notes on four other damselfishes in the Bahama Islands. Ph.D. Dissertation Univ. of Miami, Miami, FL.

de Boer, B.A. 1978. Factors influencing the distribution of the damselfish *Chromis cyanea* (Poey), Pomacentridae, on a reef at Curacao, Netherlands Antilles. Bull. Mar. Sci. 28: 550-565.

Draud, M., D.E. Itzkowitz and M. Itzkowitz. 1990. Co-defense of territory space by two species of coral reef fishes. Bull. Mar. Sci. 47: 721-724.

Gronell, Ann. 1980. Space utilization by the cocoa damselfish, *Eupomacentrus variabilis* (Pisces: Pomacentridae). Bull. Mar. Sci. 30: 237-251.

Itzkowitz, M. 1977. Group organization of a territorial damselfish, *Eupomacentrus planifrons.* Behaviour 10: 125-137.

Itzkowitz, M. 1985. Aspects of the population dynamics and reproductive success in the permanently territorial beaugregory damselfish. Mar. Behav. Physiol. 12: 57-69.

Itzkowitz, M. and T. Koch. 1991. Relationship between damselfish egg loss and brittlestars. Bull. Mar. Sci. 48: 164-167.

Kaufman, L. 1977. The threespot damselfish: effects on benthic biota of Caribbean coral reefs. Proc. 3rd Intl. Coral Reef Symp., 559–564.

Kaufman, L. 1981. There was biological disturbance on Pleistocene coral reefs. Paleobiology, 7(4): 527-532.

Knapp, R.A. 1993. The influence of egg survivorship on the subsequent nest fidelity of female bicolour damselfish, *Stegastes partitus*. Anim. Behav. 46: 111-121.

Knapp, R.A. and R.R. Warner. 1991. Male parental care and female choice in the bicolor damselfish, *Stegastes partitus:* bigger is not always better. Anim. Behav. 41: 747-756.

Knapp, R.A., P.C. Sikkel and V.T. Vredenburg. 1995. Age of clutches in nests and the within-nest spawning-site preferences of three damselfish species (Pomacentridae). Copeia. 78-88.

Lobel, P.S. 1980. Herbivory by damselfishes and their role in coral reef community ecology. Bull. Mar. Sci. 30: 273-289.

Mahoney, B.M. 1981. An examination of interspecific territoriality in the dusky damselfish, *Eupomacentrus dorsopunicans* Poey. Bull. Mar. Sci. 31: 141-146.

McDonald, G.D. 1993. Reproductive behavior and social dynamics of the yellowtail damselfish, *Microspathodon chrysurus.* Master's Thesis, Univ. of Puerto Rico.

Myrberg, A.A. 1972. Ethnology of the bicolor damselfish *Eupomacentrus partitus:* a comparative analysis of laboratory and field behaviour. Anim. Behav. Monogr. 5: 197-283.

Myrberg, A.A. 1972. Social dominance and territoriality in the bicolor damselfish, *Eupomacentrus partitus.* Behaviour. 4: 207-231.

Myrberg, A.A., B.D. Brahy and A.R. Emery. 1967. Field observations on reproduction of the damselfish, *Chromis multilineata* (Pomacentridae), with additional notes on general behavior. Copeia. 4: 819-827.

Myrberg, A.A., M. Mohler and J.D. Catala. 1986. Sound production by males of a coral reef fish (*Pomacentrus partitus*): its significance to females. Anim. Behav. 34: 913-923.

Peterson, C.W. 1990. The occurrence and dynamics of clutch loss and filial cannibalism in two Caribbean damselfishes. J. Exp. Mar. Biol. Ecol. 1990: 117-133.

Petersen, C.W. and H.C. Hess. 1991. The adaptive significance of spawning synchronization in the Caribbean damselfish, *Stegastes dorsopunicans* (Poey). J. Exp. Mar. Biol. Ecol. 151: 155-167.

Pressley, P.H. 1980. Lunar periodicity in the spawning of yellowtail damselfish, *Microspathodon chrysurus.* Env. Biol. Fish. 5: 153-159.

Robertson, D.R. 1984. Cohabitation of competing territorial damselfishes on a Caribbean coral reef. Ecology. 65: 1121-1135.

Robertson, D.R., C.W. Petersen and J.D. Brawn. 1990. Lunar reproductive cycles of benthic-brooding reef fishes: reflections of larval biology or adult biology? Ecol. Soc. Of America. 60: 311-329.

Schmale, M.C. 1981. Sexual selection and reproductive success in male of the bicolor damselfish, *Eupomacentrus partitus* (Pisces: Pomacentridae). Anim. Behav. 29: 1172-1184.

Schopmeyer, S.A and D. Lirman. 2015. Occupation Dynamics and Impacts of Damselfish Territoriality on Recovering Populations of the Threatened Staghorn Coral, *Acropora cervicornis.* PLoS One. 2015, 10(11): e0141302.

Thresher, R.E. 1984. "Reproduction in Reef Fishes" Neptune City, NJ: T.F.H. Publications.

Waldner, R.E. and D.R. Robertson. 1980. Patterns of habitat partitioning by eight species of territorial Caribbean damselfishes (Pisces: Pomacentridae). Bull. Mar. Sci. 30: 171-186.

Wellington G.M. and B.C. Victor. 1989. Planktonic larval duration of one hundred species of Pacific and Atlantic damselfishes (Pomacentiedae). Mar. Biol. 101: 557-567.

Williams, A.H. 1978. Ecology of threespot damselfish: social organization, age structure, and population stability. J. Exp. Mar. Biol. Ecol. 34: 197-213.

Dragonets

Flounders

Konstantinou, H. and D.C. Shen. 1995. The social and reproductive behavior of the eyed flounder, *Bothus ocelltus,* with notes on the spawning of *Bothus lunatus* and *Bothus ellipticus.* Env. Biol. Fish. 44: 311-324.

Moyle, P.B. and J.J.Cech, Jr. 1996. Fishes — An Introduction to Ichthyology. 3rd Ed. New Jersey: Prentice Hall.

Norman, J.R. 1934. A systematic monograph of the flatfishes (Heterosomata). London: Trustees of the British Museum. 459.

Ramachandran, V.S., C.W. Tyler, R.L. Gregory, D. Rogers-Ramachandran, S. Duensing, C. Pillsbury and C. Ramachandran. 1996. Rapid adaptive camouflage in tropical flounders. Nature. 379: 815-818.

Frogfishes

Michael, S.W. 1998. "Reef Fishes Vol. I" Shelburne, VT: Microcosm Ltd.

Pietch, T.W. and D.B. Grobeker. 1987. "Frogfishes of the World." Stanford, CA: Stanford Univ. Press.

Theodore, W.P. and D.B. Grobeker. 1990. Frogfishes. Scien. Amer. June: 96-103.

Zuberbühler, T. 2014. Frogfishes Southeast Asia, Maldives, Red Sea.http://www.critter.ch/frogfish-book.html.

References

Groupers & Sea Basses

Aguilar-Perera, A. 1994. Preliminary observations of the spawning aggregation of Nassau grouper, *Epinephelus striatus,* at Majahual, Quintana Roo, Mexico. Proc. Gulf Carib. Fish. Inst. 43: 112-122.

Aguilar-Perera, A. and W. Aguilar-Dávila. 1996. A spawning aggregation of Nassau grouper, *Epinephelus striatus* (Pisces: Serranidae) in the Mexican Caribbean. Env. Biol. Fish. 45: 351-361.

Anderson, M. 2017. Similtaneous hermaphrodites: Understanding speciation in fish called "hamlets". Smithsonian Insider. https://insider.si.edu/2017/03/qa-oscar-puebla-understanding-speciation-coral-reef-hamlets/

Archer, S.K., S.A. Heppell, B.X. Semmens, et al. 2012. Patterns of color phase indicate spawn timing at a Nassau group *Epinephelus striatus* spawning aggregation. Current Zoology. 58(1): 73-83.

Bannerot, S., W.W. Fox, Jr., and J.E. Powers. 1987. Reproductive strategies and the management of snappers and groupers in the Gulf of Mexico and Caribbean. "Tropical snappers and groupers: biology and fisheries management." (J.J. Polovina and S. Ralston, eds.) Boulder, CO: Westview Press. 561-603.

Barlow, G.W. 1975. On the sociobiology of some hermaphroditic serranid fishes, the hamlets, in Puerto Rico. Mar. Biol. 33: 295-300.

Beets, J. and A. Friedlander. 1992. Stock analysis and management strategies for red hind, *Epinephelus guttatus,* in the U.S. Virgin Islands. Proc. Gulf Carib. Fish. Inst. 42: 66-80.

Bohnsack, J.A. 1994. How marine fishery reserves can improve reef fisheries. Proc. Gulf Carib. Fish. Inst. 43: 217-241.

Carter, J., G.J. Marrow and V. Pryor. 1994. Aspects of the ecology and reproduction of Nassau grouper (*Epinephelus striatus*) off the coast of Belize, Central America. Proc. Gulf Carib. Fish. Inst. 43: 65-111.

Clark, J.A., W.E. Farrell and W.R. Peltier. 1978. Global changes in postglacial sea level: a numerical calculation. Quat. Res. 9: 265-278.

Colin, P.L. 1992. Reproduction of the Nassau grouper, *Epinephelus striatus* (Pisces: Serranidae) and its relationship to environmental conditions. Env. Biol. Fish. 34: 357-377.

Colin, P.L. 1994. Preliminary investigations of reproductive activity of the jewfish, *Epinephelus itajara.* (Pisces: Serranidae). Proc. Gulf Carib. Fish. Inst. 43: 138-147.

Colin, P.L., D.Y. Shapiro and D. Weiler. 1987. Aspects of the reproduction of two species of grouper, *Epinephelus guttatus* and E. striatus, in the West Indies. Bull. Mar. Sci. 40: 220-230.

Colin, P.L. 1994. Aggregation and spawning in the tiger grouper, *Mycteroperca tigris.* Copia. 1994(2) 511-516.

Dahlgren, C.P., K. Buch, E. Rechisky and M.A. Hixon. 2016. Multiyear tracking of Nassau grouper spawning migrations. Marine and Coastal Fisheries. (8)1: 522-535.

Domeier, M.L. 1994. Speciation in the serranid fish *Hypoplectrus.* Bull. Mar. Sci. 54(1): 103-141.

Domeier, M.L. and P.L. Colin. 1997. Tropical reef fish spawning aggregations: defined and reviewed. Bull. Mar. Sci. 60(3): 698-726.

Eggleston, D.B. 1995. Recruitment in Nassau grouper, *Epinephelus striatus*: post-settlement abundance, microhabitat features, and ontogenetic habitat shifts. Mar. Ecol. Prog. Ser. 124: 9-22.

Ellis, R., C. Koenig and F. Coleman. 2013. Spawning-related movement patterns of Goliath Grouper (*Epinephelus itajara*) off the Atlantic coast of Florida. Proceedings of the 66th Gulf and Caribbean Fisheries Institute.

Fine, J.C. 1990. Groupers in love: spawning aggregations of Nassau groupers in Honduras. Sea Frontiers. 36: 42-45.

Fischer, E.A. 1980a. The relationship between mating system and simultaneous hermaphroditism in the coral reef fish, *Hypoplectrus nigricans* (Serranidae). Anim. Behav. 28: 620-633.

Fischer, E.A. 1980b. Speciation in the hamlet (*Hypoplectrus*: Serranidae) - A continuing enigma. Copeia. 4: 649-659.

Fischer, E.A. 1981. Sexual allocation in a simultaneously hermaphroditic reef fish, *Hypoplectrus nigricans.* Amer. Nat. 117: 64-82.

Fischer, E.A. 1984a. Egg trading in the chalk bass, *Serranus tortugarum,* a simultaneous hermaphrodite. Z. Tierpsychol. 66: 143-151.

Fischer, E.A. 1984b. Local mate competition and sex allocation in simultaneous hermaphrodites. Am. Nat. 124: 590-596.

Fischer, E.A. and C.W. Petersen. 1986. Social behavior of males and simultaneous hermaphrodites in the lantern bass. Ethnology. 73: 235-246.

Fischer, E.A. and C.W. Petersen. 1987. The evolution of sexual patterns in the seabasses. BioScience. 37: 482-489.

Frias–Torres, S. 2013. Should the critically endangered Goliath grouper *Epinephelus itajara* be culled in Florida? Oryx, 47(1): 88-95.

García-Moliner, G.E. 1986. Aspects of the social spacing, reproduction and sex reversal in the red hind, *Epinephelus guttatus.* Master's Thesis, Univ. of Puerto Rico, Mayaguez.

García-Moliner, G.E. and Y. Sadovy. 2007. The Case for Regional Management of the Nassau grouper, *Epinephelus striatus.* Proceedings of the 60th Gulf and Caribbean Fisheries Institute.

Gilmore, R.G., and R.S. Jones. 1992. Color variation and associated behavior in the epinepheline groupers *Mycteroperca microlepis* (Goode and Bean) and M. phenax (Jordan and Swain). Bull. Mar. Sci. 51: 83-103.

Golden, S.K. 2000. Long-distance movement of a Nassau grouper (*Epinephelus striatus*) to a spawning aggregation in the central Bahamas. NOAA Fish. Bull. 98(3): 642-645.

Grace, M., K.R. Rademacher and M. Russell. 1994. Pictorial guide to the groupers (Teleostei: Serranidae) of the Western North Atlantic. NOAA Tech. Report NMFS 118.

Graves, J.E. and R.H. Rosenblatt. 1980. Genetic relationships of the color morphs of the serranid fish *Hypoplectrus unicolor.* Evolution. 34: 240-245.

Grover, J.J. 1993. Trophic ecology of pelagic early-juvenile Nassau grouper, *Epinephelus striatus*, during and early phase of recruitment into demersal habitats. Bull. Mar. Sci. 53(3): 1117-1125.

Heyman, W.D. and B. Kjerfve. 2008. Characterization of transient multi-species reef fish spawning aggregation at Gladden Spit, Belize. Bull. of Mar. Sci. 83(3): 531-551.

Huntsman, G.R. and W.E. Schaaf. 1994. Simulation of the impact of fishing on reproduction of a protogynous grouper, the graysby. N. Amer. J. of Fish. Management 14: 41-52.

Johnson, A.G., and L.A. Collins. 1994. Age-size structure of red grouper (*Epinephelus morio*), from the eastern Gulf of Mexico. Northeast Gulf of Mexico. Northeast Gulf Science. 13(2): 101-106.

Koenig, C., F.C. Coleman and A. Eklund. 1998. Studies of the jewfish (*Epinephelus itajara*) in the eastern Gulf of Mexico. Report to The Curtis and Edith Munson Foundation.

Koenig, C., F.C. Coleman and K. Kingon. 2011. Pattern of recovery of the Goliath grouper *Epinephelus itajara* population in the southeastern US. Bull. Mar. Sci. 87(4): 891-911.

Koenig, C., L.S. Bueno, F.C. Coleman, et al. 2016. Diel, lunar, and seasonal spawning patterns of the Atlantic goliath group, *Epinephelus itajara*, off Florida, United States.

Leonard, J.L. 1993. Sexual conflict in simultaneous hermaphrodites: Evidence from serranid fishes. Env. Biol. Fish. 36: 135-148.

Lobel, P.S. 1992. Sounds produced by spawning fishes. Env. Biol. Fish. 33: 351-358.

Peterson, C.W. 2006. Sexual selection and reproductive success in hermaphroditic seabasses. Integrative and Comparative Biology, 46(4):439-448.

Lobel, P.S. 2011. A review of the Caribbean hamlets (Serranidae, *Hypoplectrus*) with description of two new species. Zootaxa 3096: 1-17.

Mann, D.A., J.V. Locascio, F.C. Coleman and C. Koenig. 2008. Goliath grouper *Epinephelus itajara* sound production and movement patterns on aggregation sites. Endang. Species Res. 7: 229-236.

Manooch, C.S. III. 1987. Age and growth of snappers and groupers. "Tropical Snappers and Groupers: Biology and Fisheries Management." (J.J. Polovina and S. Ralston, eds.) Boulder, CO: Westview Press. 329-373.

Moe, M.A. 1969. Biology of the red grouper, *Epinephelus morio* (Valenciennes) from the eastern Gulf of Mexico. Florida Dept. Nat. Res., Res. Lab. Prof. Pap. Ser. 10: 1-95.

Mullaney, M.D. Jr. 1994. Ontogenetic shifts in the diet of gag, *Mycteroperca microlepis* (Pisces: Serranidae). Proc. Gulf Carib. Fish. Inst. 43: 432-445.

Nagelkerken, W.P. 1979. Biology of the graysby, *Epinephelus cruentatus*, of the coral reef of Curacao and other Caribbean islands. Stud. of the Fauna of Curacao and Other Caribbean Islands. 61: 1-118.

Nemeth, R.S. 2005. Population characteristics of a recovering US Virgin Islands red hind spawning aggregation following protection. Mar. Ecol. Prog. Ser. 286: 81-97.

Nemtzov, S.C., S.M. Kajiura and C.A. Lompart. 1993. Diel color phase changes in the coney, *Epinephelus fulvus* (Teleostei, Serranidae). Copeia. 3: 883-885.

Petersen, C.W. 1995. Reproductive behavior, egg trading, and correlates of male mating success in the simultaneous hermaphrodite, *Serranus tabacarius*. Env. Biol. Fish. 43: 351-361.

O'Hop, J. and J. Munyandorero. 2016. SEDAR 47 Stock Assessment Report for Goliath Grouper of the South Atlantic and Gulf of Mexico, 2016. FWC Report IHR2016-001.

Petersen, C.W. and E.A. Fischer. 1986. Mating system of the hermaphroditic coral-reef fish *Serranus baldwini*. Behav. Ecol. Sociobiol. 19: 171-178.

Powell, A.B. and J.W. Tucker. 1992. Egg and larval development of laboratory-reared Nassau grouper, *Epinephelus striatus* (Pisces, Serranidae). Bull. Mar. Sci. 50: 171-185.

Pressley, P.H. 1981. Pair formation and joint territoriality in a simultaneous hermaphrodite: the coral reef fish *Serranus tigrinus*. Z. Tierpsychol. 56: 33-46.

Randall, J.E. 1967. Food habits of reef fishes of the West Indies. Stud. Trop. Ocean. 5: 665-847.

Robertson, D.R. 2013. Who resembles whom? Mimetic and coincidental look-alikes among tropical reef fishes. PLoS ONE 8(1): e54939.

Robins, R.H., *Epinephelus itajara*: Florida Museum of Natural History. https://www.floridamuseum.ufl.edu/discover-fish/species-profiles/epinephelus-itajara/

Sadovy, Y. 1994a. Summary of papers presented at the grouper symposium of the 43rd annual meeting of the Gulf and Caribbean Fisheries Institute in Miami, Florida (Nov. 1990). Proc. Gulf Carib. Fish. Inst. 43: 482-484.

Sadovy, Y. 1994b. Grouper stocks of the western central Atlantic: the need for management and management needs. Proc. Gulf Carib. Fish. Inst. 43: 43-64.

Sadovy, Y., M. Figuerola and A. Román. 1992. Age and growth of red hind *Epinephelus guttatus*, in Puerto Rico and St. Thomas. Fish. Bull. 90: 516-528.

Sadovy, Y., P.L. Colin, and M.L. Domeier. 1994. Aggregation and spawning in the tiger grouper, *Mycteroperca tigris* (Pisces: Serranidae). Copeia. 2: 511-516.

Sadovy, Y., A. Rosario and A Román. 1994. Reproduction in an aggregating grouper, the red hind, *Epinephelus guttatus*. Env. Biol. Fish. 41: 269-286.

Sadovy, Y. and A. Eklund. 1999. Synopsis of biological information on *Epinephelus striatus* (Bloch, 1792), the Nassau grouper and E. Itajara (Lichtenstein, 1822), the jewfish. NOAA Tech. Rep. Ser.

Sadovy de Mitcheson, Y. and P.L. Colin (eds) 2012. Reef fish spawning aggregations: biology, research and management, Vol. 35: Fish and Fisheries Series. Springer, New York, NY.

Scott, D. 1962. Effect of food quality on fecundity in rainbow trout *Salmo gairdneri*. J. Fish. Res. Boars Can. 19: 715-731.

Semmens, B.X., K.E. Luke, P.G. Bush, et al. 2005. Investigating the reproductive migration and spatial ecology of Nassau grouper (*Epinephelus striatus*) on Little Cayman Island using acoustic tags – An Overview. Proceedings of the 58th Gulf and Caribbean Fisheries Institute.

Semmens, B.X. 2019. Personal communication: Tiger grouper spawning behavior during Nassau grouper annual spawning aggregation in Little Cayman.

Semmens, B.X., K.E. Luke, P.G. Bush, et al. 2007. Isopod infestation of postspawning Nassau grouper around Little Cayman Island. J. Fish. Biol. 69 (3): 933-937.

Semmens, B.X., P.G. Bush, S. Heppell, et al. 2007. Charting a course for Nassau grouper recovery in the Caribbean: What we've learned and what we still need to know. Proceedings of the 60th Gulf and Caribbean Fisheries Institute.

Shapiro, D.Y., Y. Sadovy, and M.A. McGehee. 1993a. Size, composition, and spatial structure of the annual spawning aggregation of the red hind *Epinephelus guttatus* (Pisces: Serranidae). Copeia. 1993: 367-374.

Shapiro, D.Y., Y. Sadovy and M.A. McGehee. 1993b. Periodicity of sex change and reproduction in the red hind, *Epinephelus guttatus*, a protogynous grouper. Bull. Mar. Sci. 53: 399-406.

Shapiro, D.Y., G. Garcia-Moliner and Y. Sadovy. 1994. Social system of an inshore stock of the red hind grouper, *Epinephelus guttatus* (Pisces: Serranidae). Env. Biol. Fish. 41: 415-422.

Shenker, J.M., E.D. Maddox, E. Wishinski, A. Pearl, S.R. Thorrold and N. Smith. 1993. Onshore transport of settlement-stage Nassau grouper, *Epinephelus striatus*, and other fishes in Exuma Sound, Bahamas. Mar. Ecol. Prog. Ser. 98: 31-43.

Sherman, K.D., C.P. Dahlgren, J. R. Stevens, C.R. Tyler. 2016. Integrating population biology into conservation management for endangered Nassau grouper, *Epinephelus striatus*. Mar. Ecol. Prog. Ser. Vol. 554: 263-280.

Sluka, R., M. Chiappone, K.M. Sullivan and R. Wright. 1996. Habitat and life in the Exuma Cays, the Bahamas: The status of grouper and coral reefs in the northern cays. The Nature Conservancy, Florida and Caribbean Marine Conservation Science Center. Nassau, Bahamas: Media Publishing Ltd.

Smith, C.L. 1972. A spawning aggregation of Nassau grouper, *Epinephelus striatus*. Trans. Amer. Fish. Soc., 1972, No.2.

Smith, E.L. 1972. A spawning aggregation of Nassau grouper, *Epinephelus striatus* (Bloch). Trans. Amer. Fish. Soc. 101: 257-261.

Starr, R.M., E. Ballesteros, E. Sala and J.M. Llenas. 2018. Spawning behavior of the tiger grouper (*Mycteroperca tigris*). Environ. Biol. Fish. Vol. 101, Iss. 12: 1641-1655.

Sullivan, K.M. 1994. Energetics of juvenile *Epinephelus* grouper: impact of summer temperatures and activity patterns on growth rates. Proc. Gulf Carib. Fish. Inst. 43: 148-167.

Thompson, R. and J.L. Munro. 1978. Aspects of the biology and ecology of Caribbean reef fishes: Serranidae (hinds and groupers). J. Fish Biol. 12: 115-146.

Thresher, R.E. 1978. Polymorphism, mimicry, and the evolution of the hamlets. Bull. Mar. Sci. 28: 345-353.

Tucker, J.W., P.G. Bush and S.T. Slaybaugh. 1993. Reproductive patterns of Cayman Islands Nassau grouper (*Epinephelus striatus*) populations. Bull. Mar. Sci. 52(3): 961-969.

Victor, B. 2012. *Hypoplectrus floridae* n. sp. And *Hypoplectrus ecosur* n. sp., two new barred hamlets from the Gulf of Mexico (Pisces: Serranidae): more than 3% different in COI mtDNA sequence from the Caribbean Hypoplectrus species flock. Journal of the Ocean Science Foundation. Vol. 5.

Whaylen, L., C.V. Pattengill-Semmens, B.X. Semmens, P.G. Bush and M.R. Boardman. 2004. Observations of a Nassau grouper, *Epinephelus striatus*, spawning aggregation site in Little Cayman, Cayman Islands, including multi-species spawning information. Environ. Biol. Fishes. 70: 305-313.

Whaylen, L., P. Bush, B. Johnson, et al. 2006. Aggregation dynamics and lessons learned from five years of monitoring at a Nassau grouper (*Epinephelus striatus*) spawning aggregation in Little Cayman, Cayman Islands, BWI. Proceedings of the 59th Gulf and Caribbean Fisheries Institute.

Whiteman, E.A., I.M. Côté and J.D. Reynolds. 2007. Ecological differences between hamlet (*Hypoplectrus*: Serranidae) colour morphs: between-morph variation in diet. Journal of Fish Biology 71: 235-244.

Jawfishes

Colin, P.L. 1971. Interspecific relationships of the yellowhead jawfish, *Opistognathus aurifrons* (Pisces, Opistognathidae). Copeia. 3: 469-473.

Colin, P.L. 1972. Daily activity patterns and effects of environmental conditions on the behavior of the yellowhead jawfish, *Opistognathus aurifrons*, with notes on its ecology. Zoologica. 57: 137-164.

Colin, P.L. 1973. Burrowing behavior of the yellowhead jawfish, *Opistognathus aurifrons*. Copeia. 1: 84-90.

Hess, H.C. 1993. Male mouthbrooding in jawfishes (Opistognathidae): constraints on polygamy. Bull. Mar. Sci. 52: 806-818.

Leong, D. 1967. Breeding and territorial behavior in *Opistognathus aurifrons* (Opistogathidae). Naturwissenschaften. 54: 97.

Lionfishes

Albins, M.A. 2016.Behaviours of Pacific Lionfish Facilitate Invasion of the Atlantic. from Biological Invasions and Animal Behaviour. Editors: Judith S. Weis, Daniel Sol: 309-323

Bacheler, N.M., P.E. Whitfield, R.C. Muñoz, et al. 2015. Movement of invasive adult lionfish *Pterois volitans* using telemetry: importance of controls to estimate and explain variable detection probabilities. Mar. Ecol. Prog. Ser., 527, 205–220.

Barbour, A.B., M.S. Allen, T.K. Frazer and K.D. Sherman. 2011. Evaluating the potential efficacy of invasive lionfish (*Pterois volitans*) removals. PLoS One 6, e19666.

Benkwitt, C.E. 2016. Invasive lionfish increase activity and foraging movements at greater local densities. Mar. Ecol. Prog. Ser.,558: 255–266.

Corey, E., J.M. Pitt, J. A. Morris Jr., S. Smith, et al. 2016. Diet of invasive lionfish (*Pterois volitans* and *P. miles*) in Bermuda. Mar. Ecol. Prog. Ser., 558: 193–206.

Côté, I.M. and N.S. Smith, 2018. The lionfish *Pterois* sp. invasion: Has the worst-case scenario come to pass? J. of Fish Biol. 92: 660-689.

Cure, K., C.E. Benkwitt, T.L. Kindinger, et al. 2012. Comparative behavior of red lionfish *Pterois volitans* on native Pacific versus invaded Atlantic coral reefs. Mar. Ecol. Prog. Ser. 467: 181–192.

Fishelson, L. 1997. Experiments and observations on food consumption, growth and starvation in Dendrochirus brachypterus and *Pterois volitans* (Pteroinae, Scorpaenidae). Environ. Biol. of Fishes 50 (4): 391–403.

Gardner, P.G. 2012. Reproductive biology of invasive lionfish (*Pterois volitans/miles* complex) from Little Cayman Island. Dissertation, Masters of Science, Univ. of Florida.

Gittings, S.R., A.Q. Fogg and S. Frank. 2017. Going deep for lionfish: designs for two new traps for capturing lionfish in deep water. Mar. Sanc. Con. Ser. ONMS-17-05.

Hamner, R.M., D.W. Freshwater and P.E. Whitfield. 2007. Mitochondiral cytochrome b analysis reveals two invasive species with strong founder effects in the western Atlantic. J. of Fish Biol. 71 (Sup B): 214-222.

Harris, Holden 2018. Efficacy of lionfish traps in the Northern Gulf of Mexico. Proceedings of the 71st Gulf and Caribbean Fisheries Institute. Nov. 5-9, 2018. San Andres, Colombia. Presentation.

Jud, Z.R. and C.A. Layman. 2012. Site fidelity and movement patterns of invasive lionfish, *Pterois* spp., in a Florida estuary. J. of Exp. Mar. Biol. and Ecol 414-415: 69–74.

Morris, J. A. Jr. and J. L. Akins. 2009. Feeding ecology of invasive lionfish (*Pterois volitans*) in the Bahamian archipelago. Env. Biol Fishes 86(3): 389–398.

NOAA, 2010. National Oceanic and Atmospheric Administration. National Oceanic and Atmospheric Administration., USA: United States Department of Commerce, unpaginated. http://www.noaa.gov/ (/isc/abstract/20087204983)

Peake, J., A.K. Bogdanoff, C.A. Layman, et al. 2018. Feeding ecology of invasive lionfish (*Pterois volitans* and *P. miles*) in the temperate and tropical western Atlantic. Biol. Invasions.20(9): 2567–2597.

Pitt, J.M. and T.M. Trott. 2014. Trapping lionfish in Bermuda, Part II: Lessons learned to date. Proceedings of the 67th Gulf and Caribbean Fisheries Institute. Nov. 3–7 2014, Christ Church, Barbados, pp. 221–224. Presentation.

Pusack, T.J., C.E. Benkwitt, K. Cure and T.L. Kindinger. 2016. Invasive red lionfish (*Pterois volitans*) grow faster in the Atlantic Ocean than in their native Pacific range. Env. Biol. Fishes 99: 571–579.

Rocha, L.A., C.R. Rocha, C.C. Baldwin, L.A. Weigt and M. McField. 2015. Invasive lionfish preying on critically endangered reef fish. Coral Reefs 34(3): 803-806.

Ruiz-Carus, R., R.E. Matheson, D.E. Roberts and P.E. Whitfield. 2006. The western Pacific red lionfish, *Pterois volitans* (Scorpaenidae), in Florida: evidence for reproduction and parasitism in the first exotic marine fish established in state waters. J. Biol. Con. Vol. 128: 384–390.

Schofield, P.J., D.H. Huge, T.C. Rezek, et al. 2015. Survival and growth of invasive Indo-Pacific lionfish at low salinities. Aquatic Invasions 10(3): 333–337.

Tamburello, N. and I.M. Côté. 2014. Movement ecology of Indo-Pacific lionfish on Caribbean coral reefs and its implications for invasion dynamics. Biol. Invasions, 17(6): 1639-1653.

Parrotfishes

Barlow. G.W. 1975. On the sociobiology of four Puerto Rican parrotfishes (Scaridae). Mar. Biol. 33: 281-293.

Bellwood, D.R. and J.H. Choat. 1990. A functional analysis of grazing in parrotfishes (family Scaridae): the ecological implications. Env. Biol. Fish. 28: 189-214.

Bruckner, A.W. and R.J. Bruckner. 1998a. Destruction of coral by *Sparisoma viride*. Coral Reefs 17: 350.

Bruckner, A.W. and R.J. Bruckner. 1998b. Rapid-wasting disease: pathogen or predator. Science. 279: 2023-2025.

Bruggemann, J.H., M.W.M. Kuyper and A.M. Breeman. 1994. Comparative analysis of foraging and habitat use by the sympatric Caribbean parrotfish *Scarus vetula* and *Sparisoma viride* (Scaridae). Mar. Ecol. Prog. Ser. 112: 51-66.

Bruggemann, J.H., M.J.H. van Oppen and A.M. Breeman. 1994a. Foraging by the stoplight parrotfish *Sparisoma viride*. I. Food selection in different, socially determined habitats. Mar. Ecol. Prog. Ser. 106: 41-55.

Bruggemann, J.H., M.J.H. van Oppen and A.M. Breeman. 1994b. Foraging by the stoplight parrotfish *Sparisoma* viride. II intake and assimilation of food, protein and energy. Mar. Ecol. Prog. Ser. 106: 57-71.

Bruggemann, J.H., A.M. van Kessel, J.M. van Rooij, and A.M. Breeman. 1996. Bioerosion and sediment ingestion by the Caribbean parrotfish *Scarus vetula* and *Sparisoma viride*: implications of fish size, feeding mode and habitat use. Mar. Ecol. Prog. Ser. 134: 59-71.

Buckman, N.S. and J.C. Ogden. 1972. Territorial behavior of the striped parrotfish *Scarus croicensis*. Bloch (Scaridae). Ecology. 54: 1377-1383.

Bythell, J.C. and E.H. Gladfelter. 1992. Ecological studies of Buck Island Reef National Monument, St. Croix, U.S. Virgin Islands. U.S. Dept. of the Interior, Nat. Park Serv. II: 40-61.

References

Bythell, J.C., E.H. Gladfelter and M. Bythell. 1993. Chronic and catastrophic natural mortality of three common Caribbean reef corals. Coral Reefs. 12: 143-152.

Cardwell, J.K. 1989. Behavioural endocrinology of the stoplight parrotfish, *Sparisoma viride*, a protogynous coral reef fish. Ph.D. Thesis, Univ. of British Columbia, Vancouver.

Choat, J.H. 1991. The biology of herbivorous fishes on coral reefs. "The Ecology of Fishes on Coral Reefs" (P.F. Sale, ed.) San Diego: Academic Press. 120-153.

Clavijo, I.E. 1983. Pair spawning and formation of a lek-like mating system in the parrotfish *Scarus vetula*. Copeia. 1: 253-256.

Clifton, K.E. 1989. Territory sharing by the Caribbean striped parrotfish, *Scarus iserti*: patterns of resource abundance, group size and behaviour. Anim. Behav. 37: 90-103.

Colin, P.L. 1978. Daily and summer-winter variation in mass spawning of the striped parrotfish, *Scarus croicensis*. Fish. Bull. 76: 117-124.

Dubin, R.E. 1981. Pair spawning in the princess parrotfish, *Scarus taeniopterus*. Copeia. 2: 475-477.

Dubin, R.E. and J.D. Baker. 1982. Two types of cover-seeking behavior at sunset by the princess parrotfish, *Scarus taeniopterus*, at Barbados, West Indies. Bull. Mar. Sci. 32: 572-583.

Farm, B.P. 1993. Territory dynamics in the bucktooth parrotfish (*Sparisoma radians*). Ph.D. Dissertation, Univ. of Minnesota.

Koltes, K.H. 1993. Aspects of the reproductive biology and social structure of the stoplight parrotfish, *Sparisoma viride*, at Grand Turk, Turks and Caicos Islands, B.W.I. Bull. Mar. Sci. 52: 792-805.

Lobel, P.S. and J.C. Ogden. 1981. Foraging by the herbivorous parrotfish *Sparisoma radians*. Mar. Biol. 64: 173-183.

Marconato, A. and D.Y. Shapiro. 1996. Sperm allocation, sperm production and fertilization rates in the bucktooth parrotfish. Anim. Behav. 52: 971-980.

McAfee, S.T. and S.G. Morgan. 1996. Resource use by five sympatric parrotfishes in the San Blas Archipelago, Panama. Mar. Biol. 125: 427-437.

Ogden J.C. and N.S. Buckman. 1973. Movements, foraging groups, and diurnal migrations of the striped parrotfish *Scarus croicensis* Bloch (Scaridae). Ecology. 54: 589-596.

Randall, J.E. 1967. Food habits of reef fishes of the West Indies. Stud. Trop. Ocean. 5: 665-847.

Randall, J.E. and H.A. Randall. 1963. The spawning and early development of the Atlantic parrotfish *Sparisoma rubripinne*, with notes on other Scarid and Labrid fishes. Zoologica. 48: 49-58.

Robertson, D.R. and R.R. Warner. 1978. Sexual patterns in the labroid fishes of the Western Caribbean, II: The parrotfishes (Scaridae). Smiths. Contr. Zool. 155: 1-26.

Schmitt, E.F. 1997. The influence of herbivorous fishes on coral reef communities with low sea urchin abundance: A study among reef community types and seasons in the Florida Keys. Ph.D. Dissertation, Univ. of Miami, Coral Gables, FL.

Shapiro, D.Y. 1991. Intraspecific variability in social systems of coral reef fishes. "The Ecology of Fishes on Coral Reefs" (P.F. Sale, ed.) London: Academic Press. 331-355.

Thresher, R.E. 1984."Reproduction in Reef Fishes" Neptune City, NJ: TFH Publications.

Van Rooij, J.M., E. de Jong, F. Vaandrager and J.J. Videler. 1996. Resource and habitat sharing by the stoplight parrotfish, *Sparisoma viride*, a Caribbean reef herbivore. Env. Biol. Fish. 47: 81-91.

Van Rooij, J.M., J.P. Kok and J.J. Videler. 1996. Local variability in population structure and density of the protogynous reef herbivore *Sparisoma viride*. Env. Biol. Fish. 47: 65-80.

Van Rooij, J.M., F.J. Kroon and J.J. Videler. 1996. The social and mating system of the herbivorous reef fish *Sparisoma viride*: one-male versus multi-male groups. Env. Biol. Fish. 47: 353-378.

Sharks & Rays

Pratt, H.L. and J.C.C. Carrier. 1995. Wild mating of the nurse shark. Nat. Geo. 187: 44-53.

Strong, W.R. 1990. Hammerhead shark predation on stingrays: an observation of prey handling by *Sphyrna mokarran*. Copeia. 3: 836-840.

Surgeonfishes

Alevizon, W.S. 1976. Mixed schooling and its possible significance in a tropical western Atlantic parrotfish and surgeonfish. Copeia. 4: 796-798.

Bardach, J.E. 1958. On the movements of certain Bermuda reef fishes.Ecology 39: 139-146.

Colin, P.L. 1985. Spawning of western Atlantic surgeonfishes. Natl. Geo.Soc. Rpts. 18: 243-250.

Colin, P.L. and I.E. Clavijo. 1988. Spawning activity of fishes producing pelagic eggs on a shelf edge coral reef, southwestern Puerto Rico. Bull. Mar. Sci.43: 249-279.

Clavijo, I.E. 1974. A contribution on feeding habits of three species of Acanthurids (pisces) from the West Indies. Master's Thesis. Florida Atlantic Univ. Boca Raton, FL.

Debrot, A.O. and A.A. Myrberg, Jr. 1988. Intraspecific avoidance as aproximate cause for mixed-species shoaling by juveniles of a western Atlantic surgeonfish, *Acanthurus bahianus*. Bull. Mar. Sci. 104-106.

Foster, S.A. 1985. Group foraging by a coral reef fish: a mechanism forgaining access to defended sources. Anim. Behav. 33: 782-792.

Risk, A. 1998. The effects of interactions with reef residents on the settlement and subsequent persistence of ocean surgeonfish, *Acanthurus bahianus*. Env. Biol. Fish. 51: 377-389.

Robertson, D.R. 1988. Abundances of surgeonfishes on patch-reefs in Caribbean Panama: due to settlement, or post-settlement events? Mar.Biol. 97: 495-501.

Robertson, D.R. 1991. Increase in surgeonfish populations after mass mortality of the sea urchin *Diadema antillarum* in Panama indicate food limitation. Mar. Biol. 111: 437-444.

Winterbottom, R. 1971. Movement of the caudal spine of some surgeonfishes (Acanthiurdae, Perciformes). Copeia. 3: 562-566.

Tilefishes

Baird, T.A. 1988. Female and male territoriality and mating system of the sand tilefish, *Malacanthus plumieri*. Env. Biol. Fish. 22: 101-116.

Baird. T.A. and N.R. Liley. 1989. The evolutionary significance of harem polygamy in the sand tilefish, *Malacanthus plumieri*: resource of female defence? Anim. Behav. 38: 817-829.

Clark, E., J.S. Rabin and S. Holderman. 1988. Reproductive behavior and social organization in the sand tilefish, *Malacanthus plumieri*. Env. Biol. Fish. 22: 273-286.

Wrasses

Baird, T.A. 1988. Abdominal windows in straight-tailed razorfish, *Xyrichtys martinicensis*: An unusual female sex character in a polygynous fish. Copeia. 1988(2): 497-499.

Colin, P.L. 1982. Spawning and larval development of the hogfish *Lachnolaimus maximus* (Pisces: Labridae). Fish. Bull. 80: 853-862.

Marconato, A., D.Y. Shapiro, C.W. Petersen, R.R. Warner, and T. Yoshikawa. 1997. Methodological analysis of fertilization rate in the bluehead wrasse *Thalassoma bifasciatum*: pair versus group spawns. Mar. Ecol. Prog. Ser. 161: 61-70.

Nemtzov, S.C. 1992. Intraspecific variation in the social and mating behavior of Caribbean razorfishes. (Teleostei, Labridae) Ph.D. Dissertation. State Univ. of New York, Stony Brook.

Randall, J.E. and G.L. Warmke. 1967. The food habits of the hogfish (*Lachnolaimus maximus*), a labrid fish from the western Atlantic. Carib. JSci. 7: 141-143.

Reinboth, R. 1973. Dualistic reproductive behavior in the protogynous wrasse *Thalassoma bifasciatum* and some observations on its day-night changeover. Helgolander Wiss. Meersunters. 24: 174-191.

Robertson D.R. and S.G. Hoffman. 1977. The roles of female mate choice and predation in the mating systems of some tropical Labroid fishes. Z. Tierpsychol. 45: 298-320.

Shapiro, D.Y., A. Marconato and T. Yoshikawa. 1994. Sperm economy in a coral reef fish, *Thalassoma bifasciatum*. Ecology. 75: 1334-1344.

Shen, D.C. and Eugenie Clark. 2016. Territorial and reproductive behavior of the three Caribbean Razorfishes of the Genus Xyrichtys (Labridae) at Bonaire. aqua, International Journal of Ichthyology. 22(1): 33-59.

Thresher, R.E. 1979. Social behavior and ecology of two sympatic wrasses (Labridae: *Halichoeres* spp.) off the coast of Florida. Mar. Biol. 53: 161-172.

Thresher, R.E. 1984. "Reproduction in Reef Fishes" Neptune City, NJ: T.F.H.Publications.

Victor, B.C. 1986. Duration of the planktonic larval stage of one hundred species of Pacific and Atlantic wrasses (family Labridae). Mar. Biol. 90:317-326.

Victor, B.C. 1987. The mating system of the Caribbean rosy razorfish, *Xyrichtys martinicensis*. Bull. Mar. Sci. 40: 152-160.

Warner, R.R. and E.T. Schultz. 1992. Sexual selection and male characteristics in the bluehead wrasse, *Thalassoma bifasciatum*: mating site acquisition, mating site defense, and female choice. Evol. 46: 1421-1442.

Glossary

benthic – Relating to the bottom of a body of water.

carnivore – A species that preys on animals.

chromatophores – Irregular-shaped cells in the skin of fishes that darken or lighten to modify colors.

claspers – Organs modified from the inner edge of each pelvic fin of male sharks and rays. Sperm is transferred through a groove in the clasper into a female's cloaca.

cleaners – Certain species of shrimps or young fishes that prey on tiny crustaceans infesting fishes.

cleaning station – A location continuously inhabited by cleaner fishes and/or shrimps.

cloaca – Ventral opening of sharks and rays used for reproduction and excretion.

clutch –A cluster of eggs laid in a single day.

filamentous or turf algae – Prolific, fast-growing, hair-like algae that is the primary food source of many herbivores.

gametes – Eggs and sperm.

generalist – A species that feeds on a variety of foods without a specific preference for one type.

gonochorist – An individual remaining the same sex for life.

harem – A group of two or more females and a single dominant male living together as a stable group within a defended territory.

herbivore – An animal that feeds primarily on plants.

home range – An area frequented but not defended by an individual.

initial phase – In sex-changing species, numerous, typically young, drab sexually mature individuals. Depending on the species, may include both males and females but generally females.

juvenile – The developmental stage preceding sexual maturity.

metapopulation – Widely separated populations of the same species that are isolated from one another by open sea or bottom unsuitable for larval settlement.

monogamy – Mating exclusively with a single partner.

nuclear hunting – Different species of carnivores or piscivores feeding in association with eels or octopuses.

otolith – One of three tiny calcium carbonate ear stones used for balance and sound detection, located within the semicircular canals at the base of the braincase in all bony fishes.

ontogenetic shift – Habitat change during the growth of young fishes – often due to dietary needs.

opportunist – A species that willingly eats food not included in its normal diet.

pelagic – Inhabiting the upper layers of the open sea.